SCHAUM'S OUTLINE OF

THEORY AND PROBLEMS

OF

PROJECTIVE
GEOMETRY

•

BY

FRANK AYRES, JR., Ph.D.

Formerly Professor and Head,
Department of Mathematics
Dickinson College

•

SCHAUM PUBLISHING CO.
257 Park Avenue South, New York 10010

Preface

The purpose of this book is to provide a first course in Projective Geometry for undergraduate majors in mathematics and for prospective teachers of high school geometry. For the former it will furnish an introduction to the important concept of projective spaces; for the latter it will introduce a more general geometry from which, by proper specialization, the familiar metric geometry is obtained. Since only the real geometry of one and two dimensions is considered here, every theorem may be illustrated by a diagram in the construction of which nothing more than a straight edge is required.

Chapter 1 begins with a brief survey of the geometry of Euclid and his associates. That part of this geometry which is concerned solely with the incidence of points and lines is called projective. The projective plane is then obtained by properly modifying the fundamental plane of Euclidean geometry. In Chapters 1-6 and 8-12, the reader will find the basic propositions of plane projective geometry developed entirely by synthetic methods.

Chapter 7 is concerned with an axiomatic approach. In the course of providing models for the axioms, certain finite projective geometries are introduced. Since this leads eventually to the geometry of points defined by number triples over the field of real numbers of Chapter 15, the reading of the chapter may be postponed until that time.

In Chapter 13 the procedure is reversed. Taking the projective plane as fundamental, modifications are made to obtain the affine plane in which parallel lines reappear. Additional modifications are made in Chapter 14 in order to define perpendicular lines and thus permit a return to the metric plane. Of interest here is the fact, observed perhaps for the first time by the reader, that so much of the metric geometry with which he is familiar depends on parallelism rather than on perpendicularity. Also to be noted is the great variety of metric theorems which often follow from a single projective theorem.

In Chapters 15-17 the reader is introduced to analytic methods in projective geometry. In these chapters an acquaintance with matrix algebra is assumed. For those who would wish a brief review, the Appendix will be found helpful.

The final chapter parallels Chapters 13-14. Beginning with the set of all projective transformations of the plane onto itself, the reader is led by successive steps to the familiar rigid motions of Plane Analytic Geometry.

The author wishes to take this opportunity to express his gratitude to the staff of the Schaum Publishing Company for their splendid cooperation.

FRANK AYRES, JR.

Carlisle, Penna.
August, 1967

CONTENTS

CONTENTS

Chapter 1

Introduction

EUCLIDEAN GEOMETRY

Until about 600 B.C., geometry consisted mainly of a collection of rules for finding areas and volumes. These rules, together with certain facts concerning triangles, circles, ... as developed by the Babylonians and the Egyptians, were based solely upon experience and observation. The period from 600 B.C. to 300 B.C. covers roughly the rise and decline of the classical Greek culture. This culture, with its emphasis on deductive reasoning and knowledge for its own sake, completely changed the nature of mathematics and, in particular, the role of geometry. First of all, mathematics was made abstract. For example, whereas to the Egyptians a line was a taut string, for the Greeks the words point, line, triangle, circle, ... became mental concepts suggested by appropriate physical objects. Secondly, the Greeks interpreted arithmetic and algebra in terms of geometry. A number, for example, was a length; the product of two numbers was the area of a rectangle; the product of three numbers was a volume (even today we speak of 9 as the square of 3 and 8 as the cube of 2); and geometric constructions were devised to solve equations. The debt which present day civilization owes the Greeks for making mathematics abstract is immeasurable. It must be pointed out, however, that their conversion of arithmetic and algebra into geometry was unfortunate.

The scene now shifts back to Egypt where Euclid, a professor of mathematics at the University of Alexandria, after selecting ten axioms (he also made use of other assumptions not explicitly stated) was able in his *Elements* to deduce all of the important results of the classical Greek period. Much of the material included by Euclid is quite familiar since, with only minor changes, it is the plane and solid geometry of high school. We have only to recall a few of the theorems to realize that underlying all of this geometry is the notion of measurement — the length of a line segment, the measure of an angle, ... It was then a geometry of touch but, as will be evident shortly, not always of sight.

The magnitude of the task which Euclid set for himself becomes apparent when it is realized that the theorems to be included were the product of various schools—the Ionian School established by Thales, the Pythagorean School in southern Italy, and the school established by Plato in Athens—as well as of individuals, each using axioms of his own liking. The first task then was to select an adequate set of axioms which would be universally acceptable. It was noted above that Euclid did not state explicitly all of his assumptions. Among those not stated were:

A line which contains the vertex A and an interior point P of a triangle ABC also contains a point D of the line segment BC.

In the plane determined by a point P and a line p, there exists at least one line which passes through P and is parallel to p.

The omission of the first of these was probably due to its being considered too obvious to be worthy of note. The omission of the second cannot be so simply explained and is, indeed, indicative of a characteristic of Greek thought at the time. For, although Euclid assumed in his second axiom the extension of a line segment in both directions and as far as one chooses, he used the axiom sparingly in that he extended a line segment only so far as was necessary for the problem at hand. Thus, in keeping with Greek philosophy which avoided the infinite, Euclid centered his attention on line segments and avoided consideration of a line in its entirety. As a result, his fifth axiom

> If a line p intersects two lines r and s such that the sum of the interior angles on the same side of p is less than two right angles, then the lines r and s intersect on that same side of p.

which, together with the other axioms and tacit assumptions, implies the well-known

> *Parallel Axiom*: Through any point P not on a given line p, there exists in the plane determined by P and p one and only one line parallel to p.

was to plague geometers for the next two thousand years. During this period countless attempts were made either to obtain the axiom as a consequence of the others or to replace it by a simpler one. It was not until the middle of the 18th century that mathematicians began to suspect there might exist geometries in which the Parallel Axiom did not hold. The matter was finally settled early in the last century when Bolyai and Lobatchevski independently produced a self-consistent geometry in which essentially "one and only one line" in the Parallel Axiom was replaced by "more than one line" and, a few years later, Riemann produced another in which "one and only one line" was replaced by "no line". These are the so-called non-Euclidean metric geometries.

Today it is recognized that there are many geometries; in particular, there is a non-metric geometry in which parallelism plays no role. A typical theorem of this geometry (see Fig. 1-1)

> If A_1, A_2, A_3 are distinct points on a line r and if B_1, B_2, B_3 are distinct points on another line s which meets r in the point O, then the points of intersection C_1 of A_2B_3 and A_3B_2, C_2 of A_1B_3 and A_3B_1, C_3 of A_1B_2 and A_2B_1 are *collinear* (i.e. lie on a line).

was given by Pappus in the third century A.D. Being proved by the use of Euclidean methods, the theorem was simply added to the propositions of Euclid's geometry. It is clear, however, that this theorem, concerned only with the joins of points and the intersections of lines, is of a character quite different from the typical theorem of plane geometry. Nevertheless, it was not until the 17th century that the principal theorems of this geometry were established and not until the middle of the 19th century that the geometry was freed completely of metric notions.

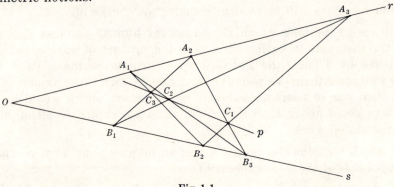

Fig. 1-1

PROJECTIVE GEOMETRY

It has been noted that Euclid's geometry is not always the geometry of sight; for example, we never see parallel lines. If one stands midway the rails of a straight railroad track, the rails appear to meet in a point on the horizon. Moreover, if a camera is used, the resulting picture will show the same phenomenon. Our concern from now on will be with the geometry, roughly that of photography, called *projective geometry*.

In producing a picture, the camera (in effect) joins each point A in its range to a point P in the lens by a straight line, that is, *projects* each point A from P, and then *sections* the resulting lines by a plane π — the film. Corresponding to any point A in the range there is a unique point A', the intersection of the line AP and π, in the section. Also, corresponding to any line q which contains A but not P there is in the section a line q' which contains A'. This follows from the fact that P forms with the line q a plane which, in turn, intersects π in the line q'. Suppose now that the subject of our picture consists of a pair of intersecting line segments and a circle drawn on a vertical wall. For practical reasons, the size of any subject is diminished in the picture. The important point is, however, that the lengths of the line segments in the picture may be varied simply by changing the distance of the lens from the wall. There are, moreover, other distortions. Although the picture shows a pair of intersecting line segments, their angles of intersection will usually be different from those in the subject; also, the circle will usually appear as an ellipse.

In Euclidean geometry, we study the lengths of line segments and measures of angles since they are invariant (unchanged) under rigid motions. It is now clear that these familiar notions will play no role in our new geometry.

We must be careful, of course, not to endow projective geometry with certain of the limitations of the camera. For instance, we shall assume that every point of space, excepting only the center of projection P, may be projected from P and that the sectioning plane may be any entire plane π which does not contain P. Usually we will have the situation that to each point $A \neq P$ of the space there will correspond a unique point A' in π. Such exceptions as there are, for example, when AP is parallel to π no correspondent A' of A is obtained, are due to certain characteristics of Euclidean space. Recall that a line p, not in the plane π, either meets π in a point or is parallel to π.

Suppose A to be a point for which there is no correspondent in π. Let π' be any plane through A which does not contain P and let r and s be any distinct lines of π' intersecting at A. Denote by r' and s' respectively the correspondents in π of r and s when projected from P. Now r' and s' must be parallel since otherwise their point of intersection would be the correspondent of A. Moreover, by interchanging the roles of π and π', we find that the correspondents in π' of the parallel lines r' and s' in π are the intersecting lines r and s.

It is now clear that so long as we deal with Euclidean space neither the property of being a point nor the property of being a pair of intersecting lines is invariant under a projection and section. It is equally clear that this state of affairs is due solely to the existence in Euclidean space of parallel lines and planes.

There are two avenues of escape from our position:

(1) Follow the pattern set by Euclid of postulating a space having precisely the properties desired.

(2) Begin with Euclidean space and somehow fashion it into another in which parallelism and metric notions are completely absent.

We shall delay the first procedure, the axiomatic approach, until we have a better idea of what projective geometry is about, that is, of what the invariants of projective geometry are. In following the second procedure, the matter of metric notions can be easily taken care of — we shall simply ignore all of them — while parallelism will be eliminated by providing 'intersections' for the parallel lines, parallel planes, etc. of Euclidean space.

PROJECTIVE SPACE

From Euclid's geometry we extract the following propositions:

(a) Any two distinct points determine one and only one line.

(b) Any three distinct non-collinear points, also any line and a point not on the line, determine one and only one plane.

(c) Two distinct *coplanar* lines, that is, two distinct lines in the same plane, either intersect in a point or are parallel.

(d) A line not in a given plane either intersects the plane in a point or is parallel to the plane.

(e) Two distinct planes either intersect in a line or are parallel.

Note that these propositions are completely free of metric notions and are concerned only with the joining of points and the intersecting of lines and planes.

In constructing a space for projective geometry, we propose to adjoin to Euclidean space certain objects. These objects will not be defined but, for the purpose of distinguishing between them, will be given names — *ideal point, ideal line* and *ideal plane*. Moreover, in order that there be no possibility of confusion the points, lines and planes of Euclidean space will now be called *ordinary points, ordinary lines* and *ordinary planes* of *ordinary space*.

The ideal elements have no inherent properties; as we proceed, we will endow them with such properties as will insure that in the newly created space, that is, in *projective space*, the following propositions hold for all possible combinations of ideal and ordinary elements:

(a') Any two distinct points determine one and only one line.

(b') Any three distinct non-collinear points, also any line and a point not on the line, determine one and only one plane.

(c') Any two distinct coplanar lines intersect in one and only one point.

(d') Any line not in a given plane intersects the plane in one and only one point.

(e') Any two distinct planes intersect in one and only one line.

Let us begin by considering an ordinary plane π and its ordinary lines. To each of these ordinary lines we adjoin an ideal point (also called a *point at infinity*) in such a manner that any two distinct intersecting ordinary lines will have distinct ideal points while any two parallel ordinary lines will have the same ideal point. Let any ordinary line together with its ideal point be called an *augmented line*. For the purpose of complete clarity, let r and s be two distinct ordinary lines of π, let R_∞ and S_∞ be their respective ideal points, and consider the augmented lines (r, R_∞) and (s, S_∞). When r and s intersect (in an ordinary point), R_∞ and S_∞ are distinct; when r and s are parallel, R_∞ and S_∞ are identical and R_∞ is then the point of intersection of the augmented lines. Thus, (c') holds for (r, R_∞) and (s, S_∞) and, hence, for any two augmented lines provided the ordinary lines are coplanar.

Define now an *augmented plane* as an ordinary plane together with the totality of ideal points adjoined to its ordinary lines. It is clear that (a') holds in an augmented plane when one of the points is an ordinary point and the other is an ideal point as well as when both points are ordinary points. There remains the case when both points are ideal points. If these ideal points are to determine a unique line, it cannot be an augmented line. (Why?) Let us then adjoin to the augmented plane containing (r, R_∞) and (s, S_∞) an *ideal line* p_∞ which contains both R_∞ and S_∞ and so can be said to be determined by these points. Let

(t, T_∞), where t is distinct from r and s, be any other augmented line of the augmented plane. When r and t are parallel, $T_\infty = R_\infty$, and then (r, R_∞) and (t, T_∞) intersect in R_∞ on p_∞. Similarly, when s and t are parallel, (s, S_∞) and (t, T_∞) intersect in S_∞ on p_∞. Suppose now that $R_\infty, S_\infty, T_\infty$ are distinct. If (c') is to hold in this case, that is, if (t, T_∞) and p_∞ are to have one and only one point in common, it is necessary that T_∞ lie on p_∞. We conclude then that the totality of ideal points adjoined to the ordinary lines of the ordinary plane π constitute the ideal line p_∞. The resulting augmented plane will be denoted by (π, p_∞).

Assume that an ideal line has been adjoined to each ordinary plane of ordinary space. We shall leave for the reader to show that (d') holds for any augmented plane and any augmented line not on the plane. Consider next two augmented planes (ρ, r_∞) and (σ, s_∞) with ρ and σ distinct ordinary planes. Clearly, (e') holds when ρ and σ intersect in an ordinary line. Suppose then that ρ and σ are parallel. In ρ take any ordinary line p and in σ any ordinary line q parallel to p. Now the common ideal point P_∞ adjoined to p and q must lie on both r_∞ and s_∞. Also if b is any ordinary line of ρ, not parallel to p, and d is any ordinary line of σ parallel to b, the common ideal point B_∞ of b and d lies on both r_∞ and s_∞. Thus, r_∞ and s_∞ must coincide and (e') holds for any two distinct augmented planes.

Although we do not give the details, it should not be surprising to find that if (b') is to hold for three non-collinear ideal points it is necessary to adjoin to ordinary space an *ideal plane* which contains these points (just as an ideal line containing the ideal points R_∞ and S_∞ was adjoined to the ordinary plane π). Also, if (e') is to hold when the two planes are any augmented plane and this ideal plane, it is necessary that the ideal plane contain all of the ideal points adjoined to all of the ordinary lines of ordinary space (just as it was found necessary that the ideal line, containing R_∞ and S_∞, contain all of the ideal points adjoined to all of the ordinary lines of the ordinary plane π). Ordinary space together with the ideal plane will be called *augmented space*.

As a final step, we drop all distinctions between ordinary and ideal points, between augmented and ideal lines, and between augmented planes and the ideal plane. For a time we shall call any point, ordinary or ideal, a *projective point*; any line, augmented or ideal, a *projective line*; and any plane, augmented or ideal, a *projective plane*. However, when there is no possibility of confusion, we shall speak simply of the points, lines and planes of projective space.

THE PRINCIPLE OF DUALITY

In the preceding section projective space was obtained by adjoining additional elements to Euclidean space. It is to be noted that while certain basic words—point, line, plane—have been retained, the meaning of each has been modified. The most obvious result of this can be seen by comparing the propositions (a)–(e) of Euclidean geometry with their counterparts (a')–(e') of projective geometry. There is a gain both in simplicity (the statements (a')–(e') are briefer) and in generality (emphasized here by the repetition of the word 'any').

A far more important gain is that the *principle of duality*, whereby for each proposition another is obtained simply by the interchange of certain key words and such other changes in notation and language as are necessary to render the statement meaningful, holds in projective geometry but not in Euclidean geometry. There is a principle of duality for the projective plane and another for projective space. Consider, first, a projective plane for whose points and lines (a') and (c') assert

Any two distinct points determine one and only one line.

and

Any two distinct lines determine one and only one point.

Either statement is obtained from the other by the simple exchange of the words 'point' and 'line'. (Note that the word 'plane', whether implied or stated explicitly, is not changed.) In addition to dual theorems in the projective plane (see Example 1.1 below) there are also dual figures. For example, (1) the dual of the figure consisting of a point P and two lines through it (see Fig. 1-2(a)) is the figure consisting of a line p and two points on it (see Fig. 1-2(b)); (2) the figure consisting of a line and a point on it (also, of a line and a point not on it) is *self-dual*, that is, the dual of either figure is another of the same type.

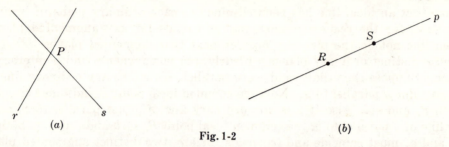

(a)   (b)

Fig. 1-2

Anticipating Example 1.1, we introduce a certain notation which will be useful throughout this book. If A, B, C, D are distinct points, everyone understands that by AB is meant the line determined by the points A and B. Let us now agree that if a, b, c, d are distinct lines and if a and b are coplanar, we shall mean by $a \cdot b$ the point of intersection of a and b. Also, assuming that all points and lines are in the same plane, let us agree that by $AB \cdot CD$ is meant the point of intersection of the lines AB and CD, by $AB \cdot a$ is meant the point of intersection of the lines AB and a, and by $(a \cdot b)(c \cdot d)$ is meant the line determined by the points $a \cdot b$ and $c \cdot d$.

Example 1.1.

Let the Theorem of Pappus (page 2) be restated as follows:

In a projective plane, let A_1, A_2, A_3 be distinct points on a line r and B_1, B_2, B_3 be distinct points on another line s; then the points $C_1 = A_2B_3 \cdot A_3B_2$, $C_2 = A_1B_3 \cdot A_3B_1$, $C_3 = A_1B_2 \cdot A_2B_1$ are collinear.

The dual of the theorem is

In a projective plane, let a_1, a_2, a_3 be distinct lines through a point R and b_1, b_2, b_3 be distinct lines through another point S; then the lines $c_1 = (a_2 \cdot b_3)(a_3 \cdot b_2)$, $c_2 = (a_1 \cdot b_3)(a_3 \cdot b_1)$, $c_3 = (a_1 \cdot b_2)(a_2 \cdot b_1)$ are *concurrent*, that is, have a common point of intersection. (See Fig. 1-3.)

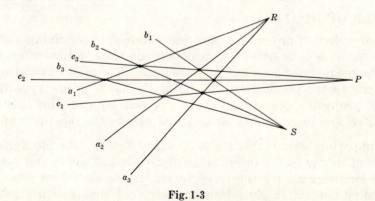

Fig. 1-3

Although we shall soon restrict our attention solely to the projective plane and thus w ll find little if any use for it, there is a principle of duality which operates in projective space. For example, (a') and (e') are dual statements, either being obtained from the other by the interchange of the words 'point' and 'plane'. (Note that here the word 'line' is not changed.) Also, the space dual of the proposition

Three distinct planes, not through the same line, determine one and only one point.

is the proposition

(b′) Three distinct points, not on the same line, determine one and only one plane.

In reality, our proposition (b′) is a postulate. Its space dual is a theorem which, when we accept the principle of duality, is automatically valid and, hence, no proof is necessary. Recall that in Euclidean geometry, a proof was required.

It is customary to say: the point P is on the line p, the line p passes through the point P, the line p lies in the plane π, the plane π passes through or contains the point P and the line p, the lines p and q intersect in the point O, etc. The task of writing the dual of a given definition, theorem, or proof of a theorem is greatly simplified by adopting the so-called 'on' language. We propose to adopt this and, thus, will write: the point P is on the line p, the line p is on the point P, the line p is on the plane π, the plane π is on the point P and the line p, the lines p and q are on the point O, etc.

THE PROJECTIVE LINE

In an ordinary plane π, take any line q and any point O not on q (see Fig. 1-4). Through O pass a line p meeting q in P. If we suppose p to rotate counterclockwise about O, then P will move along q in the direction indicated by the arrow. When p assumes the position OA, that is, when p is parallel to q, there is no point P of intersection. However, once p is beyond the position OA, the point P reappears (but on the other end of q) and, moving in the direction of the arrow, traces the remainder of q. Thus,

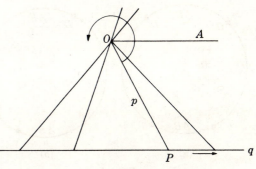

Fig. 1-4

although the motion of p about O is continuous, that is, is without jumps, the motion of P along q is not. Now had we begun with a projective plane, *every* position of p would have determined a point on q. Thus, as p rotates about O, the projective line q is traced by the continuous motion of P and we must conclude

A projective line *behaves* as if it were closed.

The reader should not attempt to form a mental picture of a closed straight line. However, for the purpose of pointing out other distinctions between ordinary and projective lines, we shall consider a circle or an ellipse as a *model* of a projective line (see Fig. 1-5(a)) and the same curve with one of its points missing as a model of an ordinary line (see Fig. 1-5(b)).

Using Fig. 1-5(b) it is clear that any point A on an ordinary line separates it into two segments, and another point B on one of the segments is sufficient to distinguish it from the other segment. Using Fig. 1-5(a) it is seen that for the projective line two distinct points A and B are necessary to separate it into two segments, and another point C on one of the segments is then sufficient to distinguish it from the other segment. A second way of describing this difference between the ordinary and projective line is as fol-

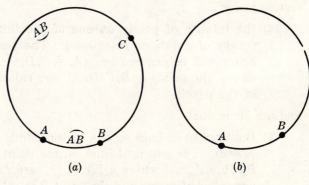

(a)   (b)

Fig. 1-5

lows: On an ordinary line there is just one path leading from one of its points A to another of
its points B; on a projective line, one may follow either of two paths AB or $\overset{\frown}{AB}$ in moving from
a point A to another point B.

Consider now a projective line and on it mark four of its points (assumed distinct)
A, B, C, D.   In considering the position of the pair C, D with respect to the pair A, B there
are two cases: (1) the pair C, D lies on one of the segments into which the pair A, B separates
the line (see Fig. 1-6(a)); (2) the points C, D lie singly on the segments into which the pair
A, B separates the line (see Fig. 1-6(b)).   In the latter case, we say that the pair of points
A, B is separated by the pair C, D.   Thus, if A, B, C, D are distinct points on a projective
line, the pair A, B is *separated* by the pair C, D provided the point C lies on one of the segments $\overset{\frown}{AB}$ or AB and the point D lies on the other segment.   Another way of putting this is
to say: If A, B, C, D are distinct points on a projective line, the pair A, B is separated by
the pair C, D provided it is not possible to move along the line from A to B without meeting
some *one* of the pair C, D.   It follows readily that if the pair C, D separates the pair A, B
then the pair A, B separates the pair C, D.

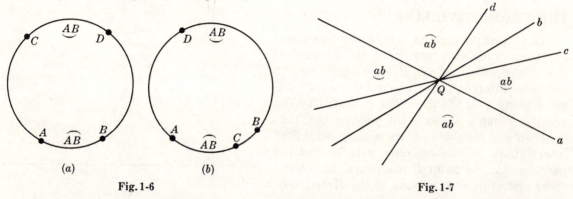

(a)   (b)

Fig. 1-6   **Fig. 1-7**

The plane dual of a projective line q with four distinct points A, B, C, D marked on it
is a point Q with four distinct coplanar lines a, b, c, d drawn through it.   Consider in Fig.
1-7 the pair a, b and denote by $\underset{\frown}{ab}$ and $\overset{\frown}{ab}$ the two sections into which they separate the plane.
Without further ado, we state: If a, b, c, d are distinct coplanar lines on a point Q, the pair
a, b will be said to be *separated* by the pair c, d provided the line c lies in one of the sections
and the line d in the other.

PERSPECTIVE PENCILS IN A PLANE

By a *figure* in projective space will be meant any collection of points, lines and planes
of the space; by a *figure* in a projective plane will be meant any collection of points and lines
of the plane.   In this section, attention will be restricted to two types of figures in a projective plane:

(a) the totality of points on one of the lines of the plane, called a *pencil of points*
(*range of points* or *point-row*).   The pencil of points on the line p of Fig. 1-8(a)
below will be denoted by $p(A, B, C, D, \ldots)$ where $A, B, C, D, \ldots$ are distinct points
on p.   The points $A, B, C, D, \ldots$ are called *elements* and the line p is called the *basis*
of the pencil.

and the plane dual

(b) the totality of lines on one of the points of the plane, called a *pencil of lines* (*flat-
pencil*).   The pencil of lines on the point P of Fig. 1-8(b) below will be denoted by
$P(a, b, c, d, \ldots)$ where $a, b, c, d, \ldots$ are distinct lines on P.   The lines $a, b, c, d, \ldots$
are called *elements* and the point P is called the *center* of the pencil.

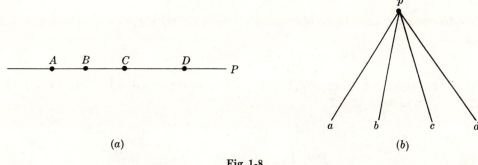

(a)   (b)

Fig. 1-8

A *one-to-one correspondence* is said to exist between the elements of two pencils provided there exists a rule which associates with each element of one pencil (the first) a unique element of the other (the second), and reciprocally, associates with each element of the second a unique element of the first. In such a correspondence between two pencils (also between two figures of any sort) each element and its associate are called *corresponding (homologous) elements*. The *identity correspondence* in which each element of a given figure is associated with itself is a somewhat trivial example.

Consider in Fig. 1-9(a) the pencil of lines $P(a, b, c, d, \ldots)$ sectioned by any line p not on P. A one-to-one correspondence between the pencil of lines and the resulting pencil of points (on p) is automatically established by the 'on' relation. For, to each line of the pencil of lines on P is thereby associated a unique point, namely, that point, of the pencil on p, which is on the line. Moreover, this association of line and point is reversible, that is, to each point of the pencil of points on p is associated a unique line, namely, that line, of the pencil of lines on P, which is on the point. Note that this correspondence is best indicated by using the same letter (small and capital) to denote any line and the point associated with it.

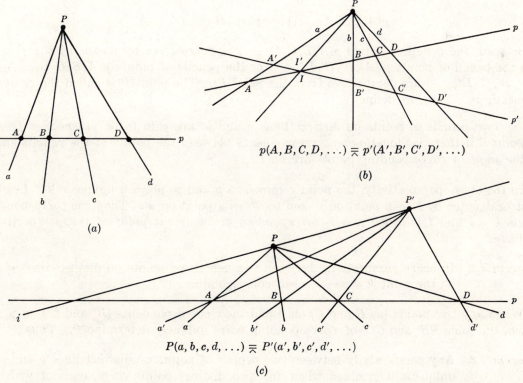

$p(A, B, C, D, \ldots) \;\overline{\overline{\barwedge}}\; p'(A', B', C', D', \ldots)$

(b)

(a)

$P(a, b, c, d, \ldots) \;\overline{\overline{\barwedge}}\; P'(a', b', c', d', \ldots)$

(c)

Fig. 1-9

The one-to-one correspondence described above is called a *perspectivity* and is indicated by writing

$$P(a,b,c,d,\ldots) \;\overline{\overline{\wedge}}\; p(A,B,C,D,\ldots)$$

We say that the pencil of lines $P(a,b,c,d,\ldots)$ is perspective with the pencil of points $p(A,B,C,D,\ldots)$ and note that

$$P(a,b,c,d,\ldots) \;\overline{\overline{\wedge}}\; p(A,B,C,D,\ldots)$$

also implies

$$p(A,B,C,D,\ldots) \;\overline{\overline{\wedge}}\; P(a,b,c,d,\ldots)$$

To distinguish this type of perspectivity from others yet to be introduced, we shall call it an *elementary perspectivity*.

Consider in Fig. 1-9(b) the pencil of lines $P(a,b,c,d,\ldots)$ sectioned by two distinct lines p and p', neither of which is on P. From the above discussion it follows that two elementary perspectivities

$$P(a,b,c,d,\ldots) \;\overline{\overline{\wedge}}\; p(A,B,C,D,\ldots)$$

and

$$P(a,b,c,d,\ldots) \;\overline{\overline{\wedge}}\; p'(A',B',C',D',\ldots)$$

are established. These perspectivities, being reversible, may be combined as follows:

$$p(A,B,C,D,\ldots) \;\overline{\overline{\wedge}}\; P(a,b,c,d,\ldots) \;\overline{\overline{\wedge}}\; p'(A',B',C',D',\ldots)$$

It is now clear that a one-to-one correspondence between the pencils of points on p and p' (A and A', B and B', C and C', ...) has been established in which each point of the pencil on p and its associate on p' determine a unique line of the pencil on P. We now replace the sequence of elementary perspectivities immediately above by

$$p(A,B,C,D,\ldots) \;\overset{P}{\overline{\overline{\wedge}}}\; p'(A',B',C',D',\ldots)$$

(to be read: the pencil of points $p(A,B,C,D,\ldots)$ is *perspective* by means of the point P with the pencil of points $p'(A',B',C',D',\ldots)$ or the pencils of points $p(A,B,C,D,\ldots)$ and $p'(A',B',C',D',\ldots)$ are perspective from the point P). The point P is called the *center of perspectivity*. Thus, we define

Two pencils of points on distinct lines p and p' are said to be *perspective from a point P* if there is a one-to-one correspondence between the points of the pencils and if the joins of corresponding points are on P.

In the above perspectivity, the point common to p and p' plays a unique role. Let this point be denoted by I, as a point on p, and by I', as a point on p'. Then the correspondent of I is $I' = I$ and I is called a *self-corresponding* or *invariant point* of the perspectivity. We have

Theorem 1.1. In every perspectivity between two pencils of points on distinct lines p and p', the point $I = p \cdot p'$ is self-corresponding.

When any two points (as B and C) on p and their correspondents (B' and C') on p' are known, the joins BB' and CC' of corresponding pairs uniquely determine P. Thus,

Theorem 1.2. Any perspectivity between two pencils of points on distinct lines p and p' is uniquely determined when any two distinct points on p, each of which is distinct from $p \cdot p'$, and their correspondents on p' are known.

The plane dual of a perspectivity between two pencils of points on distinct lines is a perspectivity between two pencils of lines on distinct points, defined as follows:

Two pencils of lines $P(a, b, c, d, \ldots)$ and $P'(a', b', c', d', \ldots)$ on distinct points are said to be *perspective from the line p* if there is a one-to-one correspondence between the two pencils and if the intersections of corresponding lines are on p.

This perspectivity, illustrated in Fig. 1-9(c), is indicated by writing

$$P(a, b, c, d, \ldots) \stackrel{p}{\overline{\wedge}} P'(a', b', c', d', \ldots)$$

Here, the line p is called the *axis of perspectivity*. For such perspectivities, we have by the principle of duality,

Theorem 1.1'. In every perspectivity between two pencils of lines on distinct points P and P', the line $i = PP'$ is self-corresponding.

and

Theorem 1.2'. Any perspectivity between two pencils of lines on distinct points P and P' is uniquely determined when any two distinct lines on P, each of which is distinct from PP', and their correspondents on P' are known.

The reader is urged to write out in full the duals of the paragraphs above concerning Fig. 1-9(b) rather than to accept, without further investigation, the theorems stated in this paragraph concerning Fig. 1-9(c).

PROJECTIVE PENCILS IN A PLANE

A perspectivity

$$p_1(A_1, B_1, C_1, D_1, \ldots) \stackrel{P_1}{\overline{\wedge}} p_2(A_2, B_2, C_2, D_2, \ldots)$$

was defined in the preceding section as establishing a one-to-one correspondence between two pencils of points on distinct lines such that the joins of corresponding points are on the point P_1. A perspectivity may also be thought of as a type of transformation, that is, a method of passing from the set of points on p_1 to the set on p_2 or a means of carrying the set of points on p_1 onto the set on p_2. Consider now a second perspectivity

$$p_2(A_2, B_2, C_2, D_2, \ldots) \stackrel{P_2}{\overline{\wedge}} p_3(A_3, B_3, C_3, D_3, \ldots)$$

by means of which we pass from the set of points on p_2 to the set of points on another line p_3. The effect of the two perspectivities carried out successively (see Fig. 1-10) is a transformation by which we pass from the set of points on p_1 to the set of points on p_3. Generally, the transformation will not be a perspectivity. To check this, it is necessary only to join corresponding points A_1 and A_3, B_1 and B_3, C_1 and C_3, D_1 and D_3, ... of Fig. 1-10 and verify that these lines are not concurrent or to verify that $I_1 = p_1 \cdot p_3$ is not invariant under the transformation. Such a

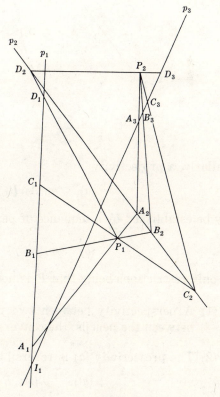

Fig. 1-10

one-to-one correspondence between the two pencils is called a *projective correspondence* or *projective transformation* and is denoted by

$$p_1(A_1, B_1, C_1, D_1, \ldots) \barwedge p_3(A_3, B_3, C_3, D_3, \ldots)$$

Thus,

A one-to-one correspondence between two pencils of points is said to be *projective* provided the correspondence is the resultant of a sequence of perspectivities. More briefly, such a correspondence is called a *projectivity*.

Fig. 1-11 illustrates a projectivity between two pencils of points on distinct lines

$$p_1(A_1, B_1, C_1, D_1, \ldots) \barwedge p_4(A_4, B_4, C_4, D_4, \ldots)$$

defined by the sequence of perspectivities

$$p_1(A_1, B_1, C_1, D_1, \ldots) \overset{P_1}{\barwedge} p_2(A_2, B_2, C_2, D_2, \ldots) \overset{P_2}{\barwedge} p_3(A_3, B_3, C_3, D_3, \ldots) \overset{P_3}{\barwedge} p_4(A_4, B_4, C_4, D_4, \ldots)$$

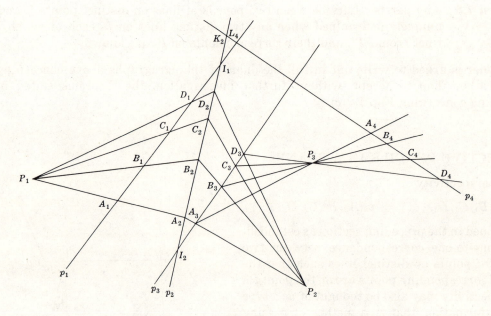

Fig. 1-11

Similarly, a projectivity

$$(a) \qquad p_1(A_1, B_1, C_1, D_1, \ldots) \barwedge p_n(A_n, B_n, C_n, D_n, \ldots)$$

may be established by a sequence of perspectivities

$$(b) \qquad p_1 \overset{P_1}{\barwedge} p_2 \overset{P_2}{\barwedge} p_3 \overset{P_3}{\barwedge} \cdots \overset{P_{n-1}}{\barwedge} p_n,$$

the only restrictions being that P_i is not on p_i and p_{i+1} is not on P_i. We note in passing that

(1) A perspectivity between two pencils of points is a special case of a projectivity between the pencils; the converse, however, is not true.

(2) The projectivity (a) is reversible, that is, (a) implies

$$(a') \qquad p_n(A_n, B_n, C_n, D_n, \ldots) \barwedge p_1(A_1, B_1, C_1, D_1, \ldots)$$

the sequence of perspectivities establishing (a') being (b) written in reverse order.

It is not necessary that all of the lines $p_1, p_2, p_3, \ldots, p_n$ above be distinct; in particular, if $p_n = p_1$ then we have

$$(c) \qquad p_1(A_1, B_1, C_1, D_1, \ldots) \; \overline{\wedge} \; p_1(A_n, B_n, C_n, D_n, \ldots)$$

For examples, the reader has only to extend the lines $P_2A_2, P_2B_2, P_2C_2, P_2D_2, \ldots$ of Fig. 1-11 to meet p_1 in $A_5, B_5, C_5, D_5, \ldots$ respectively and extend $P_3A_3, P_3B_3, P_3C_3, P_3D_3, \ldots$ of the same figure to meet p_1 in $A_6, B_6, C_6, D_6, \ldots$ respectively. Then

$$p_1(A_1, B_1, C_1, D_1, \ldots) \; \overset{P_1}{\overline{\wedge}} \; p_2(A_2, B_2, C_2, D_2, \ldots) \; \overset{P_2}{\overline{\wedge}} \; p_1(A_5, B_5, C_5, D_5, \ldots)$$

and $\qquad\qquad (d) \qquad p_1(A_1, B_1, C_1, D_1, \ldots) \; \overline{\wedge} \; p_1(A_5, B_5, C_5, D_5, \ldots)$

also

$$p_1(A_1, B_1, C_1, D_1, \ldots) \; \overset{P_1}{\overline{\wedge}} \; p_2(A_2, B_2, C_2, D_2, \ldots) \; \overset{P_2}{\overline{\wedge}} \; p_3(A_3, B_3, C_3, D_3, \ldots) \; \overset{P_3}{\overline{\wedge}} \; p_1(A_6, B_6, C_6, D_6, \ldots)$$

and $\qquad\qquad (e) \qquad p_1(A_1, B_1, C_1, D_1, \ldots) \; \overline{\wedge} \; p_1(A_6, B_6, C_6, D_6, \ldots)$

Now there is, of course, only one pencil of points on the line p_1 and the effect of each of the above projectivities is to carry this pencil of points *into itself*. Since each projectivity requires that each point of p_1 have two labels, matters will be considerably simplified if we think of each of these projectivities as between *distinct pencils on the same line*. Accordingly, we shall speak of $(c), (d), (e)$ as projectivities between *superposed* pencils of points on the line p_1.

Dually, there are projectivities of a pencil of lines into itself which, for convenience, we think of as projectivities between distinct pencils on the same point and speak of as projectivities between superposed pencils on a point.

For an example, see Problem 1.13.

In a projectivity between superposed pencils, it may happen that one or more elements are self-corresponding. (In a later chapter it is shown that if as many as three distinct elements are self-corresponding, then every element is self-corresponding.) Self-corresponding elements of superposed projectivities are more often called *double elements* (*invariant elements*) of the projectivity. Examples of superposed projectivities having two double elements are given in Problems 1.16 and 1.18.

THEOREMS ON PROJECTIVITIES

In the next two chapters answers will be found for the following questions:

(1) Given the projectivity (a) of the preceding section defined by the sequence of perspectivities (b), can this projectivity be defined by fewer than $n-1$ perspectivities?

(2) Given the projectivity (a), under what conditions will it be, in fact, a perspectivity?

(3) Given two pencils of points on distinct lines p_1 and p_n or superposed on the same line p, under what conditions can a projectivity be established between them?

It must be remembered that while these questions concern pencils of points, there is a dual of each concerning pencils of lines. Although it is possible to restrict our attention solely to pencils of points and accept without further investigation the dual of each theorem obtained, this procedure will not be followed here. In fact, we shall purposely shift back and forth from theorems concerning points to theorems concerning lines. Generally, the proof of a stated theorem and of its dual will not be given. Thus, it will be left for the reader to supply the dual of any theorem and its proof when such is not included in the text.

In Problem 1.4, we prove

Theorem 1.3. Given three distinct collinear points A_1, B_1, C_1 and another three distinct collinear points A_2, B_2, C_2 on distinct lines or on the same line, there exists at least one projectivity which carries A_1, B_1, C_1 into A_2, B_2, C_2 respectively.

In Problems 1.5 and 1.6, we prove one part and the dual of another part of

Theorem 1.4. Given four distinct points A, B, C, D on a line, there exist projectivities which carry

(a) A, B, C, D into B, A, D, C (b) A, B, C, D into D, C, B, A

(c) A, B, C, D into C, D, A, B

respectively.

Solved Problems

1.1.   In an augmented plane, let (r, R_∞) be any augmented line, let A, B, I be distinct ordinary points on this line, and let P be any ordinary point of the plane not on r. Project the points of the line from P and section by any augmented line (r', R'_∞) of the plane which passes through I, is distinct from (r, R_∞) and does not pass through P. Locate: (a) the correspondents of A, B and I; (b) the correspondent R' of R_∞; (c) the point R of (r, R_∞) whose correspondent is R'_∞.

(a)   The correspondent A' of A is the point of intersection of the projector (PA, S_∞) of A and (r', R'_∞); the correspondent B' of B is the point of intersection of (PB, T_∞) and (r', R'_∞). Since the projector (PI, U_∞) of I meets (r', R'_∞) in I, the correspondent of I is I itself.

(b)   The projector of R_∞ is the augmented line (s, R_∞), where s is the ordinary line through P and parallel to r. The correspondent R' of R_∞ is the point of intersection of r' and s.

(c) R is the point of intersection of r and the line s' passing through P and parallel to r'.

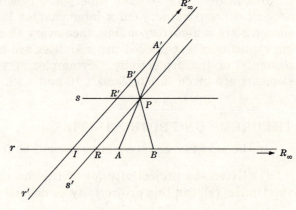

Fig. 1-12

1.2.   Assuming all points and lines are coplanar, which of the following propositions belong to plane projective geometry? For each which does, write its plane dual.

(a)   Four distinct points, no three collinear, have six joins.

(b)   The diagonals of a parallelogram bisect each other.

(c)   If A, B, C are distinct non-collinear points and if D and E are distinct points such that B, C, D are collinear and C, A, E are collinear, then there is a point F such that A, B, F are collinear and D, E, F are collinear.

(d)   The angle inscribed in a semi-circle is a right angle.

Propositions (*b*) and (*d*), concerned with lengths of line segments and the measure of an angle, do not belong to projective geometry.

The dual of (*a*) is: Four distinct lines, no three concurrent, have six points of intersection.

The dual of (*c*) is: If a, b, c are distinct non-concurrent lines and if d and e are distinct lines such that b, c, d are concurrent and c, a, e are concurrent, then there is a line f such that a, b, f are concurrent and d, e, f are concurrent.

1.3. Obtain two projectively related superposed pencils of lines.

Consider in Fig. 1-13 the pencil of lines $P_1(a_1, b_1, c_1, d_1, \ldots)$ sectioned by the line p_1, not on P_1, and the resulting pencil of points on p_1 projected from P_2, distinct from P_1 and not on p_1. In turn, the pencil of lines on P_2 are sectioned by the line p_2, distinct from p_1 and not on P_2, and the resulting pencil of points on p_2 are projected from P_1 by the lines $a_3, b_3, c_3, \ldots$ By construction,

$$P_1(a_1, b_1, c_1, d_1, \ldots) \overset{p_1}{\barwedge} P_2(a_2, b_2, c_2, d_2, \ldots) \overset{p_2}{\barwedge} P_1(a_3, b_3, c_3, d_3, \ldots)$$

and, hence,

$$P_1(a_1, b_1, c_1, d_1, \ldots) \barwedge P_1(a_3, b_3, c_3, d_3, \ldots)$$

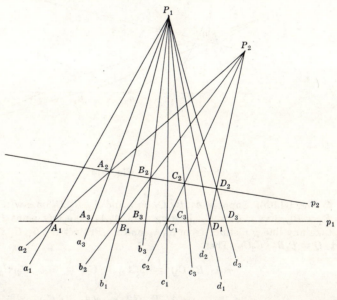

Fig. 1-13

1.4. Prove: Given three distinct collinear points A_1, B_1, C_1 and another three distinct collinear points A_2, B_2, C_2, there exists at least one projectivity which carries A_1, B_1, C_1 respectively into A_2, B_2, C_2.

In order to make perfectly clear the gist of the theorem and also to introduce a simplified notation which will hereafter be used in similar cases, suppose A_1, B_1, C_1 on a line r and A_2, B_2, C_2 on another line s. We are to prove the existence of a projectivity

$$(a) \qquad r(A_1, B_1, C_1, \ldots) \barwedge s(A_2, B_2, C_2, \ldots)$$

between the pencils of points on r and s. Note that we are not concerned with the correspondents on s of other points on r; hence it is possible that there are many different projectivities satisfying the conditions of the theorem. Since we are concerned only with given triples of points, the notation may be simplified accordingly by replacing (*a*) with

$$r(A_1, B_1, C_1) \barwedge s(A_2, B_2, C_2)$$

There are three cases to be considered:

(1) Refer to Fig. 1-14(a). Suppose A_1, B_1, C_1 on line r and A_2, B_2, C_2 on another line s with $A_1 = A_2 = r \cdot s$. Let $P = B_1 B_2 \cdot C_1 C_2$. In the perspectivity thus established, we have $r(A_1, B_1, C_1) \overset{P}{\barwedge} s(A_2, B_2, C_2)$ and, hence, $r(A_1, B_1, C_1) \barwedge s(A_2, B_2, C_2)$. The same conclusion is reached, moreover, when $r \cdot s = B_1 = B_2$ and when $r \cdot s = C_1 = C_2$.

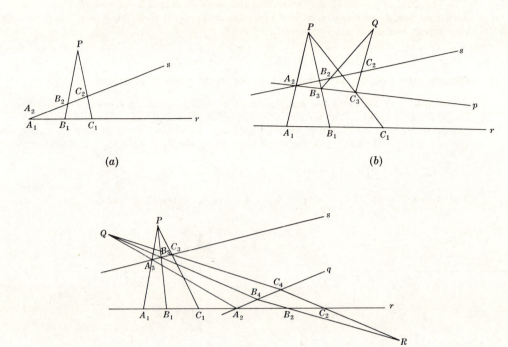

Fig. 1-14

(2) Refer to Fig. 1-14(b). Suppose A_1, B_1, C_1 on r and A_2, B_2, C_2 on s with no one of $r \cdot s = A_1 = A_2$, $r \cdot s = B_1 = B_2$, $r \cdot s = C_1 = C_2$ holding. On $A_1 A_2$ take any point P distinct from A_1 and A_2; on A_2 take any line p distinct from $A_1 A_2$ and s. Project A_1, B_1, C_1 from P into A_2, B_3, C_3 on p and let $Q = B_2 B_3 \cdot C_2 C_3$. Then

$$r(A_1, B_1, C_1) \overset{P}{\barwedge} p(A_2, B_3, C_3) \overset{Q}{\barwedge} s(A_2, B_2, C_2)$$

and $$r(A_1, B_1, C_1) \barwedge s(A_2, B_2, C_2)$$

as required.

(3) Refer to Fig. 1-14(c). Suppose A_1, B_1, C_1 and A_2, B_2, C_2 are collinear on line r. From any point P, not on r, project A_1, B_1, C_1 into A_3, B_3, C_3 on any line s, not on P, A_1, B_1 or C_1. Now, repeating the procedure in (2), we can by the use of at most two perspectivities carry A_3, B_3, C_3 into A_2, B_2, C_2 respectively. Then

$$r(A_1, B_1, C_1) \barwedge s(A_2, B_2, C_2)$$

as required.

1.5. Given four distinct points A, B, C, D on a line p, exhibit a projectivity such that

$$p(A, B, C, D) \barwedge p(C, D, A, B)$$

This is part (c) of Theorem 1-4 stated in the simpler notation of Problem 1.4. Refer to Fig. 1-15 below. Project the given points from any point P not on p, section by any line q on B distinct from p and PB, and obtain the points A', B, C', D' respectively. Note that $B' = B$. Join A' and C meeting $PD = r$ in the point D''. Then

$$p(A, B, C, D) \overset{P}{\barwedge} q(A', B, C', D') \overset{C}{\barwedge} r(D'', D, P, D') \overset{A'}{\barwedge} p(C, D, A, B)$$

and $$p(A, B, C, D) \ \barwedge \ p(C, D, A, B)$$

as required.

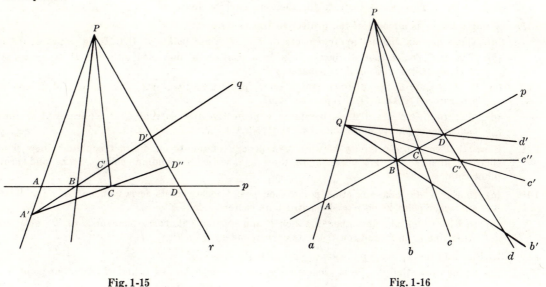

Fig. 1-15   Fig. 1-16

1.6. Given four distinct lines a, b, c, d on a point P, obtain a projectivity such that

$$P(a, b, c, d) \ \barwedge \ P(b, a, d, c)$$

Consider in Fig. 1-16 the lines a, b, c, d on P sectioned by any line p, not on P, and the resulting points A, B, C, D projected from Q, any point on the line a distinct from P and not on p.

Let $QB = b'$, $QC = c'$, $QD = d'$; $c' \cdot d = C'$; $BC' = c''$. Then

$$P(a, b, c, d) \overset{p}{\barwedge} Q(a, b', c', d') \overset{d}{\barwedge} B(b, b', c'', p) \overset{c'}{\barwedge} P(b, a, d, c)$$

and $$P(a, b, c, d) \ \barwedge \ P(b, a, d, c)$$

Supplementary Problems

1.7. Project a given circle into (a) another circle, (b) an ellipse, (c) a hyperbola, (d) a parabola.

1.8. Given Fig. 1-17(a) and (b) in a projective plane. Describe each, describe its plane dual, and illustrate by a figure. *Hint.* Fig. 1-17(a) consists of two distinct lines with two distinct points marked on one of them and three distinct points marked on the other such that none of the five points is the point of intersection of the lines. The plane dual consists of two distinct points with two distinct lines through one of them and three distinct lines through the other such that none of the five lines is the join of the points.

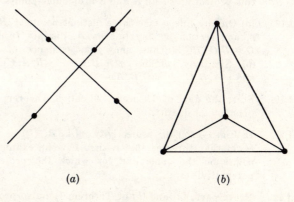

(a)   (b)

Fig. 1-17

1.9.   Draw the figures for Problem 1.2(a), 1.2(c) and for their duals.

1.10.   Write the dual of the theorem of plane projective geometry: If R_1, R_2, R_3 are distinct points on a line r, if S_1, S_2, S_3 are distinct points on another line s, and if the lines R_1S_1, R_2S_2, R_3S_3 are on a point P, then the points $R_1S_2 \cdot R_2S_1$, $R_2S_3 \cdot R_3S_2$, $R_3S_1 \cdot R_1S_3$ are on a line p which is concurrent with r and s. Draw figures to illustrate the theorem and its dual.

1.11.   Using a circle as a model of the projective line, verify:

(a)   There are six distinct arrangements of four distinct points A, B, C, D on a projective line.

(b)   If A, B, C, D are four distinct points on a projective line, then either C, D separates A, B or B, D separates A, C or A, D separates B, C.

(c)   If A, B, C, D, E are distinct points on a projective line, if C, D separates A, B and if B, E separates A, C then D, E separates A, B.

(d)   If A, B, C, D, E are distinct points on a projective line so that C, D separates A, B then D, E can never separate more than two of A, B; A, C; B, C.

1.12.   Show that in a perspectivity between two pencils of points on distinct coplanar lines p and p', there is always one but never more than one self-corresponding point. State and prove the plane dual.

1.13.   (a)   In Fig. 1-9(b) take any other point X on p and construct its correspondent X' on p'; also, take any other point Y' on p' and construct its correspondent Y on p.

(b)   In Fig. 1-9(c) take any other line x on P and construct its correspondent x' on P'; also, take any other line y' on P' and construct its correspondent y on P.

1.14.   In Fig. 1-11 let $I_1 = p_1 \cdot p_2$, $J_2 = p_2 \cdot p_3$ and $K_2 = p_2 \cdot p_4$.

(a)   In the projectivity $p_1 \overline{\wedge} p_4$, locate on p_4 the correspondent of I_1 and of any other point X_1 on p_1.

(b)   In the perspectivity $p_1 \overset{P}{\overline{\wedge}} p_2$, locate on p_1 the correspondent J_1 of J_2; then, in the projectivity $p_1 \overline{\wedge} p_4$, locate on p_4 the correspondent of J_1.

(c)   In the projectivity $p_2 \overline{\wedge} p_4$, locate the correspondent K_4 on p_4 of K_2 on p_2; also, locate on p_2 the point L_2 whose correspondent on p_4 is $L_4 = K_2$.

1.15.   In Fig. 1-13 express the projectivity $P_1 \overline{\wedge} P_2$ as a sequence of elementary perspectivities $P_1(a_1, b_1, c_1, d_1, \ldots) \overline{\overline{\wedge}} p_1(A_1, B_1, C_1, D_1, \ldots) \overline{\overline{\wedge}} \cdots$.

1.16.   In Fig. 1-13 let $P = p_1 \cdot p_2$ and $s_1 = P_1 P_2$. In the projectivity

$$P_1(a_1, b_1, c_1, d_1, \ldots) \;\overline{\wedge}\; P_1(a_3, b_3, c_3, d_3, \ldots)$$

obtain the correspondent of s_1 as s_1 itself. Thus s_1 is a double line of the projectivity. Show that $t_1 = PP_1$ is also a double line of this projectivity. When will $t_1 = s_1$ and when will $t_1 \neq s_1$?

1.17.   In Fig. 1-16 take any other line e on P and construct its correspondent f. Does it appear that the correspondent of f will be e?

1.18.   Consider in Fig. 1-16 the perspectivity

$$P(b, c, d, \ldots) \;\overset{p}{\overline{\wedge}}\; Q(b', c', d', \ldots)$$

On p take any point S distinct from the four already marked and on S take any line s distinct from p and on neither P nor Q. Let $a \cdot s = R$, $b \cdot s = B'$, $c \cdot s = C'$, $d \cdot s = D'$; $b' \cdot s = B''$, $c' \cdot s = C''$, $d' \cdot s = D''$. (a) Show that $s(B', C', D', \ldots)$ and $s(B'', C'', D'', \ldots)$ are projective. (b) Show that in this projectivity both R and S are double points.

1.19.   (a) On any line p take three distinct points A, B, C. Project these points from P, any point not on p, and section by any line on A, distinct from p and AP, to obtain the points A', B', C'. Let $Q = B'C \cdot BC'$. Thus, show the existence of a projectivity such that $p(A, B, C) \overline{\wedge} p(A, C, B)$. ($b$) Similarly, obtain $p(A, B, C) \overline{\wedge} p(B, A, C)$, $p(A, B, C) \overline{\wedge} p(B, C, A)$, $p(A, B, C) \overline{\wedge} p(C, A, B)$ and $p(A, B, C) \overline{\wedge} p(C, B, A)$.

1.20.   State the dual of Theorem 1.3; call it Theorem 1.3'. Prove this theorem without appealing to the principle of duality.

1.21.   In Fig. 1-14(c) the projectivity $p(A_1, B_1, C_1) \overline{\wedge} p(A_2, B_2, C_2)$ can be expressed as the product of no fewer than three perspectivities. Give an example of two triples A_1, B_1, C_1 and A_2, B_2, C_2 of distinct points on the same line for which the projectivity can be expressed as the product of two perspectivities.

1.22.   Prove parts (a) and (b) of Theorem 1.4; also prove parts (b) and (c) of the dual of this theorem.

Chapter 2

Cross Ratio

RATIO OF DIVISION

It will be recalled that the properties of figures studied in plane Euclidean geometry are precisely those which are invariant under the set of transformations (translations and rotations) called rigid motions. In plane projective geometry the transformations consist of projections and sections. It is natural to ask: Are there properties, other than that of being a point or a line, which are invariant under projection and section? To answer this, we return again to the Euclidean plane.

Consider in this plane any line p (see Fig. 2-1) and the segment AB determined by any two of its distinct points A and B. Let there be established on p a sense of positive direction indicated by the arrow tip. (We could, of course, have selected the opposite direction but it will not be difficult to show presently that *choosing* a direction rather than the direction chosen is significant.) Having chosen the direction, the line p will be called a *directed* or *oriented* line and the segment AB, having both *magnitude* (length) and *direction* (*always from A to B*) will be called a *directed line segment*. The segment BA has the same magnitude as AB but opposite direction; hence, $BA = -AB$, and

$$AB + BA = 0 \tag{1}$$

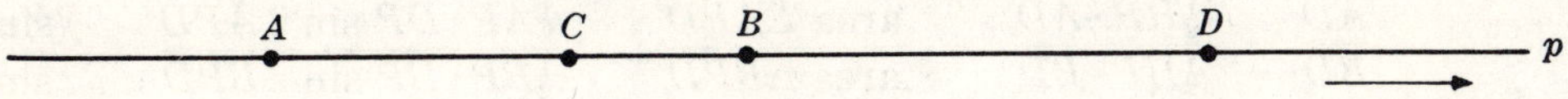

Fig. 2-1

Let C, distinct from A and B, be any other point on p. Since $AB = AC + CB$, (1) may be written in the form

$$AC + CB + BA = 0 \tag{2}$$

It will be left for the reader to verify (see Problem 2.7) that both (1) and (2) are independent of the particular orientation of p and that (2) is also independent of the position of C relative to A and B.

Finally, let D, distinct from A, B, C, be any other point on p. In Problem 2.1, we verify the relation

$$AB \cdot CD + AC \cdot DB + AD \cdot BC = 0 \tag{3}$$

When A, B, C are three distinct points on a directed line p, the *ratio of division* of the segment AB by C, denoted by (AB, C), is defined as

$$(AB, C) = AC/BC \tag{4}$$

When C is on the segment AB, as in Fig. 2-1, AC and BC have opposite signs and so $(AB, C) < 0$; when C is not on the segment AB, $(AB, C) > 0$. Again, it will be left for the reader to show that (4) is independent of the orientation of p.

19

CROSS RATIO IN THE EUCLIDEAN PLANE

Let A, B, C, D be any four distinct points (see Fig. 2-1) on a directed line p. The *cross ratio (double ratio)* of A, B with respect to C, D, denoted by $(A, B; C, D)$, is defined to be the ratio of the division ratios (AB, C) and (AB, D), that is,

$$(A, B; C, D) \;=\; \frac{AC}{BC}\Big/\frac{AD}{BD} \;=\; \frac{AC}{BC}\cdot\frac{BD}{AD} \;=\; \frac{AC}{AD}\cdot\frac{BD}{BC} \tag{5}$$

Since each of the division ratios is independent of the orientation of p, so also is (5).

When both of the points C and D are on the segment AB, both (AB, C) and (AB, D) are negative; hence $(A, B; C, D)$ is positive. When the point C is on the segment AB while D is not, then (AB, C) is negative, (AB, D) is positive, and $(A, B; C, D)$ is negative. We leave for the reader to verify that $(A, B; C, D)$ is positive when neither C nor D is on the segment AB and that $(A, B; C, D)$ is negative when D is on AB and C is not.

THE INVARIANCE OF CROSS RATIO

Consider in Fig. 2-2 the four distinct points A, B, C, D on the directed line r. From any point P, not on r, project these points onto any other directed line s, not on P, to obtain the points A', B', C', D'. The basic theorem concerning cross ratio in the Euclidean plane is

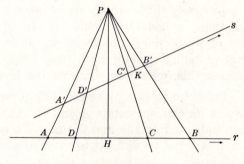

Theorem 2.1. The cross ratio of four distinct points on a line is invariant under projection and section.

Fig. 2-2

To prove this, drop from P the perpendicular to r meeting it at H. Then ignoring all signs, we have

$$(A, B; C, D) \;=\; \frac{\dfrac{AC}{BC}}{\dfrac{AD}{BD}} \;=\; \frac{\dfrac{\frac12 HP\cdot AC}{\frac12 HP\cdot BC}}{\dfrac{\frac12 HP\cdot AD}{\frac12 HP\cdot BD}} \;=\; \frac{\dfrac{\text{area }\triangle APC}{\text{area }\triangle BPC}}{\dfrac{\text{area }\triangle APD}{\text{area }\triangle BPD}} \;=\; \frac{\dfrac{\frac12 AP\cdot CP \sin\angle APC}{\frac12 BP\cdot CP \sin\angle BPC}}{\dfrac{\frac12 AP\cdot DP \sin\angle APD}{\frac12 BP\cdot DP \sin\angle BPD}} \;=\; \frac{\dfrac{\sin\angle APC}{\sin\angle BPC}}{\dfrac{\sin\angle APD}{\sin\angle BPD}}$$

What we have actually proved is

$$(A, B; C, D) \;=\; \frac{AC}{BC}\cdot\frac{BD}{AD} \;=\; \pm\frac{\sin\angle APC}{\sin\angle BPC}\cdot\frac{\sin\angle BPD}{\sin\angle APD} \tag{6}$$

Now drop the perpendicular from P to the line s meeting it in K and repeat the above argument to obtain

$$(A', B'; C', D') \;=\; \frac{A'C'}{B'C'}\cdot\frac{B'D'}{A'D'} \;=\; \pm\frac{\sin\angle APC}{\sin\angle BPC}\cdot\frac{\sin\angle BPD}{\sin\angle APD} \tag{7}$$

To complete the proof when the points are labeled as in Fig. 2-2, it is necessary to show that the sign in the right members of (6) and (7) is the same.

Let an orientation about P be fixed so that the angle APC, generated by revolving the line AP counterclockwise about P into coincidence with CP, is positive. Then angle APD is positive while angles BPC and BPD, being generated by clockwise rotation of BP about P, are negative. As a consequence, we find that $\dfrac{AC}{BC}$ and $\dfrac{\sin\angle APC}{\sin\angle BPC}$, also $\dfrac{BD}{AD}$ and $\dfrac{\sin\angle BPD}{\sin\angle APD}$, have the same sign. Moreover, this parity of sign is independent of both the orientation on p and the orientation about P. Thus the sign is $+$ in both right members of (6) and (7) and so

$$(A, B; C, D) = (A', B'; C', D')$$

as required. To show that Theorem 2.1 is independent of the order of labels, see Problem 2.11.

Dually, let us begin with any four distinct lines a, b, c, d on a point P and let these lines be cut by any lines r and s, not on P, in the sets of points A, B, C, D and A', B', C', D' respectively. By Theorem 2.1,

$$(A, B; C, D) = (A', B'; C', D')$$

This suggests the definition (see Fig. 2-3):

The cross ratio in any order of four lines a, b, c, d on a point P is the cross ratio in the same order of the four points A, B, C, D in which these lines are cut by any line not on P, i.e.,

$$(a, b; c, d) = (A, B; C, D), \quad (c, a; d, b) = (C, A; D, B), \quad \text{etc.}$$

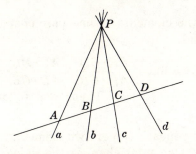

Fig. 2-3

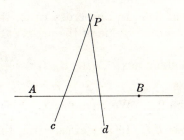

Fig. 2-4

Finally (see Fig. 2-4), let A, B be any distinct points and let c, d be any distinct lines on any point $P \neq A, B$, neither line being on A or B. Let $AB \cdot c = C$ and $AB \cdot d = D$. We define the cross ratio of any two of the given elements A, B, c, d with respect to the other two to be the corresponding cross ratio of the four points A, B, C, D; for example,

$$(A, B; c, d) = (A, B; C, D), \quad (c, A; B, d) = (C, A; B, D), \quad \text{etc.}$$

Similarly, letting $PA = a$ and $PB = b$, we define

$$(A, B; c, d) = (a, b; c, d), \quad (c, B; d, A) = (c, b; d, a), \quad \text{etc.}$$

THE TWENTY-FOUR CROSS RATIOS

Recall that we are working in the Euclidean plane. Since, in Fig. 2-1, $(A, B; C, D) < 0$ while $(A, C; B, D) > 0$, it is evident that the value of the cross ratio of four points (four lines or two points and two lines) depends upon the order in which these elements are set down in $(-, -; -, -)$. Corresponding to the twenty-four distinct permutations of four distinct symbols, there is the same number of cross ratios. The number of distinct cross ratios, however, is not twenty-four. For, taking due regard to signs, we find that

$$\frac{AC}{AD} \cdot \frac{BD}{BC} = \frac{BD}{BC} \cdot \frac{AC}{AD} = \frac{DB}{DA} \cdot \frac{CA}{CB} = \frac{CA}{CB} \cdot \frac{DB}{DA}$$

and, hence,

$$(A, B; C, D) = (B, A; D, C) = (D, C; B, A) = (C, D; A, B)$$

Thus,

Theorem 2.2. The value of a cross ratio remains unchanged when any two of the four elements are interchanged simultaneously with the other two elements.

As a result, the twenty-four cross ratios fall into six sets of four each. Setting $(A,B;C,D) = \lambda$, these sets together with their values are

$$(a) \quad (A,B;C,D) = (B,A;D,C) = (D,C;B,A) = (C,D;A,B) = \lambda$$

$$(b) \quad (A,B;D,C) = (B,A;C,D) = (C,D;B,A) = (D,C;A,B) = 1/\lambda$$

$$(c) \quad (A,C;B,D) = (C,A;D,B) = (D,B;C,A) = (B,D;A,C) = 1 - \lambda$$

$$(d) \quad (A,C;D,B) = (C,A;B,D) = (B,D;C,A) = (D,B;A,C) = 1/(1-\lambda) \qquad (8)$$

$$(e) \quad (A,D;C,B) = (D,A;B,C) = (B,C;D,A) = (C,B;A,D) = \lambda/(\lambda-1)$$

$$(f) \quad (A,D;B,C) = (D,A;C,B) = (C,B;D,A) = (B,C;A,D) = (\lambda-1)/\lambda$$

To complete the table, consider in Fig. 2-1 the four distinct points A, B, C, D on the line p and set $(A,B;C,D) = \lambda$. Then $(A,B;D,C) = \dfrac{AD}{BD} \cdot \dfrac{BC}{AC} = 1 \left/ \dfrac{AC}{AD} \cdot \dfrac{BD}{BC} \right. = 1/\lambda$ and $(8b)$ follows by Theorem 2.2. We have proved

Theorem 2.3. In any cross ratio, the interchange of the elements of one pair changes the value of the cross ratio into its reciprocal.

Consider next $(A,C;B,D) = \dfrac{AB}{CB} \cdot \dfrac{CD}{AD}$. From (3), we have $\dfrac{AB \cdot CD}{AD \cdot CB} + \dfrac{AC \cdot DB}{AD \cdot CB} = 1$ so that $(A,C;B,D) + (A,B;C,D) = 1$ and $(A,C;B,D) = 1 - \lambda$. Then $(8c)$ follows by Theorem 2.2. We have proved

Theorem 2.4. In any cross ratio, the interchange of the means (inner two elements) or the interchange of the extremes (outer two elements) changes the value of the cross ratio into its arithmetic complement.

The remainder of the table follows readily and will be left for the reader to complete.

We have now

Theorem 2.5. If the value of any one of the cross ratios of four points is known, the values of all the cross ratios are determined.

Example 2.1.

Suppose $(A,B;C,D) = 3$; then $(A,B;D,C) = 1/3$, $(A,C;B,D) = -2$, $(A,C;D,B) = -\frac{1}{2}$, $(A,D;C,B) = 3/2$, and $(A,D;B,C) = 2/3$. Now, by Theorem 2.2, the values of all twenty-four cross ratios are known.

Suppose some two of the points A, B, C, D are coincident. There are three cases:

(i) When A and B coincide, then $(A,A;C,D) = \dfrac{AC}{AD} \cdot \dfrac{AD}{AC} = 1$ and similarly when C and D coincide.

(ii) When A and C coincide, then $(A,B;A,D) = \dfrac{AA}{AD} \cdot \dfrac{BD}{BA} = 0$ and similarly when B and D coincide.

(iii) When A and D coincide, then $(A,B;C,A) = \dfrac{AC}{AA} \cdot \dfrac{BA}{BC}$ and the ratio is not defined.

We shall agree to write $(A,B;C,A) = (A,B;B,D) = \infty$ to indicate that as D approaches A, also, as C approaches B, the numerical value of $(A,B;C,D)$ increases without limit. It is not difficult to show that, conversely, if a cross ratio has one of the values $0, 1, \infty$ then some two of the points coincide.

Suppose next that three of the points, say A, B, C, are fixed while D traces out the line p. When the points have the position indicated in Fig. 2-1, then $(A,B;C,D) < 0$. As D moves to the right, the value of AC/BC remains fixed while the value of BD/AD remains > 0

and tends to 1; thus, $(A, B; C, D)$ tends to the value of AC/BC. As D moves to the left, BD/AD remains > 0 but decreases. As noted above, when D coincides with B then $(A, B; C, D) = 0$. As D continues from B to C, $(A, B; C, D)$ increases in value from 0 to 1, the latter value being attained when D coincides with C. As D continues from C to A, the value of $(A, B; C, D)$ increases from 1 to ∞, the latter denoting its value when D coincides with A. As D moves to the left from A, the value of $(A, B; C, D)$ is negative and increases to that of AC/BC.

Finally, suppose some two of the six values of the cross ratios are equal. There are two cases:

Case 1. $\lambda = 1/\lambda$. Then $\lambda^2 = 1$ and $\lambda = \pm 1$. It was noted above that when $\lambda = 1$, two of the four points coincide and the six values become only three, namely, $0, 1, \infty$. When $\lambda = -1$, the six values again reduce to three, namely, $-1, \frac{1}{2}$ and 2.

Case 2. $\lambda = 1/(1 - \lambda)$. Then λ is one of the roots of $\lambda^2 - \lambda + 1 = 0$ and the values of the cross ratios are imaginary numbers. This case does not occur here.

CROSS RATIO A PROJECTIVE PROPERTY

Consider now an augmented plane (π, p_∞). Let A, B, C, D be any four distinct ordinary points (see Fig. 2-5) on any augmented line (r, R_∞) and let P be any ordinary point not on r. Let (r', R'_∞) be another augmented line, not on P, so chosen that the correspondents A', B', C', D' respectively of the four points A, B, C, D when projected from P onto (r', R'_∞) are ordinary points. In this projection, denote by R' the correspondent of R_∞ and by R the point whose correspondent is R'_∞. By Theorem 2.1 we have

$$(A, B; C, D) = (A', B'; C', D') \tag{9}$$

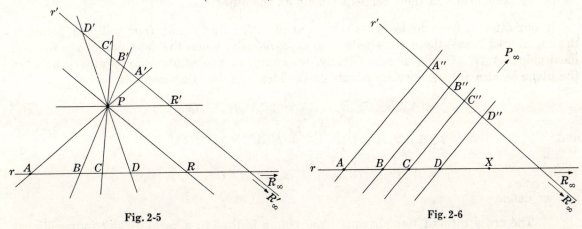

Fig. 2-5   Fig. 2-6

Next, suppose (see Fig. 2-6) that the lines (r, R_∞) and (r', R'_∞) and the points A, B, C, D are as before but that the center of projection is an ideal point P_∞, distinct from both R_∞ and R'_∞. The projectors of A, B, C, D on (r, R_∞) are parallel lines which meet (r', R'_∞) in distinct ordinary points, say A'', B'', C'', D''. It is easy to show that

$$(A, B; C, D) = (A'', B''; C'', D'') \tag{9'}$$

We now return to Fig. 2-5 and consider the case of three ordinary points and the ideal point on (r, R_∞), say A, B, C, R_∞. Their correspondents on (r', R'_∞) are the ordinary points A', B', C', R'. Now $(R', A'; B', C')$ is defined but $(R_\infty, A; B, C)$ is not. Since we are at liberty to define $(R_\infty, A; B, C)$ as best suits our purpose, we take

$$(R_\infty, A; B, C) = (R', A'; B', C') \tag{9''}$$

that is, we define

The cross ratio of three ordinary points and the ideal point on any augmented line is the cross ratio (in the same order) of their correspondents resulting from any projection onto another augmented line provided only that these correspondents be ordinary points.

When, however, the center of projection is the ideal point P_∞ of Fig. 2-6, the correspondents A'', B'', C'', R_∞' consist of three ordinary points and the ideal point of (r', R_∞'). In this case, neither $(R_\infty, A; B, C)$ nor $(R_\infty', A''; B'', C'')$ is defined. Consider $(X, A; B, C) = \dfrac{XB}{XC} \cdot \dfrac{AC}{AB}$ where X is such that B, X separates A, C. As $X \to R_\infty$, $\dfrac{XB}{XC} \to 1$. This suggests (see Problem 2.13) the definition $(R_\infty, A; B, C) = AC/AB$. Then $(R_\infty', A''; B'', C'') = A''C''/A''B''$ and, since $AC/AB = A''C''/A''B''$ (prove this), we have

$$(R_\infty, A; B, C) = (R_\infty', A''; B'', C'') \tag{9'''}$$

What we have done here is to extend the concept of cross ratio so that (9) holds when, of the four points, three are ordinary and the other is the ideal point on (r, R_∞). We are now in a position to state:

For any perspectivity between the points of two distinct augmented lines of an augmented plane, the cross ratio of any four points on one of the lines is equal to the cross ratio, in the same order, of their correspondents on the other line.

Also, since a projectivity is a sequence of perspectivities, we have

For any projectivity between the points of two distinct augmented lines of an augmented plane, the cross ratio of any four points on one line is equal to the cross ratio, in the same order, of their correspondents on the other line.

In our attempt to so define cross ratio that (9) will hold for any four collinear points in the augmented plane, there remains a final case, namely, when the four points are distinct ideal points $A_\infty, B_\infty, C_\infty, D_\infty$ of p_∞. Clearly, the center of projection is an ordinary point of the plane as also are the correspondents of the ideal points. Suppose

$$p_\infty(A_\infty, B_\infty, C_\infty, D_\infty, \ldots) \overset{P_1}{\barwedge} p_1(A', B', C', D', \ldots)$$

and

$$p_\infty(A_\infty, B_\infty, C_\infty, D_\infty, \ldots) \overset{P_2}{\barwedge} p_2(A'', B'', C'', D'', \ldots)$$

Then

$$p_1(A', B', C', D', \ldots) \barwedge p_2(A'', B'', C'', D'', \ldots)$$

and we define

The cross ratio of four distinct ideal points is the cross ratio of any four collinear ordinary points into which they may be projected.

In extending cross ratio to the augmented plane, the additional definitions needed have been made so that the following theorem be valid:

Theorem 2.6. If in the augmented plane a projectivity is established between two pencils of points on the same or on distinct augmented lines (two pencils of lines on the same or on distinct points), the cross ratio of any four elements of the first pencil is equal to the cross ratio of their correspondents, taken in the same order, in the second pencil.

This suggests the question: If A, B, C, D are any four distinct points on a line, if A', B', C', D' are four distinct points on the same or on another line, and if $(A, B; C, D) = (A', B'; C', D')$, does it necessarily follow that a projectivity of the form

$$(A, B, C, D, \ldots) \;\overline{\wedge}\; (A', B', C', D', \ldots)$$

exist or, in the simplified notation of Chapter 1, does it necessarily follow that $(A, B, C, D) \;\overline{\wedge}\;$ (A', B', C', D')? That the answer is *yes* is shown in Problem 2.2.

We are now in a position to state for the *projective plane*

Theorem 2.7. If A, B, C, D are four distinct collinear points and if A', B', C', D' are another four distinct collinear points, on the same or on distinct lines, then

$$(A, B, C, D) \;\overline{\wedge}\; (A', B', C', D') \quad \text{implies} \quad (A, B; C, D) = (A', B'; C', D')$$

and, conversely,

$$(A, B; C, D) = (A', B'; C', D') \quad \text{implies} \quad (A, B, C, D) \;\overline{\wedge}\; (A', B', C', D')$$

By Theorem 1.3, there is at least one projectivity which carries any three given collinear points A, B, C respectively into any other three given collinear points A', B', C'. Let X be any other point collinear with the first triple and denote by X' its correspondent in this projectivity. Then, by Theorem 2.7,

$$(A, B; C, X) = (A', B'; C', X')$$

and (note the query in Problem 2.3) we have

Theorem 2.8. (*The Fundamental Theorem*). Given in a projective plane three distinct collinear points (three distinct concurrent lines) and another three distinct collinear points (another three distinct concurrent lines) on the same or on distinct lines (on the same or on distinct points), there is one and only one projectivity which carries the first triple A, B, C (a, b, c) respectively into the second triple A', B', C' (a', b', c').

Note. In saying that two triples of distinct points on distinct lines p_1 and p_2 determine one and only one projectivity, we are not concerned with the various sequences of perspectivities from which the projectivity is inferred. The gist of the theorem is: If X is any other point on p_1 then, irrespective of the sequence of perspectivities which eventually carry A, B, C respectively into A', B', C', the correspondent of X is always the same point X' on p_2. As a consequence, we have

Theorem 2.9. Any projectivity $p_1 \;\overline{\wedge}\; p_2$ $(P_1 \;\overline{\wedge}\; P_2)$, with $p_1 \neq p_2$ $(P_1 \neq P_2)$, for which $P = p_1 \cdot p_2$ $(p = P_1 P_2)$ is self-corresponding, is a perspectivity.

For a proof, see Problem 2.4.

An important use of Theorem 2.9 is in proving collinearity of triples of points and concurrency of triples of lines. For an example, see Problem 2.5.

Other consequences of Theorem 2.7 are:

Theorem 2.10. If A, B, C, D are any four distinct points on a line, then $(A, B, C, D) \;\overline{\wedge}\;$ $(B, A, D, C) \;\overline{\wedge}\; (D, C, B, A) \;\overline{\wedge}\; (C, D, A, B)$.

and

Theorem 2.11. Separation is a projective property of four distinct collinear points (four distinct concurrent lines).

A valid objection to cross ratios as defined here and to theorems resulting from their use can be raised in view of the fact that these ratios were defined in terms of lengths of line segments. It will be our policy to give, whenever feasible, a proof independent of cross ratios of the more important theorems. Theorem 2.10 will be used frequently in later chapters and we point out that it was proved originally in Problems 1.5, 1.6, 1.22 of Chapter 1.

Solved Problems

2.1. For A, B, C, D distinct collinear points, verify

$$AB \cdot CD + AC \cdot DB + AD \cdot BC = 0$$

The verification is made by repeated use of (2), page 19. For the triple A, B, C we have $AC + CB + BA = 0$; whence, $AC = AB + BC$. Similarly, for the triple A, B, D we have $AB = AD + DB$ and for the triple B, C, D we have $CB = CD + DB$. Then

$$AB \cdot CD + AC \cdot DB + AD \cdot BC = AB(CD + DB) + BC(DB + AD) = AB \cdot CB + BC \cdot AB = 0$$

2.2. Given, in the augmented plane, any four distinct collinear points A, B, C, D and any other four distinct collinear points A', B', C', D' on the same line or on distinct lines. Show that

$$(A, B; C, D) = (A', B'; C', D') \quad \text{implies} \quad (A, B, C, D) \barwedge (A', B', C', D')$$

By Theorem 1.3 there always exists at least one projectivity which carries A, B, C respectively into A', B', C'. Suppose this projectivity carries D into D''. By Theorem 2.6,

$$(A, B; C, D) = (A', B'; C', D'')$$

But

$$(A, B; C, D) = (A', B'; C', D')$$

hence

$$(A', B'; C', D') = (A', B'; C', D'')$$

Then

$$(B'A', D') = (B'A', D'')$$

and we conclude that $D' = D''$.

2.3. Given, in an augmented plane, three distinct ordinary points A, B, C on an augmented line and a (real) number λ. Construct the point X on the line such that $(A, B; C, X) = \lambda$.

From $(A, B; C, X) = \dfrac{AC}{BC} \cdot \dfrac{BX}{AX} = \lambda$ it is evident that $X = B$ when $\lambda = 0$; $X = C$ when $\lambda = 1$; and $X = P_\infty$, the ideal point collinear with A, B, C, when $\lambda = AC/BC$. Excluding these values of λ, there are two cases:

(i) $\lambda > 0$. Refer to Fig. 2-7(a). Through A pass any line (q, Q_∞) distinct from $AB = (p, P_\infty)$ and orient this line so that the direction from A toward Q_∞ is positive. On (q, Q_∞) mark two points B' and C' such that $AC'/BC' = \lambda$. Let $P = BB' \cdot CC'$ and through P pass a line q' parallel to q. That $p \cdot q' = X$, the required point, follows by Theorem 2.6, that is,

$$(A, B, C, X) \overset{P}{\barwedge} (A, B', C', Q_\infty) \quad \text{and so} \quad (A, B; C, X) = (A, B'; C', Q_\infty) = (AB', C') = \lambda$$

(ii) $\lambda < 0$. Refer to Fig. 2-7(b). With due regard for signs, the procedure here is identical with that for (i) and will not be repeated. Why can we be sure there is just one point X?

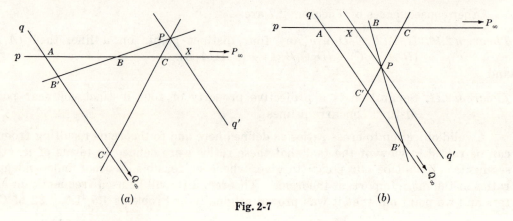

(a)   Fig. 2-7   (b)

2.4.   Prove: Any projectivity $p \barwedge p'$ $(P \barwedge P')$, where $p \neq p'$ $(P \neq P')$, for which $Q = p \cdot p'$ $(q = PP')$ is self-corresponding, is a perspectivity.

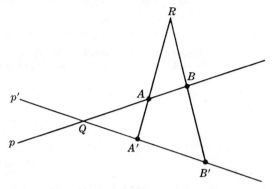

Consider in Fig. 2-8, the lines p and p' intersecting in Q. On p take two distinct points A and B, also distinct from Q. Suppose their correspondents in the projectivity are respectively A' and B' so that

$$p(Q, A, B, \ldots) \ \barwedge \ p'(Q, A', B', \ldots)$$

Let $R = AA' \cdot BB'$. Then

$$p(Q, A, B, \ldots) \ \overset{R}{\barwedge} \ p'(Q, A', B', \ldots)$$

hence, by the Fundamental Theorem the given projectivity must be this perspectivity. The second part of the theorem follows by the principle of duality.

Fig. 2-8

2.5.   Prove the plane dual of the Theorem of Pappus: In a projective plane, let a_1, a_2, a_3 be distinct lines on a point R and b_1, b_2, b_3 be distinct lines on another point S; then the lines $c_1 = (a_2 \cdot b_3)(a_3 \cdot b_2)$, $c_2 = (a_1 \cdot b_3)(a_3 \cdot b_1)$, $c_3 = (a_1 \cdot b_2)(a_2 \cdot b_1)$ are concurrent.

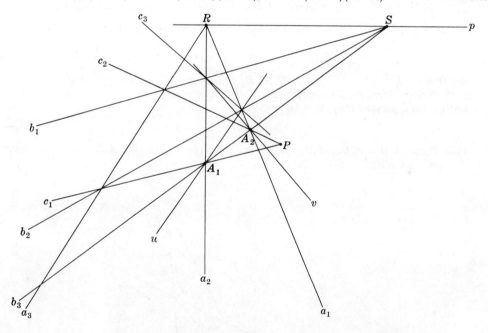

Fig. 2-9

In Fig. 2-9, let $c_1 \cdot c_2 = P$ and $RS = p$. It appears that c_3 is on P and this is what we are to prove. To construct the proof we need two perspective pencils of lines in which, say, c_1 and c_2 are corresponding elements and which include lines whose intersections determine c_3. Consider $A_1 = a_2 \cdot b_3$ on c_1 and $A_2 = a_1 \cdot b_3$ on c_2 as the centers of these pencils. On A_1 we need a line through one of the known points on c_3 and on A_2 we need a line through the other known point on c_3. Take for these lines $u = (a_1 \cdot b_2)(a_2 \cdot b_3)$ and $v = (a_2 \cdot b_1)(a_1 \cdot b_3)$ respectively. Now

$$A_1(c_1, u, b_3, a_2) \ \overset{b_2}{\barwedge} \ R(a_3, a_1, p, a_2) \quad \text{and} \quad A_2(c_2, a_1, b_3, v) \ \overset{b_1}{\barwedge} \ R(a_3, a_1, p, a_2)$$

and so

$$A_1(c_1, u, b_3, a_2) \ \barwedge \ A_2(c_2, a_1, b_3, v)$$

Since b_3 is a self-corresponding element, the projectivity is in reality the perspectivity

$$A_1(c_1, u, a_2) \ \barwedge \ A_2(c_2, a_1, v)$$

Note that this perspectivity meets the requirements listed above, that is, c_1 and c_2 are corresponding elements while $u \cdot a_1$ and $a_2 \cdot v$ are on c_3. By definition, $P = c_1 \cdot c_2$, $a_1 \cdot u = a_1 \cdot b_2$ and $a_2 \cdot v = a_2 \cdot b_1$ are collinear points. Then P is on $(a_1 \cdot b_2)(a_2 \cdot b_1) = c_3$ and so c_1, c_2, c_3 are concurrent as required.

The point P is called the *point of Pappus* for the two triples of lines on distinct points R and S.

We are now in a position to write a more tidy proof: In Fig. 2-9, let $RS = p$, $a_2 \cdot b_3 = A_1$, $a_1 \cdot b_3 = A_2$; $(a_1 \cdot b_2)(a_2 \cdot b_3) = u$, $(a_2 \cdot b_1)(a_1 \cdot b_3) = v$. Now

$$A_1(c_1, u, b_3, a_2) \overset{b_2}{\barwedge} R(a_3, a_1, p, a_2) \overset{b_1}{\barwedge} A_2(c_2, a_1, b_3, v)$$

and so
$$A_1(c_1, u, b_3, a_2) \;\barwedge\; A_2(c_2, a_1, b_3, v)$$

But this is a perspectivity; that is,
$$A_1(c_1, u, a_2) \;\overline{\barwedge}\; A_2(c_2, a_1, v)$$

Then $c_1 \cdot c_2$, $a_1 \cdot u = a_1 \cdot b_2$, $a_2 \cdot v = a_2 \cdot b_1$ are collinear and so $c_3 = (a_1 \cdot b_2)(a_2 \cdot b_1)$ is on $c_1 \cdot c_2 = P$ as required.

2.6. Prove the Theorem of Pappus: In a projective plane, let A_1, A_2, A_3 be distinct points on a line r and B_1, B_2, B_3 be distinct points on another line s; then the points $C_1 = A_2B_3 \cdot A_3B_2$, $C_2 = A_1B_3 \cdot A_3B_1$, $C_3 = A_1B_2 \cdot A_2B_1$ are collinear.

In Fig. 2-9′, let $r \cdot s = P$, $A_1B_2 \cdot A_2B_3 = U$, $A_2B_1 \cdot A_1B_3 = V$, $C_1C_2 = p$. Now

$$(C_1, U, B_3, A_2) \overset{B_2}{\barwedge} (A_3, A_1, P, A_2) \overset{B_1}{\barwedge} (C_2, A_1, B_3, V)$$

and thus
$$(C_1, U, B_3, A_2) \;\barwedge\; (C_2, A_1, B_3, V)$$

But this is a perspectivity, that is,
$$(C_1, U, A_2) \;\overline{\barwedge}\; (C_2, A_1, V)$$

Then C_1C_2, $A_1U = A_1B_2$, $A_2V = A_2B_1$ are concurrent at $A_1B_2 \cdot A_2B_1 = C_3$ and, hence, the points C_1, C_2, C_3 are collinear.

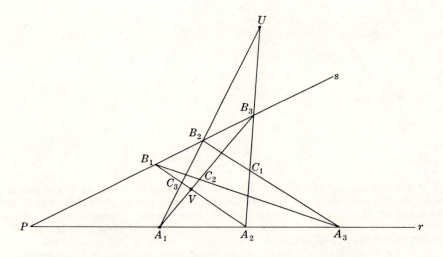

Fig. 2′-9′

The line $p = C_1C_2C_3$ is called the *line of Pappus* for the two triples of points on the distinct lines r and s.

Supplementary Problems

2.7.   (a) Show that (1) and (2), page 19, are independent of the orientation of p.

(b) Show that (2) is independent of the position of C relative to A and B.

Hint. When three distinct marks are made on the line p, there are six different ways in which the labels A, B, C may be given to them. Each must be investigated.

2.8.   Show for three distinct points A, B, C on a line that the six division ratios have values $r,\ 1/r,\ 1 - r,$ $\dfrac{1}{1-r},\ \dfrac{r}{r-1},\ \dfrac{r-1}{r}$.

2.9.   Why is (AB, C) not a projective invariant?

2.10.   Show: $(ab) + (ba) = 0,\ \ (ac) + (cb) + (ba) = 0,$
$$(ab) \cdot (cd) + (ac) \cdot (db) + (ad) \cdot (bc) = 0$$

2.11.   In Fig. 2-2, label the points (reading from left to right) B, C, D, A on r and B', C', D', A' on s and show $(A, B; C, D) = (A', B'; C', D')$.

2.12.   Complete the verification of (8), page 22.

2.13.   From the similar triangles $R'PA'$ and RAP of Fig. 2-5, obtain $R'A' = \dfrac{RP \cdot PR'}{AR}$. In the same manner obtain $R'B' = \dfrac{RP \cdot PR'}{BR}$ and $R'C' = \dfrac{RP \cdot PR'}{CR}$. Then show $(R_\infty, A; B, C) = (R', A'; B', C') = AC/AB$, $(A, B; C, R_\infty) = (A', B'; C', R') = AC/BC$, etc.

2.14.   Show for five distinct points A, B, C, D, E on a line,
$$(A, B; C, D) \cdot (A, B; D, E) \cdot (A, B; E, C) = 1$$

2.15.   Prove Theorem 2.11, page 25.

Hint. Let A, B, C, D be four collinear points such that the pair A, C is separated by the pair B, D. If $(A, B, C, D) \barwedge (A', B', C', D')$, prove that the points A', C' are separated by B', D'.

2.16.   In Figs. 2-7(a) and 2-7(b), $AB = 6$ units and $AC = 10$ units. For the same points A, B, C on p, construct X when $q \neq p$ is any other line on B and again when $q \neq p$ is any other line on C for (a) $\lambda = 5/4$ units, (b) $\lambda = -5/4$ units. (For the purpose of checking, $AX = 12$ units in (a) and $AX = 4$ units in (b).)

2.17.   Let A, B, C be distinct points on a line p such that $AB = 12$ units and $AC = 8$ units. Construct X on p such that (a) $(A, B; C, X) = -1$, (b) $(A, B; C, X) = 2$. (For the purpose of checking, $AX = 24$ units in (a) and 6 units in (b).)

2.18.   With the points A, B, C, X as in Problem 2.17(a), carry out the following construction: On A take any two distinct lines, neither of which is p; on C take any line, not p, meeting a in P and b in Q; join B and P meeting b in R; join B and Q meeting a in S. Then RS appears to be on X.

2.19.   With the points A, B, C, X as in Problem 2.17(b), carry out the following construction: On A take any two distinct lines, neither of which is p; on B take any line, not p, meeting a in P and b in Q; join C and P meeting b in R; join C and Q meeting a in S; join R and S.

2.20.   Answer the questions (2) and (3) raised on page 13.

Chapter 3

Desargues' Two-Triangle Theorem

PLANE CONFIGURATIONS

By a *plane configuration* will be meant any figure in a projective plane consisting of a_{11} points and a_{22} lines such that on each of the points are a_{12} of the lines and on each of the lines are a_{21} of the points. Such configurations will be represented symbolically by the array $\begin{pmatrix} a_{11} & a_{12} \\ a_{21} & a_{22} \end{pmatrix}$. Among these are two types, called complete:

(*a*) The *complete n-point* consisting of n points, no three collinear, and the $\frac{1}{2}n(n-1)$ lines determined by them.

and its dual

(*a′*) The *complete n-line* consisting of n lines, no three concurrent, and the $\frac{1}{2}n(n-1)$ points determined by them.

The *complete 2-point*, consisting of two distinct points and their join, and the *complete 2-line*, consisting of two distinct lines and their point of intersection, are trivial. The simplest non-trivial complete plane configurations are the *complete 3-point*, consisting of three non-collinear points and their joins, and the *complete 3-line*, consisting of three non-concurrent lines and their points of intersection. In each of these there are $a_{11} = 3$ points and $a_{22} = 3$ lines with $a_{12} = 2$ lines on each point and $a_{21} = 2$ points on each line. Each is represented symbolically by the array $\begin{pmatrix} 3 & 2 \\ 2 & 3 \end{pmatrix}$ and, hence, is self-dual. In view of this, no sharp distinction will be made between the complete 3-point and the complete 3-line; in fact, each will be called a *triangle*.

The next simplest complete plane configurations are the *complete 4-point* or *complete quadrangle* and, its dual, the *complete 4-line* or *complete quadrilateral*.

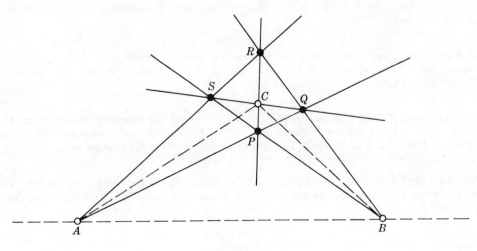

Fig. 3-1(*a*)

The *complete quadrangle*, consisting of the four distinct points P, Q, R, S, no three of which are collinear, and their six joins is illustrated in Fig. 3-1(a) above. The four given points are called the *vertices* and the six lines are called the *sides* of the complete quadrangle.

Two sides, not on the same vertex, are called a pair of *opposite sides* of the complete quadrangle. The three pairs of opposite sides determine three additional points A, B, C called the *diagonal points* of the complete quadrangle. In the figure, these three points are non-collinear and determine a triangle, the *diagonal triangle* of the complete quadrangle.

Symbolically, the complete quadrangle is represented by the array $\begin{pmatrix} 4 & 3 \\ 2 & 6 \end{pmatrix}$.

The *complete quadrilateral*, consisting of the four distinct lines p, q, r, s, no three of which are concurrent, and their six points of intersection, is illustrated in Fig. 3-1(b) below. The four given lines are called the *sides* and the six points are called the *vertices* of the complete quadrilateral.

Two vertices, not on the same side, are called a pair of opposite vertices of the complete quadrilateral. The three pairs of opposite vertices determine three additional lines a, b, c called the *diagonal lines* of the complete quadrilateral. In the figure, these three lines are non-concurrent and determine a triangle, the *diagonal triangle* of the complete quadrilateral.

Symbolically, the complete quadrilateral is represented by the array $\begin{pmatrix} 6 & 2 \\ 3 & 4 \end{pmatrix}$.

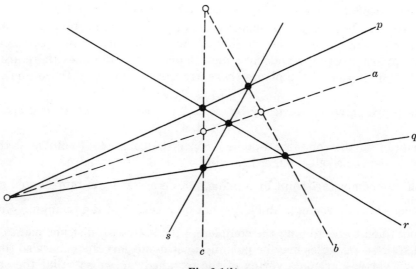

Fig. 3-1(b)

The term complete quadrangle must be used here in order to distinguish it from the *simple quadrangle* consisting of four points P, Q, R, S, no three collinear, and the four lines PQ, QR, RS, SP obtained by joining the points cyclically. Dually, there is a *simple quadrilateral* consisting of four lines p, q, r, s, no three concurrent, and the four points $p \cdot q$, $q \cdot r$, $r \cdot s$, $s \cdot p$ obtained by intersecting the lines cyclically.

We are not in a position to prove

> The diagonal points (diagonal lines) of any complete quadrangle (complete quadrilateral) are non-collinear (non-concurrent).

For the present, however, we shall work under the assumption that they are.

Two complete n-points are said to be in *perspective position* if they are in one-to-one correspondence and so situated that pairs of corresponding vertices are on concurrent lines. The point of concurrency P is called the *center of perspectivity* and we say that the two n-points are *perspective from* P.

Two complete n-lines are said to be in *perspective position* if they are in one-to-one correspondence and so situated that pairs of corresponding sides are on collinear points. The line of collinearity p is called the *axis of perspectivity* and we say that the two n-lines are *perspective from p*.

Two coplanar triangles, then, could be perspective from a point or be perspective from a line or be perspective from both a point and a line. One of the most famous theorems of plane projective geometry is

Desargues' Two-Triangle Theorem. If two coplanar triangles $A_1A_2A_3$ and $B_1B_2B_3$ are perspective from a point P, they are perspective from a line p; and, conversely.

SPACE CONFIGURATIONS

By a *space configuration* will be meant any figure in projective space consisting of a_{11} points, a_{22} lines and a_{33} planes such that on each of the points are a_{12} lines and a_{13} planes, on each of the lines are a_{21} points and a_{23} planes, and on each of the planes are a_{31} points and a_{32} lines. Symbolically, such a configuration is represented by the array $\begin{pmatrix} a_{11} & a_{12} & a_{13} \\ a_{21} & a_{22} & a_{23} \\ a_{31} & a_{32} & a_{33} \end{pmatrix}$.

The simplest complete space configurations are the *complete 4-point* and its space dual, the *complete 4-plane*. The complete 4-point consists of four non-coplanar points (*vertices*) together with the six lines (*edges*) and the four planes (*faces*) determined by them. On each of the points there are three lines and three planes, on each of the lines there are two points and two planes, and on each of the planes there are three points and three lines. The complete 4-point is represented symbolically by $\begin{pmatrix} 4 & 3 & 3 \\ 2 & 6 & 2 \\ 3 & 3 & 4 \end{pmatrix}$. We leave as an exercise the definition and description of the complete 4-plane. Since the complete 4-point and the complete 4-plane will be found to be self-dual, either will be called a *tetrahedron*.

The section of a complete 4-point by a plane, not on a vertex, is represented symbolically by $\begin{pmatrix} 6 & 2 \\ 3 & 4 \end{pmatrix}$. It consists of a triangle and a line not on a vertex of the triangle. Consider now the triangles obtained by sectioning the complete 4-point by two distinct planes, neither on a vertex. Since these triangles may be put into one-to-one correspondence so that the joins of corresponding vertices are on a vertex of the tetrahedron, we say that the triangles are perspective from a point. Also, if the sides of the two triangles can be put into one-to-one correspondence so that the pairs of corresponding sides intersect in collinear points, we say that the triangles are perspective from a line. Can you identify this line? Are the two triangles always perspective from both a point *and* a line?

The space configuration of particular interest to us is the *complete space 5-point* (see Fig. 3-2) consisting of the five points (vertices) A, B, C, D, E, no four coplanar and no three collinear, together with the ten lines (edges) and the ten planes (faces) determined by them. The symbolic representation is $\begin{pmatrix} 5 & 4 & 6 \\ 2 & 10 & 3 \\ 3 & 3 & 10 \end{pmatrix}$.

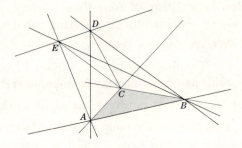

Fig. 3-2

THE DESARGUES CONFIGURATION

A section of the complete space 5-point $ABCDE$ by a plane π, not on any of the vertices, is shown in Fig. 3-3. Since the 5-point consists of ten lines and ten planes with three of the lines on each of the planes and three of the planes on each of the lines, this plane section will consist of ten points and ten lines with three of the points on each of the lines and three of the lines on each of the points. The symbolic representation of the section is $\begin{pmatrix} 10 & 3 \\ 3 & 10 \end{pmatrix}$. In order to identify these points and lines, consider the triangle ABC projected from D and again from E. The section by π consists of the triangles $A_1B_1C_1$ and $A_2B_2C_2$. Denote by P the intersection of the line DE and π. Then the planes DEA, DEB, DEC intersect π in the lines $PA_1A_2, PB_1B_2, PC_1C_2$ respectively. Having thus accounted for seven of the points and nine of the lines, we can say that, in part, the section consists of two triangles $A_1B_1C_1$ and $A_2B_2C_2$ perspective from the point P. The remaining line is p, the intersection of the plane ABC and π, on which lie the remaining points: A_3, the intersection of BC and π; B_3, the

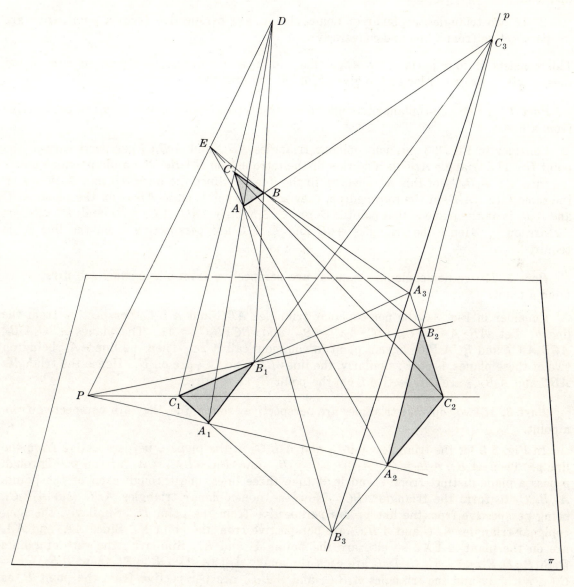

Fig. 3-3

intersection of AC and π; and C_3, the intersection of AB and π. Now the line A_1B_1, in the plane DAB, and the line A_2B_2, in the plane EAB, intersect on the line AB; thus they intersect in the point C_3. Similarly, A_1C_1 and A_2C_2 intersect in B_3 while B_1C_1 and B_2C_2 intersect in A_3. Hence the section (better known as the *Desargues configuration*) consists of the triangles $A_1B_1C_1$ and $A_2B_2C_2$ perspective from the point P and also from the line p.

DESARGUES' TWO-TRIANGLE THEOREM

The theorem, as stated on page **32**, concerns two coplanar triangles. It consists of two parts, — a theorem and its converse — which, in turn, are found to be a theorem and its plane dual. A proof, using the Fundamental Theorem, is given in Problem 3.1 and another, using the Theorem of Pappus, is indicated in Problem 3.34. Since our proof of the Fundamental Theorem and, hence, of the Theorem of Pappus employed cross ratios, we give now a third proof. However, in order to avoid cross ratios, we must prove the more general theorem:

> If two triangles, coplanar or non-coplanar, are perspective from a point, they are perspective from a line and conversely.

This consists of four parts — two theorems and their converses. We proceed now to list these parts in proper order and to give proofs of the first three.

Part 1. If two non-coplanar triangles are perspective from a point, they are perspective from a line.

Consider in Fig. 3-3 the non-coplanar triangles ABC and $A_2B_2C_2$ perspective from the point E. The triangle ABC is a section of the tetrahedron $EA_2B_2C_2$ by a plane which meets the face $\pi = A_2B_2C_2$ of this tetrahedron in the line p. Since the lines AB and A_2B_2 are on the same face EA_2B_2 of the tetrahedron, they are on a point; since AB is on the plane ABC and A_2B_2 is on the plane π, that point is C_3 on p. Similarly, $AC \cdot A_2C_2 = B_3$ and $BC \cdot B_2C_2 = A_3$ are on p. Hence the triangles ABC and $A_2B_2C_2$ are perspective from the line p, as required.

Part 2. If two non-coplanar triangles are perspective from a line, they are perspective from a point.

Consider in Fig. 3-3 the non-coplanar triangles ABC and $A_2B_2C_2$ perspective from the line p. Let $AB \cdot A_2B_2 = C_3$, $AC \cdot A_2C_2 = B_3$, and $BC \cdot B_2C_2 = A_3$. The planes $ABC_3A_2B_2$, $ACB_3A_2C_2$, and $BCA_3B_2C_2$ have a point in common; call it E. Then the line AA_2, being on two of these planes, is on E; similarly, the lines BB_2 and CC_2 are on E. Hence the triangles ABC and $A_2B_2C_2$ are perspective from the point E, as required.

Part 3. If two coplanar triangles are perspective from a line, they are perspective from a point.

In Fig. 3-3 let the triangles $A_1B_1C_1$ and $A_2B_2C_2$ in the plane π be perspective from the line p. Then $A_1B_1 \cdot A_2B_2 = C_3$, $A_1C_1 \cdot A_2C_2 = B_3$, and $B_1C_1 \cdot B_2C_2 = A_3$ are on p. Through p pass a plane distinct from π and in it take three lines, one through each of the points A_3, B_3, C_3 to form the triangle ABC. Now the non-coplanar triangles ABC and $A_1B_1C_1$, being perspective from the line p, are perspective from the point D. Similarly, the non-coplanar triangles ABC and $A_2B_2C_2$ are perspective from the point E. Since AA_1 and AA_2 are on the plane AA_1A_2, so also are the points D and E. Similarly, the sets of points B, B_1, B_2, D, E and C, C_1, C_2, D, E are each coplanar. Let $A_1A_2 \cdot DE = P$; then $B_1B_2 \cdot DE = C_1C_2 \cdot DE = P$ and the triangles $A_1B_1C_1$ and $A_2B_2C_2$ are perspective from the point P, as required.

Part 4. If two coplanar triangles are perspective from a point, they are perspective from a line.

Since the triangles of Parts 3 and 4 are coplanar, the theorem of Part 4 is the plane dual of that of Part 3 and hence may be accepted as valid. On the other hand, the proof of Part 3 is made in projective space and thus cannot be dualized in the plane. It is suggested that the reader provide an independent proof and compare it with that given in Problem 3.2.

The Two-Triangle Theorem can be used to prove a number of interesting constructions and theorems. As expected, the procedure in each is to locate a pair of triangles perspective from a point (line) and hence perspective from a line (point). Although some of the theorems may be valid both in space and in the plane, our concern from now on will be limited strictly to a single plane. It will be understood then that when, for example, two lines are given, they are coplanar; when four or more points are given, they are coplanar; and when the dual of a construction or theorem is called for, the plane dual is meant.

Example 3.1.

Let there be given a line p and two distinct points A and B not on p. Without joining A and B (assume that this is impossible for some reason), obtain the intersection of p and AB.

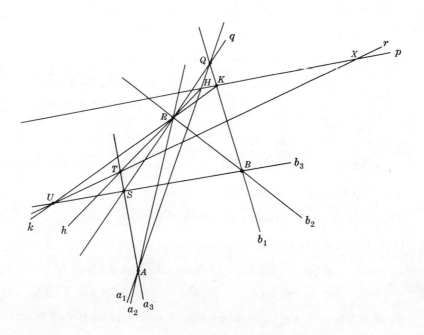

Fig. 3-4

In Fig. 3-4 we begin with the line p, the two points A, B and, assuming the construction possible, a point X somewhere on p. In order to have perspective triangles we must add, in the most general way possible, additional points and lines. Let q, distinct from p be any line not on A or B. On q take any three distinct points Q, R, S and, joining them to A and B, obtain the lines a_1, a_2, a_3 and b_1, b_2, b_3. Now we look for two triangles which by construction are perspective from a point and hence will be perspective from a line. We would like, if possible, this line to be AB which, recall, is also to contain X. Since a_1 and a_3 are on A, b_1 and b_3 are on B, and X is on p, we consider the triangle $a_1 b_1 p$ and the triangle $a_3 b_3 r$ with r to be determined. Let $p \cdot a_1 = H$ and $p \cdot b_1 = K$; then the triangle $a_1 b_1 p$ is the triangle QHK. Since QS is on R, we try R as the center of perspectivity. Let $h = HR$, $k = KR$; $T = h \cdot a_3$, $U = k \cdot b_3$. Then $TU = r$ is the missing line. Now the triangles $a_1 b_1 p$ and $a_3 b_3 r$ (QHK and STU), being perspective from R, are perspective from a line. This line is on $A = a_1 \cdot a_3$, $B = b_1 \cdot b_3$, and locates $X = p \cdot r$ as the intersection of p and AB.

See also Problems 3.3, 3.4.

SUCCESSIVE PERSPECTIVITIES

Suppose the projectivity

$$p_1(A_1, B_1, C_1, D_1, \ldots) \; \overline{\wedge} \; p_{n+1}(A_{n+1}, B_{n+1}, C_{n+1}, D_{n+1}, \ldots)$$

follows from the given sequence of n successive perspectivities

(i) $p_1(A_1, B_1, C_1, D_1, \ldots) \; \overset{P_1}{\overline{\wedge}} \; p_2(A_2, B_2, C_2, D_2, \ldots) \; \overset{P_2}{\overline{\wedge}} \; p_3(A_3, B_3, C_3, D_3, \ldots)$

        $\overset{P_3}{\overline{\wedge}} \; p_4(A_4, B_4, C_4, D_4, \ldots) \; \overset{P_4}{\overline{\wedge}} \; \cdots \; \overset{P_n}{\overline{\wedge}} \; p_{n+1}(A_{n+1}, B_{n+1}, C_{n+1}, D_{n+1}, \ldots).$

In Problem 1.4 it was shown that every projectivity between two pencils of points (pencils of lines) on distinct lines (points) can be expressed as the product of at most two perspectivities and every projectivity between two pencils of points (pencils of lines) on the same line (point) can be expressed as the product of at most three perspectivities. Another proof of this theorem will be indicated by providing a systematic procedure for reducing (i) to one or at most two perspectivities when $p_1 \neq p_{n+1}$ and a succession of two or at most three perspectivities when $p_1 = p_{n+1}$. To effect such a reduction, it is necessary to know under what conditions two consecutive perspectivities can be reduced to a single perspectivity and under what conditions three successive perspectivities can be reduced to two. These questions are answered in the following theorems:

Theorem 3.1.   If the distinct lines p_1, p_2, p_3 are on a point O and if

$$p_1(A_1, B_1, C_1, D_1, \ldots) \; \overset{P}{\overline{\wedge}} \; p_2(A_2, B_2, C_2, D_2, \ldots) \; \overset{Q}{\overline{\wedge}} \; p_3(A_3, B_3, C_3, D_3, \ldots)$$

there exists a point R on the line PQ such that

$$p_1(A_1, B_1, C_1, D_1, \ldots) \; \overset{R}{\overline{\wedge}} \; p_3(A_3, B_3, C_3, D_3, \ldots)$$

For a proof, see Problem 3.5.

Corollary.   If p_1, p_2, p_3 are distinct lines or if $p_1 = p_3$, if p_2' is any line on $p_1 \cdot p_2$, and if

$$P_1(A_1, B_1, C_1, D_1, \ldots) \; \overset{P}{\overline{\wedge}} \; p_2(A_2, B_2, C_2, D_2, \ldots) \; \overset{Q}{\overline{\wedge}} \; p_3(A_3, B_3, C_3, D_3, \ldots)$$

there exists a point X on PQ such that

$$P_1(A_1, B_1, C_1, D_1, \ldots) \; \overset{X}{\overline{\wedge}} \; p_2'(A_2', B_2', C_2', D_2', \ldots) \; \overset{Q}{\overline{\wedge}} \; p_3(A_3, B_3, C_3, D_3, \ldots)$$

Theorem 3.2.   If no three of the distinct lines p_1, p_2, p_3, p_4 are concurrent and if

$$p_1(A_1, B_1, C_1, D_1, \ldots) \; \overset{P}{\overline{\wedge}} \; p_2(A_2, B_2, C_2, D_2, \ldots)$$
$$\overset{Q}{\overline{\wedge}} \; p_3(A_3, B_3, C_3, D_3, \ldots) \; \overset{R}{\overline{\wedge}} \; p_4(A_4, B_4, C_4, D_4, \ldots)$$

there exists a line p_*, distinct from p_1 and p_4, and points S and T such that

$$p_1(A_1, B_1, C_1, D_1, \ldots) \; \overset{S}{\overline{\wedge}} \; p_*(A_*, B_*, C_*, D_*, \ldots) \; \overset{T}{\overline{\wedge}} \; p_4(A_4, B_4, C_4, D_4, \ldots)$$

For a proof, see Problem 3.6.

Theorem 3.3.   If p_1, p_2, p_3, p_4 are distinct lines of which p_1, p_2, p_4 are on a point O while p_3 is not (p_1, p_3, p_4 are on a point O while p_2 is not) and if

$$p_1(A_1, B_1, C_1, D_1, \ldots) \; \overset{P}{\overline{\wedge}} \; p_2(A_2, B_2, C_2, D_2, \ldots)$$
$$\overset{Q}{\overline{\wedge}} \; p_3(A_3, B_3, C_3, D_3, \ldots) \; \overset{R}{\overline{\wedge}} \; p_4(A_4, B_4, C_4, D_4, \ldots)$$

there exists a line p_*, distinct from p_1 and p_4, and points S and T such that

$$p_1(A_1, B_1, C_1, D_1, \ldots) \; \overset{S}{\overline{\wedge}} \; p_*(A_*, B_*, C_*, D_*, \ldots) \; \overset{T}{\overline{\wedge}} \; p_4(A_4, B_4, C_4, D_4, \ldots)$$

For a proof, see Problem 3.7.

After applying the above theorems to (i) and the sequences which result after each application of a theorem, it is clear:

(ii) For the case $p_1 \neq p_{n+1}$, we have $p_1 \stackrel{U}{\barwedge} p_{**} \stackrel{V}{\barwedge} p_{n+1}$
which further reduces to $p_1 \stackrel{W}{\barwedge} p_{n+1}$ when p_1, p_{**}, p_{n+1} are concurrent.

(iii) For the case $p_1 = p_{n+1}$, we have

$$p_1(A_1, B_1, C_1, D_1, \ldots) \stackrel{U}{\barwedge} p_{**}(A_{**}, B_{**}, C_{**}, D_{**}, \ldots)$$
$$\stackrel{V}{\barwedge} p_{***}(A_{***}, B_{***}, C_{***}, D_{***}, \ldots) \stackrel{W}{\barwedge} p_1(A_{n+1}, B_{n+1}, C_{n+1}, D_{n+1}, \ldots)$$

which further reduces to $p_1(A_1, \ldots) \stackrel{X}{\barwedge} p_{**} \stackrel{W}{\barwedge} p_1(A_{n+1}, \ldots)$ or to $p_1(A_1, \ldots) \stackrel{U}{\barwedge} p_{***} \stackrel{Y}{\barwedge}$
$p_1(A_{n+1}, \ldots)$ when p_1, p_{**}, p_{***} are concurrent.

PERSPECTIVE QUADRANGLES

Consider in Fig. 3-5 the complete quadrangles $P_1Q_1R_1S_1$ and $P_2Q_2R_2S_2$ so situated that the intersections $A = P_1Q_1 \cdot P_2Q_2$, $D = P_1R_1 \cdot P_2R_2$, $B = P_1S_1 \cdot P_2S_2$, $G = Q_1R_1 \cdot Q_2R_2$, $F = R_1S_1 \cdot R_2S_2$ are on a line o. Let $E = Q_1S_1 \cdot Q_2S_2$. The triangles $P_1Q_1R_1$ and $P_2Q_2R_2$, being perspective from the line $ADG = o$, are perspective from O the common point on P_1P_2, Q_1Q_2 and R_1R_2. Similarly, the triangles $P_1R_1S_1$ and $P_2R_2S_2$ are perspective from the line $BDF = o$ and so are perspective from $O = P_1P_2 \cdot R_1R_2$. Then S_1S_2 is on O. Finally, the triangles $P_1Q_1S_1$ and $P_2Q_2S_2$, being perspective from O, are perspective from the line $AB = o$; hence E is on o. We have proved

Theorem 3.4. Given two complete quadrangles so situated that the intersections of five pairs of corresponding sides are on a line, then (i) the intersection of the sixth pair of corresponding sides is on this line and (ii) the two quadrangles are perspective both from a point and from a line.

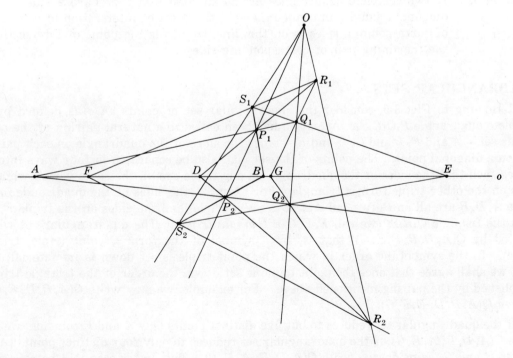

Fig. 3-5

The set of six points in which a complete quadrangle meets a line, not on a vertex, is called a *quadrangular set of points*. There are three cases: (*a*) the six points are distinct as in Fig. 3-5; (*b*) two of the points A and F (B and G) coincide, in which case the line o is on the diagonal point A (B); (*c*) each of the two pairs of points A, F and B, G coincide, in which case the line o is a side of the diagonal triangle of the quadrangle. (Our assumption concerning the diagonal points of a complete quadrangle excludes the possibility of the six points coinciding further.)

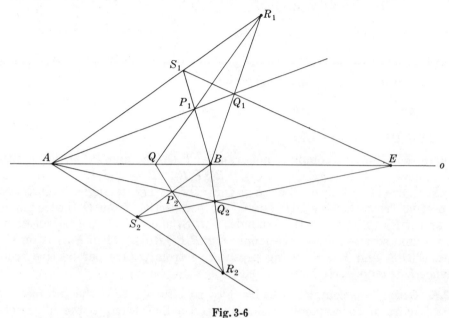

Fig. 3-6

For the case (*c*), Theorem 3.4 may be restated as

Theorem 3.5. If two complete quadrangles are so situated that they have a side of their diagonal triangles in common, and if the point of intersection of a fifth pair of corresponding sides is on this line, so also is the point of intersection of the remaining pair of corresponding sides.

QUADRANGULAR SETS

Returning to Fig. 3-5, consider the quadrangular set of points on o as defined by the complete quadrangle $P_1Q_1R_1S_1$. Attention has been called to a natural pairing of the points of this set — A and F, B and G, D and E — since the sides of the quadrangle on each pair are also on a diagonal point. The points of the set may also be separated in four ways into two triples such that the sides of the quadrangle on one triple are on the same vertex while the sides on the other triple form a triangle. For example, the sides of the quadrangle on the triple A, D, B are all on P_1 (we call A, D, B a *point* triple) and the sides on the triple F, E, G form the triangle $Q_1R_1S_1$ (we call F, E, G a *triangle triple*). The quadrangular set will be denoted by $Q(A, D, B; F, E, G)$, that is, by a symbol of the form Q(point triple; triangle triple). In the symbol the order in which the point triple is set down is immaterial; however, we shall agree that once the point triple is set down, the order of the triangle triple is established by the pairing mentioned above. For example, we may write $Q(A, B, D; F, G, E)$ but not $Q(A, B, D; E, F, G)$.

If the quadrangular set reduces to but five distinct points (say A and F coincide) we shall write $Q(A, D, B; A, E, G)$; if the quadrangular set reduces to only four distinct points (A and F, also B and G, coincide) we write $Q(A, D, B; A, E, B)$. This latter case will be considered in detail in the next chapter. There a more useful symbol will be introduced.

Solved Problems

3.1. Prove, using the Fundamental Theorem: If two coplanar triangles $A_1A_2A_3$ and $B_1B_2B_3$ are perspective from a point P, they are perspective from a line p.

Let $A_1B_1 = c_1$, $A_2B_2 = c_2$, $A_3B_3 = c_3$; we are given c_1, c_2, c_3 concurrent at P. Let $A_1A_2 = a_3$, $A_1A_3 = a_2$, $A_2A_3 = a_1$; $B_1B_2 = b_3$, $B_1B_3 = b_2$, $B_2B_3 = b_1$; $a_1 \cdot b_1 = C_1$, $a_2 \cdot b_2 = C_2$, $a_3 \cdot b_3 = C_3$; $C_1C_3 = p$. We are to prove C_2 is on p.

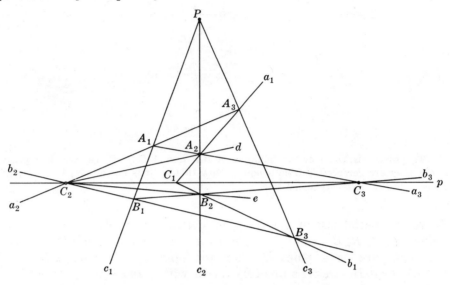

Fig. 3-7

Let $A_2C_2 = d$ and $B_2C_2 = e$.   Then

$$A_2(a_1, a_3, c_2, d) \ \overset{a_2}{\underset{\wedge}{=}} \ P(c_3, c_1, c_2, PC_2) \ \overset{b_2}{\underset{\wedge}{=}} \ B_2(b_1, b_3, c_2, e)$$

and
$$A_2(a_1, a_3, c_2, d) \ \overline{\wedge} \ B_2(b_1, b_3, c_2, e)$$

This projectivity is a perspectivity (why?) and the axis of perspectivity is on $a_1 \cdot b_1 = C_1$, $a_3 \cdot b_3 = C_3$ and $d \cdot e = C_2$. Hence, $a_2 \cdot b_2 = C_2$ is on $p = C_1C_3$, as required.

3.2. Prove without using the Fundamental Theorem: If two coplanar triangles are perspective from a point, they are perspective from a line.

Refer to Fig. 3-3. Let the triangles $A_1B_1C_1$ and $A_2B_2C_2$ in the plane π be perspective from the point P. Let $C_3 = A_1B_1 \cdot A_2B_2$, $B_3 = A_1C_1 \cdot A_2C_2$, and $A_3 = B_1C_1 \cdot B_2C_2$.

On P take any line, not on π, and on it take distinct points D and E. In the plane DEA_1A_2, let $A = DA_1 \cdot EA_2$; in the plane DEB_1B_2, let $B = DB_1 \cdot EB_2$; in the plane DEC_1C_2, let $C = DC_1 \cdot EC_2$. By construction, the triangles ABC and $A_1B_1C_1$ are perspective from D and hence are perspective from the line of intersection p of their planes. Similarly, the triangles ABC and $A_2B_2C_2$ are perspective from E and hence are also perspective from p. Now $AB \cdot A_1B_1$ and $AB \cdot A_2B_2$ are on p; hence they are the same point C_3. Similarly, B_3 and A_3 are on p, as required.

3.3. State and prove the dual of Example 3.1, page 35.

Let there be given a point P and two distinct lines a and b, not on P. Without intersecting a and b (assume that this is impossible for some reason), obtain the join of P and the point of intersection of a and b.

In Fig. 3-8 below let Q, distinct from P, be any point not on a or b. On Q take distinct lines q, r, s and, intersecting them with a and b, obtain the points A_1, A_2, A_3 and B_1, B_2, B_3. Let $h = PA_1$, $k = PB_1$; $H = h \cdot r$; $K = k \cdot r$; $t = A_3H$, $u = B_3K$; $R = t \cdot u$; $x = PR$. We assert that x is the required line.

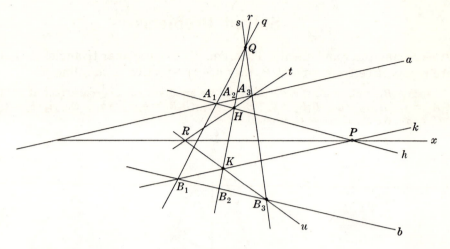

Fig. 3-8

To prove this, consider the triangles A_1B_1P and A_3B_3R. Since they are perspective from the line r, they are perspective from a point. Now this point is on $a = A_1A_3$, $b = B_1B_3$ and $x = PR$. Thus $x = PR$ is the required line.

3.4. Prove the special case of the Theorem of Pappus: If A_1, A_2, A_3 are distinct points on a line r, if B_1, B_2, B_3 are distinct points on another line s, and if $c_1 = A_1B_1$, $c_2 = A_2B_2$, $c_3 = A_3B_3$ are on a point O, then the points $C_1 = A_2B_3 \cdot A_3B_2$, $C_2 = A_1B_3 \cdot A_3B_1$, $C_3 = A_1B_2 \cdot A_2B_1$ are on a line concurrent with r and s.

Refer to Fig. 3-9. Let $r \cdot s = P$. The triangles $C_1A_2B_2$ and $C_2A_1B_1$, being perspective from the line OA_3B_3 (check this), are perspective from P and so C_1, C_2, P are collinear. Similarly, the triangles $C_1A_3B_3$ and $C_3A_1B_1$, being perspective from the line OA_2B_2, are perspective from P and so C_1, C_3, P are collinear. Then C_1, C_2, C_3, P are collinear as required.

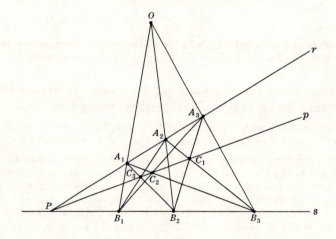

Fig. 3-9

3.5. Prove: If the distinct lines p_1, p_2, p_3 are on a point O and if

$$p_1(A_1, B_1, C_1, D_1, \ldots) \overset{P}{\barwedge} p_2(A_2, B_2, C_2, D_2, \ldots) \overset{Q}{\barwedge} p_3(A_3, B_3, C_3, D_3, \ldots)$$

there exists a point R on PQ such that

(a) $$p_1(A_1, B_1, C_1, D_1, \ldots) \overset{R}{\barwedge} p_3(A_3, B_3, C_3, D_3, \ldots)$$

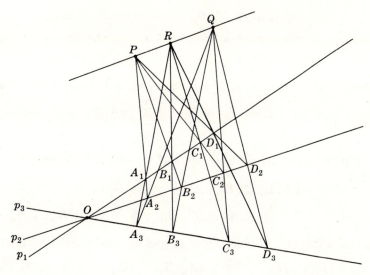

Fig. 3-10

Refer to Fig. 3-10. The triangles $A_1A_2A_3$ and $B_1B_2B_3$, being perspective from O, are perspective from the line PQ on which is the point $R = A_1A_3 \cdot B_1B_3$. Similarly, the triangles $A_1A_2A_3$ and $C_1C_2C_3$ are perspective from PQ on which is the point $R' = A_1A_3 \cdot C_1C_3$. But, since R and R' are on both A_1A_3 and PQ, they must coincide. Thus we have (a) as required.

3.6. Prove: If no three of the distinct lines p_1, p_2, p_3, p_4 are concurrent and if

$$p_1(A_1, B_1, C_1, D_1, \ldots) \overset{P}{\barwedge} p_2(A_2, B_2, C_2, D_2, \ldots)$$

$$\overset{Q}{\barwedge} p_3(A_3, B_3, C_3, D_3, \ldots) \overset{R}{\barwedge} p_4(A_4, B_4, C_4, D_4, \ldots)$$

there exists a line p_* and points S, T such that

(b) $p_1(A_1, B_1, C_1, D_1, \ldots) \overset{S}{\barwedge} p_*(A_*, B_*, C_*, D_*, \ldots) \overset{T}{\barwedge} p_4(A_4, B_4, C_4, D_4, \ldots)$

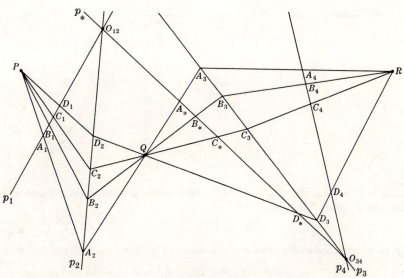

Fig. 3-11

Refer to Fig. 3-11. Let $O_{12} = p_1 \cdot p_2$ and $O_{34} = p_3 \cdot p_4$. Take $p_* = O_{12}O_{34}$ and on it obtain $p_*(A_*, B_*, C_*, D_*, \ldots)$ such that

$$p_*(A_*, B_*, C_*, D_*, \ldots) \overset{Q}{\barwedge} p_2(A_2, B_2, C_2, D_2, \ldots)$$

Now
$$p_2(A_2, B_2, C_2, D_2, \ldots) \overset{P}{\doteq} p_1(A_1, B_1, C_1, D_1, \ldots)$$

hence, since p_1, p_2, p_* are concurrent, Theorem 3.1 applies, and there exists a point S on PQ such that

$$p_1(A_1, B_1, C_1, D_1, \ldots) \overset{S}{\doteq} p_*(A_*, B_*, C_*, D_*, \ldots)$$

Similarly, since the lines p_3, p_4, p_* are concurrent, there exists a point T on QR such that

$$p_*(A_*, B_*, C_*, D_*, \ldots) \overset{T}{\doteq} p_4(A_4, B_4, C_4, D_4, \ldots)$$

and we have (b).

3.7. Prove: If p_1, p_2, p_3, p_4 are distinct lines of which p_1, p_2, p_4 are on a point O while p_3 is not, and if

$$p_1(A_1, B_1, C_1, D_1, \ldots) \overset{P}{\doteq} p_2(A_2, B_2, C_2, D_2, \ldots)$$

$$\overset{Q}{\doteq} p_3(A_3, B_3, C_3, D_3, \ldots) \overset{R}{\doteq} p_4(A_4, B_4, C_4, D_4, \ldots)$$

there exists a line p_*, distinct from p_1 and p_4, and points S and T such that

$$p_1(A_1, B_1, C_1, D_1, \ldots) \overset{S}{\doteq} p_*(A_*, B_*, C_*, D_*, \ldots) \overset{T}{\doteq} p_4(A_4, B_4, C_4, D_4, \ldots)$$

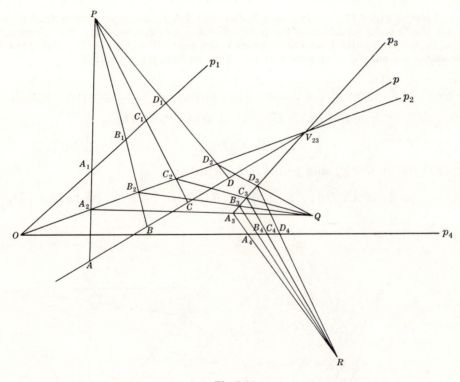

Fig. 3-12

Refer to Fig. 3-12. On $V_{23} = p_2 \cdot p_3$ take any line p distinct from p_2, p_3 and not on P. Project the pencil on p_2 from P onto p. Then

$$p(A, B, C, D, \ldots) \overset{P}{\doteq} p_2(A_2, B_2, C_2, D_2, \ldots) \overset{Q}{\doteq} p_3(A_3, B_3, C_3, D_3, \ldots)$$

Since p, p_2, p_3 are concurrent, Theorem 3.1 assures the existence of a point U on PQ such that

$$p(A, B, C, D, \ldots) \overset{U}{\doteq} p_3(A_3, B_3, C_3, D_3, \ldots)$$

In the sequence $p_1 \overset{P}{\doteq} p \overset{U}{\doteq} p_3 \overset{R}{\doteq} p_4$, the lines p_1, p, p_3, p_4 are such that no three are concurrent. Then Theorem 3.2 assures the existence of a line p_* and points S, T such that

$$p_1 \overset{S}{\doteq} p_* \overset{T}{\doteq} p_4$$

as required.

Supplementary Problems

3.8. (a) Four distinct points A_1, A_2, A_3, A_4, no three on the same line, determine a unique complete quadrangle. Show that these same points determine exactly three different simple quadrangles: $A_1A_2A_3A_4$, $A_1A_2A_4A_3$, $A_1A_3A_2A_4$.

  (b) State and verify the dual of (a).

  (c) Describe the complete plane 5-point and its plane dual.

  (d) Discuss as in (a) the simple plane 5-point and its dual. Enumerate several simple pentagons determined by the five distinct points.

  (e) Same as (d) for the simple plane 6-point (hexagon) and its dual.

3.9. In obtaining the Desargues configuration, the triangle ABC (see Fig. 3-2) was projected from both D and E. Repeat the discussion when the triangle BCD is projected from both A and E.

3.10. In the Desargues configuration (see Fig. 3-3) show:

  (a) There are ten pairs of perspective triangles,

  (b) The complete quadrangle $A_1B_1C_1P$ and the complete quadrilateral $p, A_2B_2, B_2C_2, C_2A_2$ are so situated that each side of the quadrangle is on a vertex of the quadrilateral. List four other such pairs of quadrangles and quadrilaterals.

3.11. Let there be given in the same plane a triangle ABC and a point P. Let the joins of P and the vertices meet the opposite sides of the triangle in A', B', C' respectively. Construct the axis of perspectivity p of the triangles ABC and $A'B'C'$.   Dualize.

  The axis p is called the *polar line* of P while P is called the *pole* of p *with respect to the triangle* ABC. Is p ever on P?

3.12. (a) Establish the Corollary to Theorem 3.1, page 36.

  (b) Reduce $p_1(A_1, B_1, \ldots) \overset{P_1}{\underset{\wedge}{=}} p_2(A_2, B_2, \ldots) \overset{P_2}{\underset{\wedge}{=}} p_1(A_3, B_3, \ldots) \overset{P_3}{\underset{\wedge}{=}} p_2(A_4, B_4, \ldots)$ to $p_1(A_1, B_1, \ldots) \overset{X}{\underset{\wedge}{=}} p_2(A_4, B_4, \ldots)$.

3.13. State and prove the dual of Problem 3.4.

3.14. Prove: If three triangles are perspective in pairs from a common point P, their axes of perspectivity are concurrent.

3.15. State and prove the converse of Problem 3.14.

3.16. In Fig. 1-1, take A_4 any other point on r; let $B_1A_4 \cdot p = C_4$ and $A_1C_4 \cdot s = B_4$. Prove $(A_1, A_2, A_3, A_4) \overline{\wedge} (B_1, B_2, B_3, B_4)$.

3.17. On any line r take four distinct points A_1, A_2, A_3, A_4 and project them from any point P, not on r, into B_1, B_2, B_3, B_4 respectively on any other line s. Construct the line of Pappus p for the triples A_1, A_2, A_3 and B_2, B_1, B_4. Show that $T = A_4B_1 \cdot B_3A_2$ is on p and $(A_1, A_2, A_3, A_4) \overline{\wedge} (B_2, B_1, B_4, B_3)$.

3.18. Prove for the points of Problem 3.17:

  (a) $(A_1, A_2, A_3, A_4) \overline{\wedge} (B_4, B_3, B_2, B_1)$ (b) $(A_1, A_2, A_3, A_4) \overline{\wedge} (B_3, B_4, B_1, B_2)$

3.19. Prove: If a complete quadrangle $PQRS$ and a complete quadrilateral $pqrs$ are so situated that the sides PQ, PR, PS, QR, QS of the quadrangle are on the vertices $r \cdot s, q \cdot s, q \cdot r, p \cdot s, p \cdot r$ respectively of the quadrilateral, then the side RS of the quadrangle is on the vertex $p \cdot q$ of the quadrilateral.

3.20. Given five points of a quadrangular set, construct the sixth point when (a) the given points are distinct, (b) two of the five points are coincident.

3.21. State and construct the dual of Problem 3.20.

3.22. Prove: If two triangles ABC and $A'B'C'$ are perspective from a point, the points $R = AB \cdot A'B'$, $S = BC \cdot B'C'$, $T = CA \cdot C'A'$, $U = AB' \cdot A'B$, $V = BC' \cdot B'C$, $W = CA' \cdot AC'$ lie by threes on four lines.

3.23. Prove the surmise in Problem 1.17, page 18. *Hint.* In the figure let $p \cdot e = E$, $p \cdot f = F$; $QE = e'$, $QF = f'$; $e' \cdot d = E'$, $f' \cdot d = F'$; $BE' = e''$, $BF' = f''$. Prove e, c', f'' concurrent.

3.24. (*a*) Prove: The diagonal triangle of a complete quadrangle is perspective with each of the four triangles whose vertices are three of the vertices of the quadrangle. (*b*) Show that the complete quadrilateral whose sides are the four axes of perspectivity obtained in (*a*) has the same diagonal triangle as the complete quadrangle.

3.25. State and prove the dual of Problem 3.24.

3.26. In Fig. 3-9, let $A = A_1B_2 \cdot B_1A_3$, $B = A_1B_3 \cdot B_1A_2$, $C = A_2B_1 \cdot B_2A_3$, $D = A_2B_3 \cdot B_2A_1$, $E = A_3B_1 \cdot B_3A_2$, $F = A_3B_2 \cdot B_3A_1$; $R = BE \cdot AF$, $S = BD \cdot AC$, $T = CE \cdot DF$. Prove: (*a*) the lines AB, CD, EF are on O, (*b*) the points R, S, T are on p.

3.27. Given, as in Fig. 3-1(*b*), the quadrilateral $pqrs$ having abc as diagonal triangle. Let $(a \cdot b)(p \cdot r) = d$, $(a \cdot b)(q \cdot s) = e$, $(b \cdot c)(p \cdot q) = f$, $(b \cdot c)(r \cdot s) = g$, $(c \cdot a)(p \cdot s) = i$, $(c \cdot a)(q \cdot r) = j$. Prove each of the triples of lines d, g, i; d, f, j; e, f, i; e, g, j concurrent.
Hint. Consider the triangles abc and prs perspective from q.

3.28. State and prove the dual of Problem 3.27.

3.29. Given a triangle abc and a line p not on a vertex. Construct a quadrilateral $pqrs$ having abc as diagonal triangle. Is the quadrilateral unique?
Hint. Let $(a \cdot b)(c \cdot p) = d$, $(a \cdot c)(b \cdot p) = g$, $(b \cdot c)(d \cdot g) = h$, then $(a \cdot h)(b \cdot g) = s$, $(a \cdot h)(c \cdot d) = r$, $(b \cdot r)(c \cdot s) = q$.

3.30. Show that the quadrilateral of Problem 3.29 is determined when the diagonal triangle and any side is given.

3.31. State and prove the dual of Problem 3.29; also, the dual of Problem 3.30.

3.32. The sides of a variable triangle pass through three fixed collinear points while two of the vertices move along fixed lines r and s. Prove that the third vertex describes a line concurrent with r and s.

3.33. Through each vertex of a given triangle a line is drawn. Show that these lines are concurrent if and only if their intersections with a given line x, together with the intersections of the sides of the triangle with x form a quadrangular set.

3.34. Two triangles $A_1A_2A_3$ and $B_1B_2B_3$ such that A_1B_1, A_2B_2, A_3B_3 are concurrent in a point C_1 while A_1B_2, A_2B_3, A_3B_1 are concurrent in a point C_2 are said to be *doubly perspective*; if, in addition A_1B_3, A_2B_1, A_3B_2 are concurrent in a point C_3, the triangles are said to be *triply perspective*.
(*a*) Given a triangle $A_1A_2A_3$ and two distinct points B_1, B_2, locate B_3 so that the triangles $A_1A_2A_3$ and $B_1B_2B_3$ are doubly perspective.
(*b*) Verify that the two triangles of (*a*) are triply perspective.
(*c*) Prove: If two triangles $A_1A_2A_3$ and $B_1B_2B_3$ are perspective in any two of the orders

$$A_1, A_2, A_3 \; \overline{\overline{\wedge}} \; B_1, B_2, B_3 \qquad A_1, A_2, A_3 \; \overline{\overline{\wedge}} \; B_2, B_3, B_1 \qquad A_1, A_2, A_3 \; \overline{\overline{\wedge}} \; B_3, B_1, B_2$$

they are also perspective in the third order.
(*d*) Show that any two of the triangles $A_1A_2A_3, B_1B_2B_3, C_1, C_2, C_3$ are triply perspective with the vertices of the third triangle as centers of perspectivity.

3.35. Show that the vertices of the three triangles of Problem 3.34(*d*) lie on three lines in accordance with the Theorem of Pappus.

3.36. Prove the theorem of Problem 3.1 using the Theorem of Pappus.
Hint. In Fig. 3-7 let $A_2A_3 \cdot B_1B_3 = Q$, $A_2B_1 \cdot A_3B_3 = R$, $A_1A_2 \cdot PQ = S$, $B_1B_2 \cdot PQ = T$. Then, applying the Theorem of Pappus to the collinear triples:
(*a*) A_1, P, B_1 and A_3, Q, A_2,   find C_2, R, S collinear;
(*b*) B_2, P, A_2 and B_3, Q, B_1,   find C_1, R, T collinear;
(*c*) B_1, A_2, R and Q, S, T,   find C_2, C_1, C_3 collinear.

Chapter 4

Harmonic Sets

HARMONIC SETS OF POINTS AND LINES

Consider in Fig. 4-1 the set A, B, D, E of four points obtained when the sides of the complete quadrangle $PQRS$ are sectioned by the side $c = AB$ of its diagonal triangle. Fig. 4-1', the dual of Fig. 4-1, exhibits the set a, b, d, e of four lines obtained when the vertices of the complete quadrilateral $pqrs$ are projected from the vertex $C = a \cdot b$ of its diagonal triangle. An examination of these figures suggests the following dual definitions.

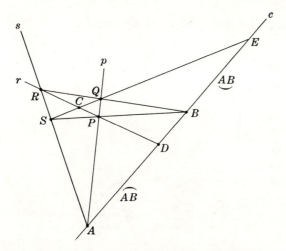

Fig. 4-1

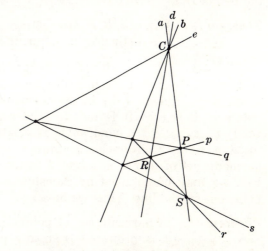

Fig. 4-1'

Four collinear points A, B, D, E are said to form a *harmonic set (harmonic tetrad) of points* when there exists a complete quadrangle having two opposite sides on A, another two opposite sides on B, while the third pair of opposite sides are singly on D and E. The point D (E) is called the *harmonic conjugate* of E (D) with respect to A and B. This relation will be indicated by writing $H(A, B; D, E)$. Notice that in $H(A, B; D, E)$ A and B play equal roles; hence we might have written $H(B, A; D, E)$. Also, D and E play equal roles; hence we might have used $H(A, B; E, D)$ or $H(B, A; E, D)$. In any case, we say that the pair A, B is *separated harmonically* by the pair D, E.

When A, B, D are distinct points on a line c, the harmonic conjugate of D with

Four concurrent lines a, b, d, e are said to form a *harmonic set (harmonic tetrad) of lines* when there exists a complete quadrilateral having two opposite vertices on a, another two opposite vertices on b, while the third pair of opposite vertices are singly on d and e. The line d (e) is called the *harmonic conjugate* of e (d) with respect to a and b. This relation will be indicated by writing $H(a, b; d, e)$. Notice that in $H(a, b; d, e)$ a and b play equal roles; hence we might have written $H(b, a; d, e)$. Also, d and e play equal roles; hence we might have used $H(a, b; e, d)$ or $H(b, a; e, d)$. In any case, we say that the pair a, b is *separated harmonically* by the pair d, e.

When a, b, d are distinct lines on a point C, the harmonic conjugate of d with respect

respect to A and B may be constructed as follows: On A take any two distinct lines $p \neq c$ and $s \neq c$; on D take any line $r \neq c$. Let $p \cdot r = P$, $s \cdot r = R$; $BP \cdot s = S$, $BR \cdot p = Q$. Then $SQ \cdot c = E$ is the required point.

to a and b may be constructed as follows: On a take any two distinct points $P \neq C$ and $S \neq C$; on d take any point $R \neq C$. Let $PR = p$, $RS = r$; $(b \cdot p)S = s$, $(b \cdot r)P = q$. Then $(s \cdot q)C = e$ is the required line.

First, a word about the symbol $H(A,B;D,E)$ adopted here. From Fig. 4-1, we find

$$(i) \qquad (A,B,D,E) \overset{R}{\barwedge} (S,Q,C,E) \overset{P}{\barwedge} (B,A,D,E)$$

so that

$$(A,B;D,E) = (B,A;D,E)$$

Now in view of the assumption concerning the diagonal points of complete quadrangles, we conclude $(A,B;D,E) = -1$ and so may take $H(A,B;D,E)$ as equivalent to this. However, to avoid cross ratios, we will consider $H(A,B;D,E)$ as merely shorthand for "the collinear points A,B,D,E form a harmonic set" or "the points A,B are separated harmonically by the points D,E". We have

Theorem 4.1. When A,B,D are distinct collinear points, $H(A,B;D,E)$ implies A,B,D,E are distinct collinear points.

Fig. 4-2 is Fig. 4-1′ with such changes in labels as to make clear that the line s sections the harmonic set of lines a,b,d,e in the harmonic set of points A,B,D,E. (It is left for the reader to verify that $PQRS$ is a quadrangle associated with these points.) Conversely, beginning with any harmonic set of points on a line, a figure such as Fig. 4-2 may be obtained by choosing as R any point not on the line. Thus we have

Theorem 4.2. A harmonic set of points on a line s is projected from any point R, not on s, by a harmonic set of lines.

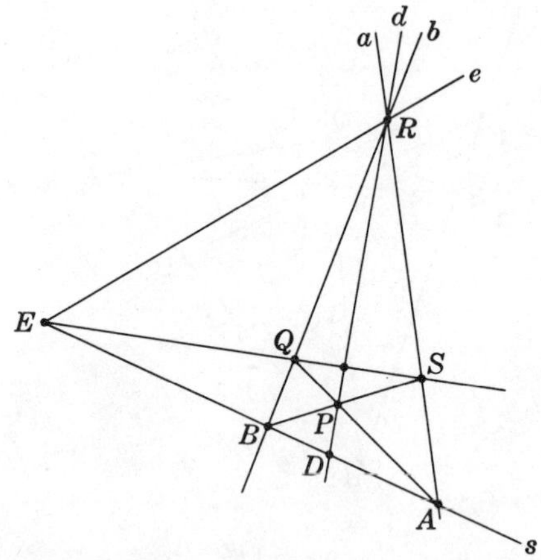

and its dual

Fig. 4-2

Theorem 4.2′. A harmonic set of lines on a point R is sectioned by any line, not on R, in a harmonic set of points.

Let four concurrent lines a,b,d,e be sectioned by the distinct lines r and s in the points A,B,D,E and A',B',D',E' respectively. By definition $(A,B,D,E) \overset{}{\barwedge} (A',B',D',E')$ and Theorems 4.2-4.2′ imply: if $H(A,B;D,E)$, then $H(A',B';D',E')$. Since a projectivity is a sequence of perspectivities, we have for any two sets of collinear points A,B,D,E and A'',B'',D'',E'',

Theorem 4.3. If $(A,B,D,E) \overset{}{\barwedge} (A'',B'',D'',E'')$, then

$$H(A,B;D,E) \quad \text{implies} \quad H(A'',B'';D'',E'')$$

and conversely.

Let A,B,D,E be four collinear points such that $H(A,B;D,E)$. By Theorem 2.10, page 25, $(A,B,D,E) \overset{}{\barwedge} (D,E,A,B)$. Then $H(D,E;A,B)$ and we have

Theorem 4.4. If A,B,D,E are four collinear points such that the pair A,B is separated harmonically by the pair D,E, then the pair D,E is separated harmonically by the pair A,B.

Suppose, as in Fig. 4-1, that D is taken on the segment $\widehat{AB}$ of c; clearly, when $D \neq A$ and $D \neq B$, then E is on the segment $\underset{\frown}{AB}$. In Problem 4.6 the reader is asked to verify that when D is near A then E is near A and when D is near B then E is near B. Let D describe the segment $\widehat{AB}$, moving from A to B. This motion of D does not disturb the points R, Q and, hence, does not disturb the lines AR, AQ, BR. On the contrary, the line CR revolves counterclockwise about R, the point P moves from A to Q over the segment shown in the figure, the line BP revolves clockwise about B, the point S moves from A to R over the segment shown, and E moves from A to B over $\underset{\frown}{AB}$. We leave for the reader to investigate the effect of reversing the direction of motion of D and state

Theorem 4.5. If D describes $\widehat{AB}$ in a given sense, the harmonic conjugate E of D with respect to A, B describes $\underset{\frown}{AB}$ in the opposite sense.

The directions for constructing the harmonic conjugate of D with respect to A, B allow the two lines on A and the line on D to be any whatever, except that all be distinct from $o = AB$ and that the two on A be distinct from each other. This freedom of choice follows from Theorem 3.5, page 38, which states in effect that the location of E is fixed by that of the other three points and is independent of the particular quadrangle used. Thus Theorem 3.5 may be restated here as

Theorem 4.6. If A, B, D are collinear points, the harmonic conjugate of D with respect to A, B is unique.

and its dual as

Theorem 4.6'. If a, b, d are concurrent lines, the harmonic conjugate of d with respect to a, b is unique.

In Problem 4.1, we prove

Theorem 4.7. If A and B are two vertices of a complete quadrangle, E is its diagonal point on AB, and D is on AB and the join of the other diagonal points, then $H(A, B; D, E)$.

In Problem 4.2, we prove

Theorem 4.8. If A', B', C' are three distinct collinear points, one on each side of a given triangle ABC, the lines which join the harmonic conjugate of each of these points, with respect to the two vertices on the side of the triangle, to the third vertex are concurrent.

Theorem 4.8 provides yet another procedure for proving three distinct lines concurrent. Its dual is

Theorem 4.8'. If a', b', c' are three distinct concurrent lines, one on each vertex of a triangle ABC, the points in which the harmonic conjugate of each of these lines, with respect to the two sides on the vertex of the triangle, meets the third side are collinear.

HARMONIC PROPERTIES OF A COMPLETE QUADRANGLE

In a complete quadrangle, any two of its vertices are collinear with one vertex of its diagonal triangle. Thus (see Fig. 4-3 below) each side of the quadrangle meets its diagonal triangle in two points — a diagonal point and an additional point which will be called a *harmonic point* associated with the quadrangle. For example, the side PQ of the complete quadrangle meets the diagonal triangle in the diagonal point A and in the harmonic point $F = PQ \cdot BC$. In addition to F, the harmonic points for the complete quadrangle of Fig. 4-3 are $D = PR \cdot AB$, $E = QS \cdot AB$, $G = RS \cdot BC$, $I = PS \cdot AC$, $J = QR \cdot AC$. We have

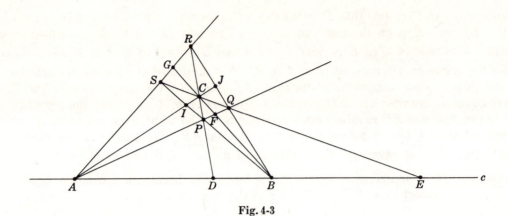

Fig. 4-3

Theorem 4.9. On any side of a complete quadrangle are two vertices of the quadrangle, a vertex of its diagonal triangle, and one of the harmonic points.

Theorem 4.10. On any side of the diagonal triangle of a complete quadrangle are two vertices of the diagonal triangle and two harmonic points.

By projecting A, B, D, E from C onto RS, we obtain A, G, R, S respectively. Thus,

Theorem 4.11. The vertex of the diagonal triangle and the harmonic point on any side of a complete quadrangle separate harmonically the vertices of the quadrangle on that side.

Now the lines BR, BS, BA, BG are a harmonic set (Theorem 4.2) and their section J, I, A, C by the line AC is a harmonic set (Theorem 4.2'). Thus we have

Theorem 4.12. The two sides of a complete quadrangle on any vertex of its diagonal triangle are separated harmonically by the sides of the diagonal triangle on that vertex.

and

Theorem 4.13. The diagonal points on any side of the diagonal triangle of a complete quadrangle are separated harmonically by the harmonic points (associated with the quadrangle) on that side.

From Problem 3.28, page 44, we have

Theorem 4.14. The six harmonic points associated with a complete quadrangle lie by threes on four lines.

HARMONIC NET ON A PROJECTIVE LINE

In Fig. 4-4, A, B, C are three distinct points on the line o. On A take any line $q \neq o$ and on q take distinct points $R \neq A$ and $S \neq A$. Let $BR \cdot CS = D$, $AD = t$; $CR \cdot t = E_1$, $SE_1 \cdot o = F_1$; $F_1R \cdot t = E_2$, $SE_2 \cdot o = F_2$; $F_2R \cdot t = E_3$, $SE_3 \cdot o = F_3$; .... By construction, we have $H(A, C; B, F_1)$, $H(A, F_1; C, F_2)$, $H(A, F_2; F_1, F_3)$, ..., $H(A, F_{n-1}; F_{n-2}, F_n)$, ... The set of points

$$A, B, C, F_1, F_2, F_3, F_4, \ldots \tag{1}$$

are said to constitute a *harmonic sequence* generated by the three given points A, B, C. Since the elements of any harmonic set are distinct, it seems reasonable to suppose that all points of (*1*) are distinct. (A proof will be given later.) Assuming this true, it is clear, since the construction may be continued indefinitely, that the number of points in the harmonic sequence is infinite.

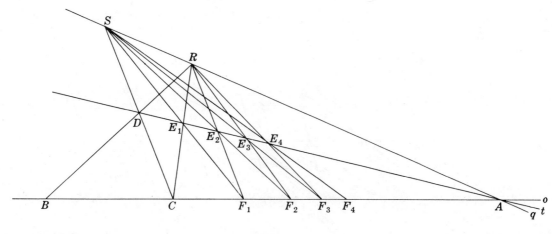

Fig. 4-4

Let us begin once more with the points A, B, C on o. First construct F_1 as before and label it G_1; then construct additional points on o to obtain a set

$$A, B, C, G_1, G_2, G_3, G_4, G_5, \ldots \tag{2}$$

such that for $W = G_n$, where n is any positive integer, we have $H(X, Y; Z, W)$, that is, W is the harmonic conjugate of Z with respect to X and Y, for *some* triple of points X, Y, Z which precede G_n in (2). For example, G_4 might be such that $H(A, G_3; G_2, G_4)$ or $H(B, C; G_1, G_4)$ or $H(G_1, G_3; G_2, G_4)$. In any event $W = G_n$ is said to be *harmonically related* to the given points A, B, C. The set of all points harmonically related to A, B, C is called a *harmonic net* (*net of rationality*) on o and is denoted by $R(A, B, C)$. We note first of all that there is no systematic procedure, that is no rule of order, for locating the points $G_2, G_3, G_4, G_5, \ldots$. Thus it will not be expected that all the G's are necessarily distinct. Since the harmonic sequence generated by A, B, C is included in $R(A, B, C)$, the harmonic net is also an infinite set of points. The two sets, however, are never identical; for example, G_4 defined by $H(B, C; G_1, G_4)$ is a member of the harmonic net but not of the harmonic sequence.

In Problem 4.4, we prove

Theorem 4.15. A harmonic net is determined by any three distinct points of the net.

HARMONIC NET ON AN AUGMENTED LINE

We go back for a moment to the augmented plane of Chapter 1 which, it will be recalled, consists of the Euclidean plane together with a line of ideal points. Our purpose is to construct the harmonic net $R(P_\infty, P_0, P_1)$ where P_0 and P_1 are any two distinct ordinary points of the augmented line (p, P_∞). Since this may be done, as will be seen, in a systematic manner, we shall gain thereby a clearer picture of the harmonic net of the preceding section. In the construction (see Fig. 4-5 below) use will be made of the distinct ideal points $R_\infty \neq P_\infty$ and $S_\infty \neq P_\infty$ on the ideal line l_∞. Let $P_0 R_\infty \cdot P_1 S_\infty = E_0$, $E_0 P_\infty = t$; $P_1 R_\infty \cdot t = E_1$, $E_1 S_\infty \cdot p = P_2$; $P_2 R_\infty \cdot t = E_2$, $E_2 S_\infty \cdot p = P_3; \ldots$. That P_2 is the harmonic conjugate of P_0 with respect to P_∞, P_1, and thus $H(P_\infty, P_1; P_0, P_2)$, is verified by the following considerations: On P_∞ are the lines t and l_∞; on P_0 is the line $P_0 R_\infty$ meeting t in E_0 and l_∞ in R_∞. Now $P_1 E_0$ meets l_∞ in S_∞, $P_1 R_\infty$ meets t in E_1 and $E_1 S_\infty$ meets p in P_2. Consider now a coordinate system established on p in which P_0 is the *origin* (i.e. the point with coordinate 0) and P_1 is the *unit point* (i.e. the point with coordinate 1). Since

$$P_0 P_2 = P_0 P_1 + P_1 P_2 = P_0 P_1 + E_0 E_1$$

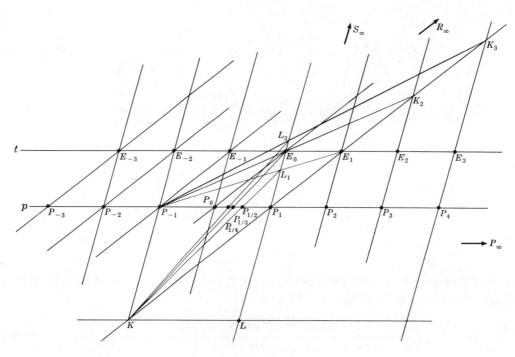

Fig. 4-5

and $E_0E_1 = P_0P_1$, it follows that P_2 is the point on p with coordinate 2. Thus by successive steps using $H(P_\infty, P_{n-1}; P_{n-2}, P_n)$, we can construct P_n, having coordinate n, for $n = 2, 3, 4, \ldots$.

Let $P_0S_\infty \cdot t = E_{-1}$, $E_{-1}R_\infty \cdot p = P_{-1}$; $P_{-1}S_\infty \cdot t = E_{-2}$, $E_{-2}R_\infty \cdot p = P_{-2}$; $\ldots$. Now P_{-1} is the point on p having -1 as coordinate, P_{-2} is the point on p having -2 as coordinate, $\ldots$. Thus by successive steps using $H(P_\infty, P_{1-n}; P_{2-n}, P_{-n})$, we can construct P_{-n}, having $-n$ as coordinate, for $n = 1, 2, 3, 4, \ldots$.

Next, let P_x be defined by $H(P_1, P_{-1}; P_2, P_x)$. In the construction (see Fig. 4-5) take on P_1 the lines P_1R_∞ and P_1S_∞ and take on P_2 the line P_2S_∞ meeting P_1R_∞ in $K_1 = E_1$ and P_1S_∞ in S_∞. Let $P_{-1}K_1 \cdot P_1S_\infty = L_1$ and $P_{-1}S_\infty \cdot P_1R_\infty = K$; then $P_x = KL_1 \cdot p$. We show that $P_x = P_{1/2}$, the point on p having $\frac{1}{2}$ as coordinate. Let $KP_\infty \cdot P_1S = L$. From the similar triangles $L_1P_xP_1$ and L_1KL, we have

$$\frac{P_xP_1}{KL} = \frac{P_1L_1}{LL_1} = \frac{\frac{2}{3}P_1E_0}{\frac{8}{3}P_1E_0} = \frac{1}{4}$$

Then $P_xP_1 = \frac{1}{4}KL = \frac{1}{4} \cdot 2 = \frac{1}{2}$; $P_0P_x = \frac{1}{2}$ and $P_x = P_{1/2}$. Similarly, (see Problem 4.28), we can construct $P_{1/n}$, having $1/n$ as coordinate, for $n = 3, 4, 5, \ldots$. Finally, beginning anew with any triple of points $P_\infty, P_0, P_{1/n}$ and repeating the constructions of the two paragraphs above, we locate the points

$$\ldots, P_{-3/n}, P_{-2/n}, P_{-1/n}, P_0, P_{1/n}, P_{2/n}, P_{3/n}, \ldots.$$

Thus far a systematic procedure has been outlined for locating on an augmented line a set of points (i.e. a subset of the ordinary points of the line) in one-to-one correspondence with the set of all rational numbers. Let P_a, P_b, P_c be any three distinct points of the subset and consider the point P_x defined by $H(P_a, P_b; P_c, P_x)$. Then (see Chapter 2)

$$(P_a, P_b; P_c, P_x) = \frac{a-c}{a-x} \cdot \frac{b-x}{b-c} = -1$$

and, since a, b, c are rational numbers, so also is x. Thus P_x is a point of the subset and so the subset is precisely $R(P_\infty, P_0, P_1)$.

Returning to Fig. 4-4 in the projective plane, let us relabel A as P_∞, B as P_0, C as P_1, D as E_0, $F_1 = G_1$ as P_2 and repeat systematically the constructions outlined above. The resulting subset of points on the projective line (i.e. the harmonic net) is again in one-to-one correspondence with the set of all rational numbers. Here, however, we are not concerned with any possible interpretation of the subscript a of the point P_a of the net; the subscript serves simply as a convenient label and nothing more.

Solved Problems

4.1. Prove: If A and B are two vertices of a complete quadrangle, E is its diagonal point on AB, and D is on AB and the join of the other diagonal points, then $H(A, B; D, E)$.

The theorem follows immediately from Fig. 4-1 by taking $ABQS$ as the complete quadrangle. Give directions for constructing D when A, B, E (in that order) are given on a line; also, when A, E, B (in that order) are given.

4.2. Prove: If A', B', C' are three distinct collinear points, one on each side of a given triangle ABC, the lines which join the harmonic conjugates of each of these points, with respect to the two vertices on the side of the triangle, to the third vertex are concurrent.

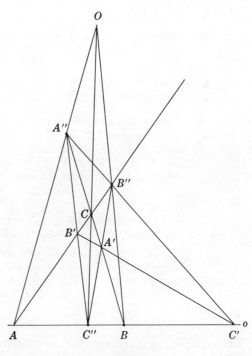

Fig. 4-6

Refer to Fig. 4-6. Construct the harmonic conjugate C'' of C' with respect to A, B; then $H(A, B; C', C'')$. Let $A'' = BC \cdot B'C''$ and $B'' = AC \cdot A'C''$. That A'' and B'' are the remaining harmonic conjugates of the theorem follows from

$$(A, B, C', C'') \overset{B'}{\doublebarwedge} (C, B, A', A'')$$

and $$(A, B, C', C'') \overset{A'}{\doublebarwedge} (A, C, B', B'')$$

Now

$$(A, C, B', B'') \overset{A'}{\doublebarwedge} (A, B, C', C'') \overset{B'}{\doublebarwedge} (C, B, A', A'')$$

and so $$(A, C, B', B'') \barwedge (C, B, A', A'')$$

From (i), page 46,

$$(C, B, A', A'') \barwedge (B, C, A', A'')$$

and so $$(A, C, B', B'') \barwedge (B, C, A', A'')$$

But this is a perspectivity; hence $AB, A'B', A''B''$ are on C'. Similarly, $BC, B'C', B''C''$ are on A' and $CA, C'A', C''A''$ are on B'. Then the triangles ABC and $A''B''C''$ are perspective from the line $A'B'C'$. Finally, they are perspective from a point, that is, AA'', BB'', CC'' are concurrent as required.

Note. There follows from $A'' = BC \cdot B'C''$, $B'' = AC \cdot A'C''$, $C'' = AB \cdot A'B''$: If the sides of a triangle are cut by any line not on a vertex, the harmonic conjugates of any two of these intersections, each with respect to the vertices on the side of the triangle, are collinear with the third intersection.

4.3. Consider in Fig. 4-7 the complete quadrangle $PQRS$, its diagonal triangle ABC, and the associated harmonic points D, E, F, G, I, J. Let $DI \cdot PQ = K$, $DJ \cdot RS = L$, $EI \cdot RS = M$, $EJ \cdot PQ = N$. Prove the triples of points (a) B, K, M and (b) B, N, L are collinear.

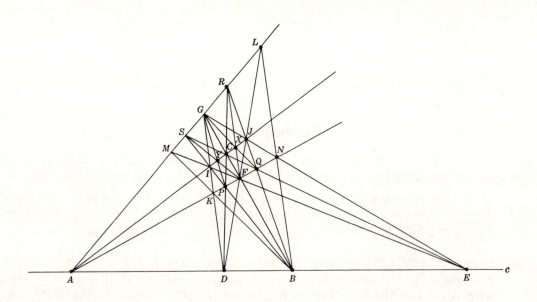

Fig. 4-7

(a) We have $(A, C, I, J) \overset{E}{\underset{\wedge}{=}} (A, S, M, G)$ and

$$(A, C, I, J) \overset{D}{\underset{\wedge}{=}} (A, P, K, F) \overset{B}{\underset{\wedge}{=}} (A, S, M_1, G)$$

where $M_1 = BK \cdot RS$. Since $H(A, C; I, J)$ we have $H(A, S; M, G)$ and $H(A, S; M_1, G)$. Then, by Theorem 4.6, $M_1 = M$ and so M is on BK.

(b) A proof similar to the above may be given. We vary the procedure here by attempting a proof using Theorem 4.8'. First, we need a triangle having B, L, N singly on its sides; try triangle AQR. Next, we need a point X such that the harmonic conjugate of AX with respect to AQ and AR meets RQ in B, the harmonic conjugate of QX with respect to QA and QR meets AR in L, and the harmonic conjugate of RX with respect to RA and RQ meets AQ in N. Since

$$(G, F, C, B) \overset{A}{\underset{\wedge}{=}} (R, Q, J, B)$$

we have $H(R, Q; J, B)$; then X must lie on AJ. Similarly, $H(A, R; G, L)$ requires X to lie on GQ and $H(A, Q; F, N)$ requires X to lie on FR.

The proof then consists in showing that AJ, GQ, FR are concurrent. Consider on RS the distinct points S, G, R and on PQ the distinct points P, F, Q. By the theorem of Problem 3.4, page 40, the points $SF \cdot PQ = Y$, $SQ \cdot PR = C$, $GQ \cdot FR = X$ are collinear with A. Thus X lies on AC; the lines AJ, GQ, FR are concurrent; and the points B, L, N are collinear.

4.4. Prove: A harmonic net is determined by any three distinct points of the net.

Let X, Y, Z be three distinct points of (2), the harmonic net $R(A, B, C)$. Then, beginning with A, B, C we obtain X, Y, Z after constructing a finite number of harmonic sets, the first of which is $H(A, C; B, G_1)$. Now $H(A, C; B, G_1)$ implies $H(A, C; G_1, B)$. Hence the points harmonically related to A, B, C are also harmonically related to A, G_1, C, that is, $R(A, B, C) = R(A, G_1, C)$. Since Y is harmonically related to A, G_1, C we have, after a finite number of steps, $R(A, B, C) = R(A, Y, C)$ and, similarly, $R(A, B, C) = R(X, Y, C) = R(X, Y, Z)$.

Supplementary Problems

4.5. If a, b, d, e are four distinct concurrent lines such that $H(a, b; d, e)$, there are seven other arrangements of these lines which are also harmonic sets. List them.
Partial Ans. $H(a, b; e, d)$, $H(d, e; a, b)$.

4.6. On a line take three distinct points A, D, B in that order. Construct the harmonic conjugate of D with respect to A and B when *(a)* D is near A, *(b)* D is near B, *(c)* D is equidistant from A and B.

4.7. State the dual of Problem 4.6 and make the constructions.

4.8. What would you consider a natural choice for the harmonic conjugate of D with respect to A and B of Problem 4.6 when *(a)* $D = A$, *(b)* $D = B$?

4.9. Prove: $H(A, B; C, D)$ and $H(A', B'; C', D)$ on distinct lines implies AA', BB', CC' concurrent.
Hint. $(A, B, C, D) \barwedge (A', B', C', D)$.

4.10. Prove: $H(A, B; D, E)$ if and only if $(A, B, D, E) \barwedge (A, B, E, D)$.

4.11. Call the line of Pappus p in Fig. 3-9, page 40, the *polar line of O with respect to r and s* and call O the *pole of the line p with respect to r and s*. Let $A_1 B_1 \cdot p = Q_1$. Show that Q_1 is the harmonic conjugate of O with respect to A_1 and B_1. Thus prove: As a line revolves about a point O meeting two distinct lines r and s (neither on O) in the points R and S respectively, the harmonic conjugate of O with respect to R and S describes the polar line of O with respect to r and s.

4.12. Discuss the dual of Problem 4.11.

4.13. For any position of the revolving line in Problem 4.11, let $RS \cdot p = Q$. Show that the polar line of Q with respect to r and s passes through O. Thus, prove: If, with respect to two distinct lines r and s, the polar line of a point O is on a point Q, then the polar line of Q with respect to r and s is on O.

4.14. Discuss the dual of Problem 4.13.

4.15. On a line p take four distinct points A, B, C, D such that $H(A, B; C, D)$. Project these points from any point P, not on p, and section by any other line q on A, but not on P, obtaining the points A, B', C', D', respectively. Show that $Q = CD' \cdot C'D$ is on the line PB.

4.16. State and prove the dual of Problem 4.15.

4.17. Using Fig. 4-3, page 48, show that the complete quadrilateral whose vertices are the six harmonic points of the complete quadrangle $PQRS$ has the same diagonal triangle as the quadrangle.

4.18. Use the result of Problem 4.17 to obtain an alternate construction to that of Problem 3.29, page 44, of the complete quadrilateral when its diagonal triangle and one side are given.

4.19. Give an alternate construction to that of Problem 3.31, page 44, of the complete quadrangle when its diagonal triangle and one vertex are given.

4.20. Prove Theorem 4.2, page 46, without using the notion of cross ratio. *Hint.* In Fig. 4-1, let $QA = a$, $QB = b$, $QD = d$, $QE = e$; $AB = p$, $BS = q$, $AR = r$, $PR = s$ and obtain $H(a, b; d, e)$ from the complete quadrilateral $pqrs$.

4.21. Using Fig. 4-6, prove:
(a) Theorem 4.8′.
  Hint. Take $a' = OA$, $b' = OB$, $c' = OC$.
(b) Let the lines joining the vertices of the triangle ABC to a fixed point O, not on a side, meet the opposite sides in A'', B'', C'', then the points in which each side of the given triangle meets the join of the two points (of A'', B'', C'') not on that side are collinear.

4.22. *(a)* In Fig. 4-7 there are four sets of harmonic points on RS; list them. Show that $(A, G; M, R) = (A, G; L, S) = -\frac{1}{2}$.
(b) Excluding the points X, Y from Fig. 4.7, there are in all twelve sets of harmonic lines; list them.

4.23. Using Fig. 4-7 prove: The triples of points K, C, L and M, C, N are collinear.

4.24. Let r and s be two distinct lines intersecting at O and let P be any point not on either of these lines. Construct the harmonic conjugate of OP with respect to r and s. (a) Show that this line is the line of Pappus for any triple of points A_1, A_2, A_3 on r and B_1, B_2, B_3 on s so selected that A_1B_1, A_2B_2, A_3B_3 are concurrent at P, that is, show that the line is the polar line of P with respect to r and s. (b) Show that any line on P which meets the lines r, s in distinct points R, S respectively meets the polar line of P with respect to r and s in a point T such that $H(R, S; P, T)$.

4.25. Given a triangle ABC and a point P distinct from the vertices, construct the polar lines c, b, a of P with respect to the pairs of lines AC, CB; AB, BC; BA, AC respectively. Let $c \cdot b = A'$, $c \cdot a = B'$, $b \cdot a = C'$. (a) Show that the triangles ABC and $A'B'C'$ are perspective. Label the axis of perspectivity p. (b) Show that p is the polar line of P with respect to the triangle ABC as defined in Problem 3.11, page 43. (c) Construct the polar line of Q, any point on p, with respect to each of the triangles ABC and $A'B'C'$ and show that these lines meet on p.

4.26. In Fig. 4-8, A, B, D are any three distinct points on a line while A', B', D' are the harmonic conjugates of each of these points with respect to the other two, that is, $H(A, B; D, D')$, $H(B, D; A, A')$, $H(D, A; B, B')$. The points $C = PR \cdot QS$, $E = PQ \cdot ST$, $F = PS \cdot QU$ are diagonal points of complete quadrangles associated with the harmonic sets and $G = PR \cdot ST$. Prove:

(a) $G = PR \cdot QU$.

    Hint. Suppose $G_1 = PR \cdot QU$. Then $(A, B, D, D') \overset{P}{\overline{\wedge}} (Q, F, U, G_1) \overset{S}{\overline{\wedge}} (D, B, A, A_1')$.

(b) The triples A, C, F; B, C, E; D, E, F are collinear.

    Hint. Suppose $AC \cdot UQ = F_1$. Then $(A, B, D, D') \overline{\overline{\wedge}} (S, Q, D, C) \overline{\overline{\wedge}} (U, Q, B', F_1)$ and $(D, A, B, B') \overline{\overline{\wedge}} (Q, U, F, B')$.

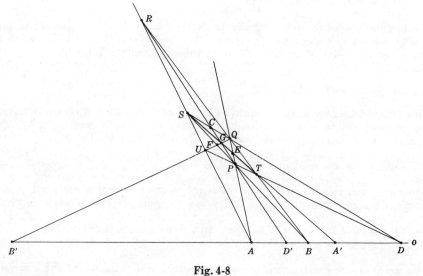

Fig. 4-8

4.27. Using Fig. 4-5, verify:

(a) $H(P_\infty, P_2; P_1, P_3)$ and $H(P_\infty, P_3; P_2, P_4)$.

    Hint. Take, as before, the lines t and l_∞ on P_∞ and on P_1 take the line P_1R_∞ meeting t in E_1 and l_∞ in R_∞. Then $P_2E_1 \cdot l_\infty = S_\infty$, $P_2R_\infty \cdot t = E_2$, and $E_2S_\infty \cdot p = P_3$.

(b) $H(P_\infty, P_{-1}; P_0, P_{-2})$ and $H(P_\infty, P_{-2}; P_{-1}, P_{-3})$.

(c) $H(P_1, P_{-1}; P_3, P_{1/3})$ and $H(P_1, P_{-1}; P_4, P_{1/4})$.

    Hint. On P_1 take P_1R_∞ and P_1S_∞, on P_3 take P_3S_∞ meeting P_1R_∞ in K_2 and P_1S_∞ in S_∞. Let $P_{-1}K_2 \cdot P_1S_\infty = L_2 = E_0$; then $L_2K \cdot p = P_{1/3}$.

(d) $H(P_\infty, P_0; P_n, P_{-n})$ for $n = 1, 2, 3$.

    Hint. On P_∞ take t and l_∞; on P_n take P_nR_∞ meeting t in E_n and l_∞ in R_∞. Then $P_0R_\infty \cdot t = E_0$ and $[E_0(P_0E_n \cdot l_\infty)] \cdot p = E_{-n}$.

4.28. Using Fig. 4-5, construct: (a) P_5; (b) $P_{1/5}, P_{2/5}, P_{3/5}$; (c) $P_{-1/5}, P_{-2/5}, P_{-3/5}$.

4.29. On a projective line take three distinct points and label them in order P_0, P_1, P_∞. Construct and label a number of points of $R(P_\infty, P_0, P_1)$.

Chapter 5

Projectivities

PROJECTIVITIES AND THE PAPPUS CONFIGURATION

In Chapter 1 a projectivity, $p_1 \overline{\wedge} p_n$, between two distinct pencils of points on distinct lines, $p_1 \neq p_n$, or on the same line, $p_1 = p_n$, was said to be established whenever the two pencils in question were members of a sequence of perspectivities

$$p_1 \ \overset{P_1}{\overline{\wedge}} \ p_2 \ \overset{P_2}{\overline{\wedge}} \ p_3 \ \overset{P_3}{\overline{\wedge}} \ \cdots \ \overset{P_{m-2}}{\overline{\wedge}} \ p_{m-1} \ \overset{P_{m-1}}{\overline{\wedge}} \ p_m$$

Dually, a projectivity, $P_1 \overline{\wedge} P_n$, between two distinct pencils of lines on distinct points, $P_1 \neq P_n$, or on the same point, $P_1 = P_n$, was said to be established whenever the two pencils in question were members of a sequence of perspectivities

$$P_1 \ \overset{p_1}{\overline{\wedge}} \ P_2 \ \overset{p_2}{\overline{\wedge}} \ P_3 \ \overset{p_3}{\overline{\wedge}} \ \cdots \ \overset{p_{m-2}}{\overline{\wedge}} \ P_{m-1} \ \overset{p_{m-1}}{\overline{\wedge}} \ P_m$$

By the Fundamental Theorem, every projectivity between two pencils of like forms is completely determined when three elements of one pencil and their correspondents in the other pencil are given. Combining this with the results of Problem 1.4, page 15, we have

Theorem 5.1. A projectivity between two pencils of points on distinct lines (two pencils of lines on distinct points), which is not a perspectivity, can always be expressed as a product of exactly two perspectivities. A projectivity between two superposed distinct pencils can always be expressed as a product of not less than two and not more than three perspectivities.

The problem of constructing the projectivity when three distinct elements of one pencil and their correspondents in the other pencil are given, is essentially that of locating the correspondent of any fourth element of either pencil. Precisely:

If distinct points A, B, C, D on one line and distinct points A', B', C' on another line or if distinct points A, B, C, D; A', B', C' on the same line are given, locate the point D' which, in the projectivity established by the triples of points A, B, C and A', B', C', is the correspondent of D. One procedure for solving this problem is as follows:

(i) Establish the projectivity $(A, B, C) \overline{\wedge} (A', B', C')$ by a minimum sequence of perspectivities,

(ii) Note the effect of this sequence of perspectivities on D.

A somewhat less tedious procedure will now be given. For this purpose, consider in Fig. 5-1 below the two projective pencils of points $r(A, B, C, D, \ldots)$ and $s(A', B', C', D', \ldots)$ on distinct lines. Let $AA' = a$, $BA' = b$, $CA' = c$, $DA' = d$, $\ldots$; $AB' = b'$, $AC' = c'$, $AD' = d', \ldots$. Since

$$A'(a, b, c, d, \ldots) \ \overline{\overline{\wedge}} \ r(A, B, C, D, \ldots) \ \overline{\wedge} \ s(A', B', C', D', \ldots) \ \overline{\overline{\wedge}} \ A(a, b', c', d', \ldots)$$

we have
$$A'(a, b, c, d, \ldots) \ \overline{\wedge} \ A(a, b', c', d', \ldots)$$

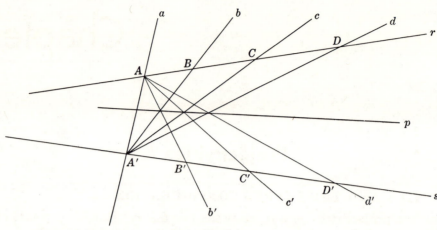

Fig. 5-1

But this is a perspectivity (why?) and so the points $b \cdot b'$, $c \cdot c'$, $d \cdot d'$, ... are on a line p, the axis of the perspectivity $A' \stackrel{p}{=} A$. This line p will now be called the *axis of projectivity* of $r \barwedge s$. Since p contains $b \cdot b' = BA' \cdot B'A$ and $c \cdot c' = CA' \cdot C'A$, it follows that p is the line of Pappus for the triples A, B, C and A', B', C'. Now the line of Pappus q for the triples A, B, D and A', B', D' is on both $b \cdot b'$ and $d \cdot d'$; hence, $q = p$. We have proved

Theorem 5.2. If two pencils of points $r(A, B, C, D, \ldots)$ and $s(A', B', C', D', \ldots)$ on distinct lines are projective, the axis of projectivity is the line of Pappus for any triple of distinct points of one pencil and their correspondents in the other pencil.

Its dual is

Theorem 5.2'. If two pencils of lines $R(a, b, c, d, \ldots)$ and $S(a', b', c', d', \ldots)$ on distinct points are projective, the *center of projectivity* is the point of Pappus for *any* triple of distinct lines of one pencil and their correspondents in the other pencil.

We return to our problem and state the following partial solution:

If A, B, C, D are distinct points on a line r (see Fig. 5-2) and A', B', C' are distinct points on another line s, the correspondent D' of D on r in the non-perspective projectivity established by the triples A, B, C and A', B', C' is located by

(a) Constructing the line of Pappus p for the two triples,

(b) Joining D to any of the points A', B', C' (say C') and marking its intersection X'' with p,

(c) Joining X'' and C and marking its intersection D' with s.

In this construction $C'D \cdot CD'$ is on p and by Theorem 5.2, D and D' are correspondents in the projectivity $(A, B, C) \barwedge (A', B', C')$, as required.

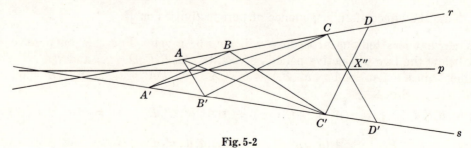

Fig. 5-2

We leave for the reader to show that the above construction holds without change when the projectivity $r \barwedge s$ is in reality a perspectivity. For the case of a projectivity between superposed pencils of points, see Problem 5.1.

PROJECTIVITIES BETWEEN SUPERPOSED PENCILS

In the identity projectivity $r(A, B, C, D, \ldots) \barwedge r(A, B, C, D, \ldots)$, every point is a self-corresponding point (*double point*). Since a projectivity is uniquely determined by any three of its elements and their correspondents, we have

Theorem 5.3.   If a projectivity between two superposed pencils of points (pencils of lines) has three double points (double lines), the projectivity is the identity, that is, every point (line) is a double point (double line).

It follows that a projectivity (not the identity) between two superposed pencils could, perhaps, have no double element, one double element, or two distinct double elements. The existence of projectivities having just one double element and of projectivities having two distinct double elements was established in Problem 1.16, page 18. Another example of a projectivity with two double elements is given in Problem 5.2. The projectivity of Problem 1.6, page 17, has no double element; however, we are not at the moment in a position to prove this.

In Problem 5.3, page 60, we prove

Theorem 5.4.   Every non-identity projectivity between superposed pencils having a given double element can be constructed as the product of two perspectivities.

Problem 5.3 also provides constructions of projectivities for which one double element and two distinct pairs of distinct corresponding elements are known. Two constructions, which appear to be different, are given since each may be found in the literature. In the next section it will be shown that they are, however, essentially the same.

From Problem 5.3, there also follows

Theorem 5.5.   If a non-identity projectivity between two superposed pencils has one double element it has a second which may, however, coincide with the first.

With respect to its double elements, non-identity projectivities between superposed pencils are called *elliptic* when they have no double element, are called *parabolic* when they have just one double element, i.e., two double elements which are coincident, and are called *hyperbolic* when they have two distinct double elements.

We shall hereafter restrict M, N (m, n) to denote double points (double lines) of projectivities having double elements. By the Fundamental Theorem,

(1) A hyperbolic projectivity is defined by giving

   (i)   Its double elements and any pair of distinct corresponding elements.

   (ii) One of its double elements and any two distinct pairs of distinct corresponding elements.

   (iii) Any three distinct pairs of distinct corresponding elements.

(2) A parabolic projectivity is defined by giving

   (i)   Its double element and any pair of distinct corresponding elements.

   (ii) Any three distinct pairs of distinct corresponding elements.

Thus, $(M, N, A) \barwedge (M, N, A')$ is a hyperbolic projectivity. Since a parabolic projectivity is a hyperbolic projectivity with coincident double elements, a parabolic projectivity with double point M may be indicated by $(M, M, A) \barwedge (M, M, A')$.

On a line o take any four distinct points M, N, X, X' such that $H(M, N; X, X')$. By Problem 5.2 we know that the projectivity $(M, N, X) \barwedge (M, N, X')$ is hyperbolic with M, N the double points. From the discussion in Chapter 4, we also know that as X describes the line o in either direction, its correspondent X' describes the same line in the opposite direction. We say, in this case, that the projectivity is *opposite*.

On the other hand, in the hyperbolic projectivity of Fig. 5-6(a), we conclude with the help of Problem 5.12, that as x revolves about O in either direction, its correspondent x' revolves about O in the same direction. In this case, we say that the projectivity is *direct*.

Using a circle as a model of the projective line, it is not difficult to see that if a variable point X describes the line moving in one direction while its correspondent X' describes the line moving in the opposite direction, then the projectivity $X \barwedge X'$ will always have two distinct double points. Thus, with the existence of elliptic projectivities yet to be established, we state

Theorem 5.6.   Every opposite projectivity is hyperbolic; every elliptic and every parabolic projectivity is direct.

See Problem 5.4

HYPERBOLIC AND PARABOLIC PROJECTIVITIES

An examination of Fig. 5-5(a) reveals:

(a) The points M, N, A, A', B, B' constitute the quadrangular set $Q(M, A, B; N, B', A')$ determined on the line o by the complete quadrangle RSA_1B_1.

(b) The complete quadrangle RSA_1B_1 assures the existence of the hyperbolic projectivity $(M, N, A, B) \barwedge (M, N, A', B')$ as shown in Solution 1, Problem 5.3. Conversely, the projectivity determines the quadrangle which, in turn, yields the quadrangular set. We have

Theorem 5.7.   If M, N, A, A', B, B' are any six distinct collinear points, then $(M, N, A, B) \barwedge (M, N, A', B')$ implies $Q(M, A, B; N, B', A')$ and conversely.

In Solution 1, Problem 5.3, N is located simply by the construction of a complete quadrangle having one pair of opposite sides singly on A and B', another pair of opposite sides singly on A' and B, and a fifth side on M. If now in Fig. 5-5(a) the labels M and N, R and A_1, S and B_1, p and q are interchanged, there results the dual of Fig. 5-6(a). It follows then that the two solutions of Problem 5.3 are essentially one and the same.

In Problem 5.5, we prove

Theorem 5.8.   The product of two parabolic projectivities having the same double point M is either the identity or another parabolic projectivity having M as double point.

and in Problem 5.6, we prove

Theorem 5.9.   The projectivity $(M, A, A') \barwedge (M, A', A'')$ is parabolic if and only if $H(M, A'; A, A'')$.

QUADRANGULAR SETS

In Problem 5.7, we prove the first of the three parts of

Theorem 5.10. Six distinct collinear points X, Y, Z, X', Y', Z' form the quadrangular set $Q(X, Y, Z; X', Y', Z')$ if and only if

$$(X, Y, Z, X') \;\overline{\wedge}\; (X', Y', Z', X) \qquad\qquad (1)$$

or $$(X, Y, Z, Y') \;\overline{\wedge}\; (X', Y', Z', Y) \qquad\qquad (2)$$

or $$(X, Y, Z, Z') \;\overline{\wedge}\; (X', Y', Z', Z) \qquad\qquad (3)$$

In the next chapter it will be shown that the quadrangular set $Q(X, Y, Z; X', Y', Z')$ on a line implies the special projectivity $(X, X', Y, Y', Z, Z') \;\overline{\wedge}\; (X', X, Y', Y, Z', Z)$ between two superposed pencils on the line; in other words, any one of *(1)*, *(2)*, *(3)* of Theorem 5.10 implies the others.

Solved Problems

5.1. Given the distinct points $A, B, C, D; A', B', C'$ on a line o, construct the correspondent of D in the projectivity $(A, B, C) \;\overline{\wedge}\; (A', B', C')$.

Refer to Fig. 5-3. Project the points A', B', C' from any point P, not on o, into the points A'', B'', C'' respectively on any line s, not on any of the points P, A', B', C'. Construct the line of Pappus p for the triples A, B, C and A'', B'', C''. Let $DA'' \cdot p = X$, $AX \cdot s = D''$; then D'' is the correspondent of D in the projectivity $(A, B, C) \;\overline{\wedge}\; (A'', B'', C'')$. That $PD'' \cdot o = D'$ is the required correspondent of D follows from

$$o(A', B', C', D') \;\overset{P}{\underset{\wedge}{=}}\; s(A'', B'', C'', D'') \;\overline{\wedge}\; o(A, B, C, D)$$

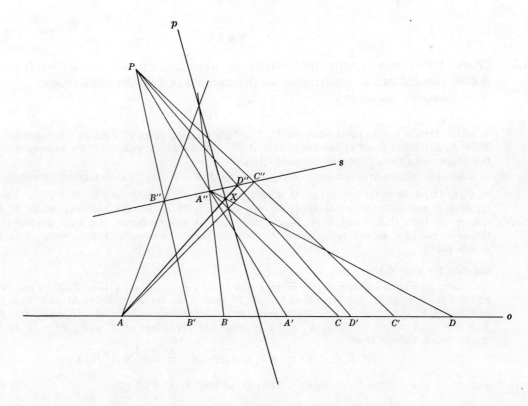

Fig. 5-3

5.2. Show that the correspondence between the points of a line and their harmonic conjugates with respect to two distinct fixed points of the line is a projectivity having the fixed points as double points.

Refer to Fig. 5-4. Let o be the line, let A and B be the fixed points, and Z_1 be any other point on o. On A take any two distinct lines $r \neq o$, $s \neq o$ and on Z_1 take any line, not o, meeting r in P_1 and s in R. Let $BP_1 \cdot s = S_1$ and $BR \cdot r = t \cdot r = Q$. Then $S_1Q \cdot o = Z_1'$ is the harmonic conjugate of Z_1 with respect to A and B. (This, to be sure, is the usual construction; it is repeated here in order to introduce more useful designations to the vertices and sides of the complete quadrangle.) Now take any other point Z_i on o and carry through the above construction using, however, the same lines r, s on A and t on B as before. Let $Z_iR \cdot r = P_i$, $BP_i \cdot s = S_i$ and $S_iQ \cdot o = Z_i'$. From

$$o(A, B, Z_i) \stackrel{R}{\overline{\wedge}} r(A, Q, P_i) \stackrel{B}{\overline{\wedge}} s(A, R, S_i) \stackrel{Q}{\overline{\wedge}} o(A, B, Z_i')$$

follows

$$(A, B, Z_i) \;\overline{\wedge}\; (A, B, Z_i')$$

in which A and B are exhibited as double points.

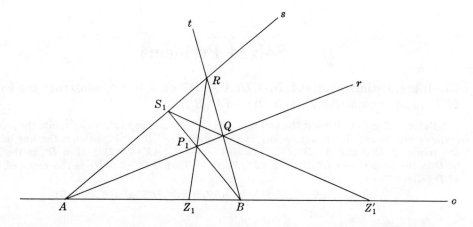

Fig. 5-4

5.3. Prove: Every non-identity projectivity between superposed pencils having a given double element can be constructed as the product of two perspectivities.

Consider the projectivity

$$o(M, A, B, \ldots) \;\overline{\wedge}\; o(M, A', B', \ldots)$$

in which M is a double point while A, A'; B, B' are distinct pairs of distinct corresponding points. With X, any other point of the pencil $o(M, A, B, \ldots)$, we are to construct its correspondent X' in the projectivity using just two perspectivities.

Assume the construction made and denote by R and S the two centers of perspectivity used so that $(M, A, B) \stackrel{R}{\overline{\wedge}} (M', A_1, B_1) \stackrel{S}{\overline{\wedge}} (M, A', B')$ or $(M, A, B) \stackrel{R}{\overline{\wedge}} (M, A_1, B_1) \stackrel{S}{\overline{\wedge}} (M, A', B')$. Then either (1) M, R, S are collinear while M, A_1, B_1 are not, (2) M, A_1, B_1 are collinear while M, R, S are not, or (3) both M, R, S and M, A_1, B_1 are collinear. It will be found that both (1) and (2) insure the existence of a second double point N distinct from M while (3) obtains when M is the only double point.

Solution for Case (1).

Refer to Fig. 5-5(a) below. On M take any line $q \neq o$ and on q take distinct points $R \neq M$ and $S \neq M$. Project M, A, B from R and M, A', B' from S, locate $A_1 = RA \cdot SA'$ and $B_1 = RB \cdot SB'$, and let $A_1B_1 = p$. For X, any other point on o, let $RX \cdot p = X_1$ and $SX_1 \cdot o = X'$. Also, let $p \cdot o = N$ and $p \cdot q = T$. That X' is the required correspondent of X while $N \neq M$ is a second double point, follows from

$$o(M, N, A, B, X) \stackrel{R}{\overline{\wedge}} p(T, N, A_1, B_1, X_1) \stackrel{S}{\overline{\wedge}} o(M, N, A', B', X')$$

and

$$o(M, N, A, B, X) \;\overline{\wedge}\; o(M, N, A', B', X')$$

For the case when p is also on M, that is, when (3) obtains, see Fig. 5-5(b) below.

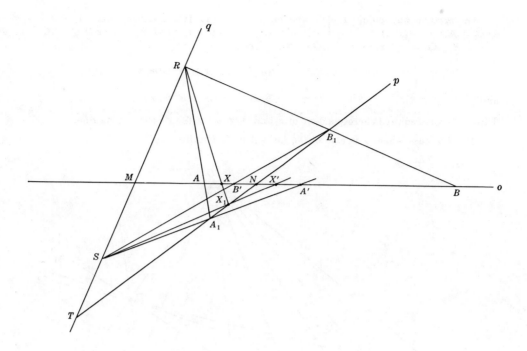

Fig. 5-5(a)

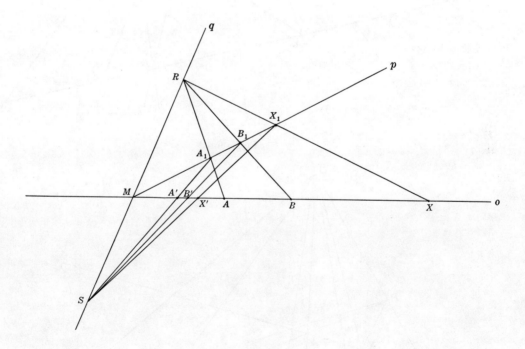

Fig. 5-5(b)

Solution for dual of Case (2).

For the sake of variety we treat two superposed pencils of lines. Consider, then, in Fig. 5-6(a) below the projectivity

$$O(m, a, b, \ldots) \; \overline{\wedge} \; O(m, a', b', \ldots)$$

where m is a double line and a, a'; b, b' are distinct pairs of distinct corresponding lines. Let x be any other line of the pencil $O(m, a, b, \ldots)$.

On m take any point $P \neq O$ and on P take any two distinct lines $a_1 \neq m$, $b_1 \neq m$. Let $a \cdot a_1 = A$, $a' \cdot a_1 = A'$, $b \cdot b_1 = B$, $b' \cdot b_1 = B'$; $AB = r$, $A'B' = s$, $r \cdot s = Q$, $OQ = n$, $PQ = t$; $x \cdot r = X$, $XP = x_1$, $x_1 \cdot s = X'$, $OX' = x'$. Then

$$O(m, n, a, b, x) \ \overset{r}{\wedge} \ P(m, t, a_1, b_1, x_1) \ \overset{s}{\wedge} \ O(m, n, a', b', x')$$

and

$$O(m, n, a, b, x) \ \overline{\wedge} \ O(m, n, a', b', x')$$

Thus x' is the required correspondent of x and $n \neq m$ is the second double line.

For the case when Q is also on m, i.e. when (3) obtains, see Fig. 5-6(b).

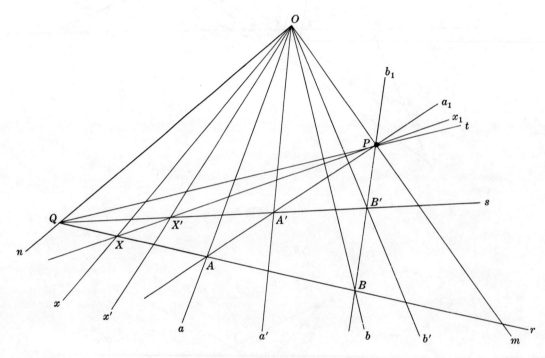

Fig. 5-6(a)

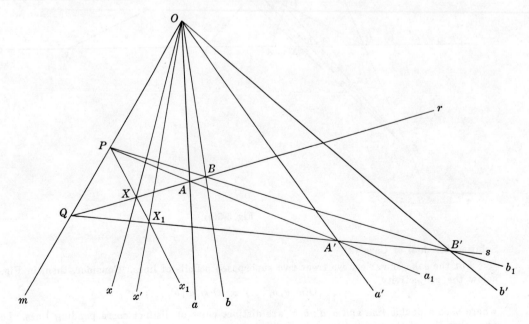

Fig. 5-6(b)

5.4. Prove: If M, N, A, A', B, B' are distinct collinear points such that $H(M, N; A, A')$ and $H(M, N; B, B')$, the pair A, A' does not separate the pair B, B'.

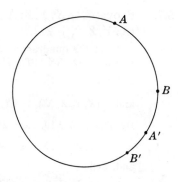

From Problem 5.2, we have that A, A' and B, B' are pairs of corresponding points in the opposite hyperbolic projectivity $(M, N, A, B) \barpi (M, N, A', B')$. Since A, A' and B, B' are pairs which separate M, N harmonically, this projectivity may also be defined as $(A, B, A') \barpi (A', B', A)$.

Suppose the pair A, A' separates the pair B, B' so that the distribution of these points on the line is as indicated in Fig. 5-7. Then the directions of ABA' and $A'B'A$ being the same, the projectivity $(A, B, A') \barpi (A', B', A)$ is direct. But this is a contradiction; hence A, A' do not separate B, B'.

Fig. 5-7

5.5. Prove: The product of two parabolic projectivities having the same double point M is either the identity or another parabolic projectivity having M as double point.

Consider the product of $(M, M, A) \barpi (M, M, A')$ and $(M, M, A') \barpi (M, M, A'')$. Clearly M is a double point of this product. Suppose $B \neq M$ is a second double point. Since the first projectivity cannot carry B into B, it must be of the form $(M, M, B) \barpi (M, M, B')$ with $B \neq B'$. Now the second projectivity must carry B' into B, that is, must be of the form $(M, M, B') \barpi (M, M, B)$. But then the product is the identity projectivity. Thus, unless the product of the two given parabolic projectivities is the identity, it too is parabolic with double point M.

5.6. Prove: The projectivity $(M, A, A') \barpi (M, A', A'')$ is parabolic if and only if $H(M, A'; A, A'')$.

Consider the parabolic projectivity $(M, A, A') \barpi (M, A', A'')$ of Fig. 5-8. By construction, the complete quadrangle $D'D''RS$ has one pair of opposite sides on M, another pair of opposite sides on A' and the third pair singly on A and A''; hence, $H(M, A'; A, A'')$. Conversely, if the pairs of opposite sides of the quadrangle determine the points M, A', A, A'' as in the figure, then

$$(M, A, A') \overset{R}{\barwedge} (M, D', D'') \overset{S}{\barwedge} (M, A', A'')$$

and

$$(M, A, A') \barpi (M, A', A'')$$

Since the axis of projectivity is on M, this projectivity is parabolic.

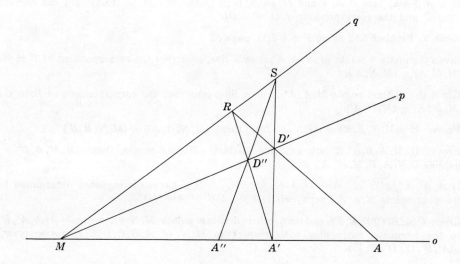

Fig. 5-8

5.7.   Prove: $Q(X, Y, Z; X', Y', Z')$ implies $(X, Y, Z, X') \barwedge (X', Y', Z', X)$ and, conversely, $(X, Y, Z, X') \barwedge (X', Y', Z', X)$ implies $Q(X, Y, Z; X', Y', Z')$.

Let the quadrangular set be determined on o by the complete quadrangle $PQRS$ as in Fig. 5-9 and let $PQ \cdot RS = T$. Then

$$(X, Y, Z, X') \overset{R}{\barwedge} (T, P, Q, X') \overset{S}{\barwedge} (X, Z', Y', X')$$

and $(X, Y, Z, X') \barwedge (X, Z', Y', X')$

By Theorem 2.10, page 25,

$(X, Z', Y', X') \barwedge (X', Y', Z', X)$

and, hence,

$(X, Y, Z, X') \barwedge (X', Y', Z', X)$

For the converse, suppose

$(X, Y, Z, X') \barwedge (X', Y', Z', X)$

but $Q(X, Y, Z; X', Y', Z'')$. Then, as above,

$(X, Y, Z, X') \barwedge (X', Y', Z'', X)$

and so

$(X', Y', Z', X) \barwedge (X', Y', Z'', X)$

But this projectivity, having three double points, is the identity. Hence, $Z' = Z''$ and $Q(X, Y, Z; X', Y', Z')$ as required.

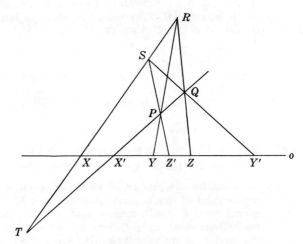

Fig. 5-9

Supplementary Problems

5.8.   In Fig. 5-2, page 56, let $p \cdot r = R$, $p \cdot s = S'$; $r \cdot s = P$, as a point on r and $r \cdot s = Q'$ as a point on s. Construct the correspondent of each of these points.

5.9.   In the perspectivity $r(A, B, C, \ldots) \overset{P}{\barwedge} s(A', B', C', \ldots)$ take any other point D on r and construct, using the line of Pappus of the triples A, B, C and A', B', C', the correspondent D' of D. What is the essential difference between your figure and Fig. 5-2?

5.10.   In Fig. 5-3 for Problem 5.1: (a) let $G = p \cdot o$ and locate its correspondent G', (b) let $o \cdot s = O$, as a point on o, and locate its correspondent O'.

5.11.   State and prove the dual of Problem 5.1.

5.12.   In Fig. 5-6(a) take C on r and D' on S, both to the left of Q. Construct the correspondent c' of $c = OC$ and the correspondent d of $d' = OD'$.

5.13.   Same as Problem 5.12 using Fig. 5-6(b), page 62.

5.14.   Given the distinct points M, N, A, A', B on a line, construct the correspondent of B in the projectivity $(M, N, A) \barwedge (M, N, A')$.

5.15.   Given the distinct points M, A, A', B on a line, construct the correspondent of B in the projectivity $(M, M, A) \barwedge (M, M, A')$.

5.16.   Prove: If $(M, N, A, B) \barwedge (M, N, A', B')$, then $(M, N, A, A') \barwedge (M, N, B, B')$.

5.17.   Prove: If M, A, B, A', B' are any five distinct collinear points, then $(M, M, A, B) \barwedge (M, M, A', B')$ implies $Q(M, A, B; M, B', A')$ and conversely.

5.18.   If $A, A', A'', A''', \ldots, A^{(m-1)}, A^{(m)}, A^{(m+1)}, \ldots$ is a harmonic sequence determined by the distinct collinear points M, A, A', verify $H(M, A^{(m)}; A^{(m-1)}, A^{(m+1)})$.

5.19.   Given $Q(A, B, C; D, E, F)$ and two other distinct points M, N on a line o, let A', B', C', D', E', F' be the harmonic conjugates with respect to M, N of A, B, C, D, E, F respectively. Show that $Q(A', B', C'; D', E', F')$.

Chapter 6

Involutions

DEFINITION

Consider the non-identity projectivity

$$\text{(i)} \qquad O(a, b, c, d, \ldots) \; \overline{\wedge} \; O(a', b', c', d', \ldots)$$

Think of each line on O as having two names or labels — a given name (say a) and an alias (say d') or an unprimed label (say c) and a primed label (say a'). The projectivity pairs given names and aliases but, in general, the line with given name a and its correspondent, the line with alias a', are different lines.

Suppose now that (i) is expressed in terms of given names only, say,

$$\text{(i')} \qquad O(a, b, c, d, \ldots) \; \overline{\wedge} \; O(c, m, q, a, \ldots)$$

and, in turn, each element on the right is replaced by its correspondent in the projectivity and then is expressed in terms of given names. As the effect of the second application of the projectivity, we would have, say,

$$\text{(i'')} \qquad O(a, b, c, d, \ldots) \; \overline{\wedge} \; O(c', m', q', a', \ldots) = O(q, r, u, c, \ldots)$$

In general, if this procedure were continued, the right member of (i') would never again appear.

In Problem 1.6, page 17, the existence of a special projectivity was established in which the correspondent of a was $a' = b$ and the correspondent of b was $b' = a$. Moreover, it was found in Problem 3.23, page 43, that this is also true of every pair of corresponding lines. In such a projectivity, the pair of elements a and a' are said to correspond *doubly* (we prefer *reciprocally*) and the projectivity is called an *involution*. An involution then is a projectivity among the lines on a point (the points on a line) such that the correspondence x with x' (X with X') is reciprocal.

Think of the right and left members of (i') as distinct systems and denote them by S_1 and S_2 respectively; also, denote the one-to-one correspondence between their elements by τ. Then (i') may be expressed by $\tau(S_1) = S_2$ and (i'') by $\tau(\tau(S_1)) = \tau(S_2) = S_3$, where S_3 denotes $O(q, r, u, c, \ldots)$. The correspondence τ is an involution if and only if $\tau(\tau(S_1)) = \tau^2(S_1) = S_1$. More generally, a non-identity one-to-one correspondence τ between the lines on a point (the points on a line) is an involution if and only if τ^2 is the identity.

In Problem 6.1, we prove

Theorem 6.1. If, in a projectivity among the lines on a point, a line a and its correspondent $a' \neq a$ correspond reciprocally, then any other line x and its correspondent x' also correspond reciprocally.

As a consequence, we have

Theorem 6.2. An involution is determined by any two of its reciprocal pairs.

For a construction, see Problem 6.2.

The reciprocal pairs of an involution are sometimes called conjugate pairs. We reserve this term for later use and so will continue to speak of reciprocal pairs.

In Problem 6.3, we prove

Theorem 6.3. The three pairs of opposite sides of a complete quadrangle are sectioned by any line o, not on a vertex, in three reciprocal pairs of an involution among the points of o. If the line o is not on a diagonal point of the quadrangle, each reciprocal pair consists of distinct points.

This theorem and its dual

Theorem 6.3'. The three pairs of opposite vertices of a complete quadrilateral are joined to any point O, not on a side, in three reciprocal pairs of an involution among the lines on O. If the point O is not on a diagonal line of the quadrilateral, each reciprocal pair consists of distinct lines.

provide alternate constructions of the involution determined by two distinct pairs of reciprocal elements.

Since the six points of Theorem 6.3 constitute a quadrangular set, we have proved the statement made at the end of the last section in Chapter 5.

A proof of

Theorem 6.4. If a line o meets the sides PQ, PS, SQ of a triangle PQS in the points A', B', D' which, together with three other points A, B, C on o, are three reciprocal pairs of an involution among the points of o, the lines AS, BQ, DP are on a point.

is immediate since the six points constitute the quadrangular set $Q(A, B, D; A', B', D')$. An independent proof is given in Problem 6.4.

DOUBLE ELEMENTS

On a line o take three distinct points M, N, X and construct the harmonic conjugate X' of X with respect to M and N. If M, N are kept fixed while X describes the line, the resulting projectivity $X \barwedge X'$ is an involution having M and N as double points. This involution is naturally called *hyperbolic*. Since the involution is uniquely determined by its double points while $(M, N, X, X') \barwedge (M, N, X', X)$ implies $H(M, N; X, X')$, we have proved

Theorem 6.5. Every hyperbolic involution is merely the correspondence between pairs of harmonic conjugates with respect to the double points.

There follows

Theorem 6.6. Every hyperbolic involution is opposite.

In Problem 6.5, we prove

Theorem 6.7. If an involution has one double element, it has a second which is (a) distinct from the first and (b) the harmonic conjugate of the first with respect to any reciprocal pair.

As an immediate consequence, we have

Theorem 6.8. There is no parabolic involution.

By Theorem 6.6, no two reciprocal pairs of a hyperbolic involution separate each other. Thus the involution determined by two reciprocal pairs A, A' and B, B' which separate each other (see Problem 1.5 for an example) is *elliptic*. Since an involution is a special type of projectivity, this establishes the existence of elliptic projectivities. We have also

Theorem 6.9. Every direct involution is elliptic.

PAIRS OF INVOLUTIONS

Consider two involutions I_1 and I_2 among the points on a line. Our purpose here is to inquire into the possibility of these involutions having a reciprocal pair A, A' in common.

First, suppose that the involutions are hyperbolic: I_1 with M_1, N_1 as double points and I_2 with M_2, N_2 as double points. Denote by I_3 the involution having M_1, N_1 and M_2, N_2 as reciprocal pairs. If I_3 is hyperbolic, there exist points M_3, N_3 (the double points of I_3) such that $H(M_3, N_3; M_1, N_1)$ and $H(M_3, N_3; M_2, N_2)$. Then $M_3, N_3 = M_3'$ are a reciprocal pair of both I_1 and I_2. If, however, I_3 is elliptic then no points M_3, N_3 exist. Thus,

Theorem 6.10. Two hyperbolic involutions among the points on a line have a reciprocal pair A, A' in common if and only if their double points do not separate each other.

Next, suppose that I_1 is elliptic and I_2 is hyperbolic. Denote by $\pi_3 = I_1 I_2$ the projectivity among the points of a line which results when I_1 and I_2 are applied successively in that order. For example, let I_1 carry the point X into X' and I_2 carry X' into X''; then $I_1 I_2$ carries X into X''. If I_1 and I_2 should have a reciprocal pair A, A' in common, then I_1 interchanges them and I_2 restores them to the original order. Thus A, A' would be the double points of the projectivity π_3. The problem is then reduced to determining whether or not π_3 is hyperbolic. Since I_1 is direct and I_2 is opposite, it is clear that π_3 is opposite. Then the double points of π_3 are a common pair and we have

Theorem 6.11. If, of two involutions among the points on a line, one is elliptic and the other is hyperbolic, they always have a reciprocal pair in common.

Finally, suppose both I_1 and I_2 are elliptic. Let the reciprocal pair A, A' of I_1 be carried into A'', A''' by I_2. These points may be assumed distributed over the line as follows:

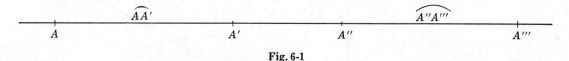

Fig. 6-1

As a variable point X moves over AA' from A to A', its correspondent under the projectivity $\pi_3 = I_1 I_2$ moves over $A'' A'''$ in the same direction, that is, from A''' to A''. Now $\widehat{AA'}$ is included in $\widehat{A''A'''}$; hence X and X' must coincide in some point of $\widehat{AA'}$ which, being a double point of π_3 is a point of the reciprocal pair common to I_1 and I_2. Thus,

Theorem 6.12. Two elliptic involutions among the points on a line always have a reciprocal pair in common.

In Problem 6.6, we prove

Theorem 6.13. If the involutions

$$I_1 : (A, A', B, B') \barwedge (B', B, A', A) \quad \text{and} \quad I_2 : (A, A', B, B') \barwedge (A', A, B', B)$$

have M, N as a common reciprocal pair, then $I = I_1 \cdot I_2$ is an involution having M, N as double points.

Solved Problems

6.1. Prove: If, in a projectivity among the lines on a point, any line a and its correspondent a' correspond reciprocally, then any other line x and its correspondent x' also correspond reciprocally.

We have given a projectivity $x \barwedge x'$ among the lines on a point O with the property that for the corresponding pair of lines a, a', we have $(a, a', x) \barwedge (a', a, x')$. We are to prove $(a, a', x, x') \barwedge (a', a, x', x)$.

As in Fig. 6-2, take any line r not on O and take on a any point A distinct from O and from $a \cdot r$. Construct on A the lines a_1', x_1, x_1' such that

$$(a_1', x_1, x_1') \stackrel{r}{\barwedge} (a', x, x')$$

Let $a' \cdot a_1' = A'$, $x' \cdot x_1 = V$, and $A'V = v$. Then

$$(a, a', x, x') \stackrel{r}{\barwedge} (a, a_1', x_1, x_1') \stackrel{x'}{\barwedge} (a', a_1', v, r) \stackrel{x_1}{\barwedge} (a', a, x', x)$$

and

$$(a, a', x, x') \barwedge (a', a, x', x)$$

as required.

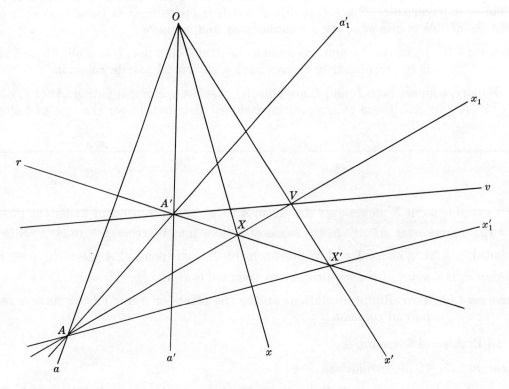

Fig. 6-2

6.2. Construct the involution on a line given two reciprocal pairs of points of that involution.

Let the reciprocal pairs be A, A' and B, B' on the line o of Fig. 6-3 below and let C be any other point on o. We are to construct the mate C' of C in the involution. Take any point R, not on o, and on A take any line a distinct from o and not on R. Locate on a the points A_1', B_1, B_1', C_1 such that

$$(A_1', B_1, B_1', C_1) \stackrel{R}{\barwedge} (A', B, B', C)$$

Let $B'B_1 \cdot A'R = V$, $B'C_1 \cdot A'R = W$ and $B_1W \cdot o = C'$. Then C' is the required point since

$$(A, A', B, B', C) \stackrel{R}{\barwedge} (A, A_1', B_1, B_1', C_1) \stackrel{B'}{\barwedge} (A', A_1', V, R, W) \stackrel{B_1}{\barwedge} (A', A, B', B, C')$$

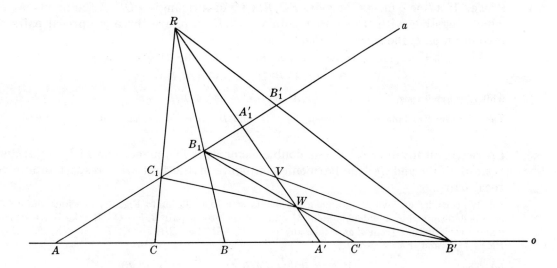

Fig. 6-3

6.3. Prove: The three pairs of opposite sides of a complete quadrangle are sectioned by any line o, not on a vertex, in three reciprocal pairs of an involution among the points of o.

Using Fig. 6-4, we have

$$(A, A', B', D') \;\overset{S}{\barwedge}\; (A'', A', P, Q) \;\overset{R}{\barwedge}\; (A, A', D, B)$$

and so $$(A, A', B', D') \;\barwedge\; (A, A', D, B)$$

By Theorem 2.10, page 25, $$(A, A', D, B) \;\barwedge\; (A', A, B, D)$$

hence, $$(A, A', B', D') \;\barwedge\; (A', A, B, D)$$

Since A, A' is a reciprocal pair, the projectivity is an involution having B, B' and D, D' as reciprocal pairs by the dual of Theorem 6.1, page 65.

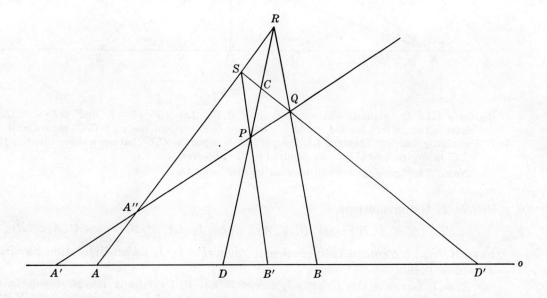

Fig. 6-4

6.4. Prove: If a line o meets the sides PQ, PS, SQ of a triangle PQS in the points A', B', D' which, together with three other points A, B, D on o, are three reciprocal pairs of an involution on o, the lines AS, BQ, DP are on a point.

In Fig. 6-4, suppose $AS \cdot BQ = R$ but $RP \cdot o = D_1$. By Theorem 6.3, we have

$$(A, A', B', D') \,\barwedge\, (A', A, B, D_1)$$

while, by hypothesis, $(A, A', B', D') \,\barwedge\, (A', A, B, D)$

Then, by the Fundamental Theorem, $D_1 = D$ and DP is on R as required.

6.5. Prove: If an involution has one double element, it has a second which is (a) distinct from the first and (b) the harmonic conjugate of the first with respect to any reciprocal pair.

We consider an involution among the points of a line. Refer to Fig. 6-5 in which the involution on o is determined by the double point M and any reciprocal pair B, B'. On M take two distinct lines r and s each distinct from o; on r take any point $P \neq M$. Let $BP \cdot s = Q$, $B'P \cdot s = S$, $BS \cdot B'Q = R$, $PR \cdot s = K$ and $PR \cdot o = N$.

(a) Then $(M, N, B, B') \overset{P}{\,\doublebarwedge\,} (M, K, Q, S) \overset{R}{\,\doublebarwedge\,} (M, N, B', B)$

exhibits $N \neq M$ as the second double point.

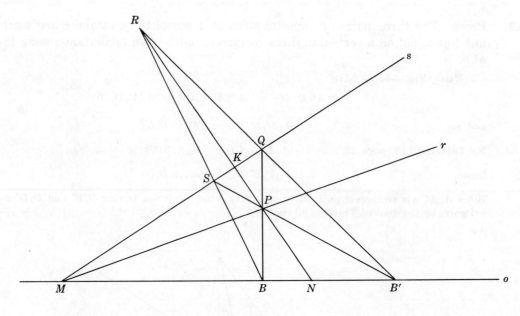

Fig. 6-5

(b) On o take any point C distinct from M, N, B, B'. Let $CP \cdot s = T$ and $RT \cdot o = C'$. (This construction is left for the reader.) Now, by construction, the pair C, C' separate M, N harmonically and, by Theorem 4.4, the pair M, N separate C, C' harmonically. Clearly, the pair C, C' is any reciprocal pair as required by the theorem.

(*Note*. The figure for (b) will be used in Problem 6.16.)

6.6. Prove: If the involutions

$$I_1: (A, A', B, B') \,\barwedge\, (B', B, A', A) \quad \text{and} \quad I_2: (A, A', B, B') \,\barwedge\, (A', A, B', B)$$

have M, N as a common reciprocal pair, then $I = I_1 \cdot I_2$ is an involution having M, N as double points.

Since I_1 carries A into B' while I_2 carries B' into B, I carries A into B. Similarly, we find $I: (A, A', B, B') \,\barwedge\, (B, B', A, A')$ is an involution. Since each of I_1 and I_2 interchanges M and N, I carries M into M and N into N. Thus, M and N are double points.

Supplementary Problems

6.7. Prove: If in a hyperbolic projectivity there exists one pair of corresponding elements which are harmonic conjugates with respect to the double points, the projectivity is an involution.

6.8. Construct the involution among the lines on a point, given: (*a*) the double lines m, n; (*b*) a double line m and a reciprocal pair a, a'.

6.9. Prove: If two pairs of points on a line do not separate each other, there exists only one pair of points which separate each of the given pairs harmonically.

6.10. Prove: If $H(A, B; D, D')$, $H(B, D; A, A')$, $H(D, A; B, B')$, then $Q(A', B', D'; A, B, D)$.

    Hint. In Fig. 4-8, page 54, consider the quadrangle *CEFG*.

6.11. Prove: The six collinear points A, A'; B, B'; D, D' are three reciprocal pairs of an involution if and only if

$$(A, B, D, A') \barpi (A', B', D', A)$$

or

$$(A, B, D, B') \barpi (A', B', D', B)$$

or

$$(A, B, D, D') \barpi (A', B', D', D)$$

6.12. Given the triangle of Fig. 6-6, let a', b', c' be distinct lines lying singly on the vertices A, B, C respectively and let p be any line distinct from a, b, c; a', b', c'. Let $a \cdot p = A'$, $b \cdot p = B'$, $c \cdot p = C'$; $a' \cdot p = A''$, $b' \cdot p = B''$, $c' \cdot p = C''$. Show that a', b', c' are concurrent if and only if $Q(A', B', C'; A'', B'', C'')$.

    Hint. Suppose a', b', c' are concurrent at P and consider the complete quadrangle *ABCP*. Suppose a', b', c' non-concurrent; let $b' \cdot c' = P$, $AP = a''$, and $a'' \cdot p = A'''$.

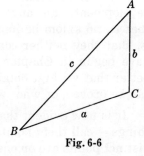

Fig. 6-6

6.13. State and prove the dual of Problem 6.12.

6.14. Prove: If p is the line of Pappus of the projectivity

$$r(A_1, B_1, C_1, D_1, \ldots) \barpi s(A_2, B_2, C_2, D_2, \ldots)$$

and if

$$s(A_2, B_2, C_2, D_2, \ldots) \overset{P}{\barpi} r(A_3, B_3, C_3, D_3, \ldots)$$

show that

$$r(A_1, B_1, C_1, D_1, \ldots) \barpi r(A_3, B_3, C_3, D_3, \ldots)$$

is an involution if and only if P is on p. State and prove the dual.

6.15. Prove: Any projectivity among the points on a line (lines on a point) may be expressed as the product of two involutions.

    Hint. Let the projectivity π carry A into A' and A' into A''. Define I_1: $(A', A, A'') \barpi (A', A'', A)$ and $I_2 = I_1 \cdot \pi$. Then $I_1 \cdot I_2 = \pi$.

6.16. Prove: If $H(M, N; B, B')$ and $H(M, N; C, C')$, then M, N is a reciprocal pair of the involution $(B, C, B', C') \barpi (C, B, C', B')$.

    Hint. See note in Problem 6.5.

6.17. An involution $(A, A', B) \barpi (A', A, B')$ is given on a line o. (*a*) Describe the construction of the mate C' of any other point C on o suggested by Theorem 6.3. (*b*) Let D be yet another point on o and construct its mate by the above procedure; also, by using the figure of Problem 6.2.

Chapter 7

Axioms for Plane Projective Geometry

INTRODUCTION

Plane projective geometry is concerned solely with those properties of plane figures which remain invariant under one or more central projections and sections. Our presentation thus far, while following roughly the historical development, tends to suggest that the content of this geometry is that which remains of Euclidean geometry when it is stripped of all theorems which in any way have to do with measurements. One way of negating this idea would be to establish a foundation for projective geometry, that is, set down a system of axioms, completely independent of any notions of Euclidean geometry. This we now propose to do using eleven axioms in all.

The construction of systems of axioms for projective geometry is a relatively new development. For any such system it is desirable that the axioms be simple and independent, that is, no axiom be implied by the others. It is urgent only that they be consistent, that is, that they neither contradict each other nor imply contradictory theorems. We might have begun in Chapter 1 with some one of the available systems. However, it seemed better that we first obtain some idea of what we wished to prove and, since not everything can be proved, of what we would have to assume.

It is now evident that for our plane projective geometry we need two sets of objects or things — call them points and lines, respectively — and a relation "on" such that every two distinct points are on one and only one line and every two lines are on one and only one point. Also, we must assume or be able to prove the Two-Triangle Theorem of Desargues (page 32), the Theorem of Pappus (page 2), the special case of the Theorem of Pappus (page 40), the non-collinearity of the diagonal points of a complete quadrangle and the Fundamental Theorem (Theorem 2.8, page 25).

INITIAL AXIOMS

We begin with two non-empty sets of "things" (called respectively points and lines) and a relation "on" such that:

Axiom 1. If A and B are distinct points, then there is at least one line on both.

Axiom 2. If A and B are distinct points, then there is at most one line on both.

Axiom 3. If p and q are two distinct lines, then there is at least one point on both.

Axiom 4. There are at least three distinct points on any line.

Axiom 5. Not all points are on the same line.

It is to be noted here that no attempt is made to define the "things" called points and lines nor to define the relation "on". Since the sets are non-empty, we know that there is at least one point and one line. Apart from this, our only other information is that given by the axioms and such facts (theorems) as these axioms imply.

THE PRINCIPLE OF DUALITY

In Problems 7.1-7.3, we prove

Theorem 7.1.   If p and q are distinct lines, then there is at most one point on both.

Theorem 7.2.   Not all lines are on the same point.

Theorem 7.3.   There are at least three distinct lines on any point.

We now have eight facts which when properly paired — Axioms 1 and 3, Axiom 2 and Theorem 7.1, Axiom 4 and Theorem 7.3, Axiom 5 and Theorem 7.2 — consist of four statements and their duals. Thus far, then, the principle of duality holds in our geometry. We must be careful to retain this property as we proceed.

THE NUMBER OF POINTS ON A LINE

By Axiom 4 and Theorem 7.3 there are at least three points on every line and at least three lines on every point. Could there then be more points on some one line than on another and, consequently, more lines on some one point than on another? Such a state of anarchy is ruled out by (see Problems 7.4-7.5 for proofs):

Theorem 7.4.   If there are exactly $n \geqq 3$ distinct points on some line p, there are exactly n distinct points on every line.

and

Theorem 7.5.   If there are exactly n distinct points on any line, there are exactly $n^2 - n + 1$ distinct points in all.

The duals of these theorems:

Theorem 7.4′.   If there are exactly $n \geqq 3$ distinct lines on some point P, there are exactly n distinct lines on every point.

and

Theorem 7.5′.   If there are exactly n distinct lines on any point, there are exactly $n^2 - n + 1$ distinct lines in all.

are, of course, valid by the principle of duality. The reader is urged, however, to provide an independent proof of each.

CONSISTENCY OF THE AXIOMS

It might be wise at this point to see whether or not there exists a model of our system. To build a model, we take two non-empty sets of concrete objects, identify the elements of one set as points and of the other as lines, and so define the relation "on" that our axioms become provable statements (theorems) concerning the proposed model.

Example 7.1.

Consider in Fig. 7-1 an ornament consisting of seven beads

$$B_1, B_2, B_3, \ldots, B_7$$

attached by three's to seven wires

$$w_1, w_2, w_3, \ldots, w_7$$

Identify the beads as points, the wires as lines, and let B_i on w_j (also, w_j on B_i) mean the bead B_i is attached to the wire w_j. There remains now the job of proving all of the axioms or of disproving at least one of them.

Proof of Axioms 1 and 2. There are $_7C_2 = 21$ selections of two distinct points — $B_1, B_2; B_1, B_3; \ldots; B_1, B_7: B_2, B_3; B_2, B_4; \ldots; B_2, B_7: \ldots: B_6, B_7.$

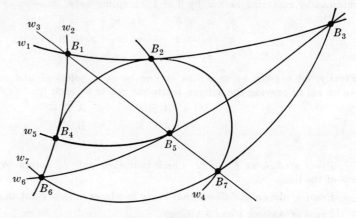

Fig. 7-1

First, we check to see that B_1 and B_2 are on some line; they are on line w_1. Then we check to see that B_1 and B_2 are not on some other line; they are not. Continuing in this manner with each of the selections, we finally conclude that if X and Y are distinct points, they are on at least one and at most one line, as required.

Proof of Axiom 4.   There are seven lines on each of which are exactly three points and, hence, at least three points, as follows:

$$w_1: B_1, B_2, B_3 \qquad w_3: B_1, B_5, B_7 \qquad w_5: B_3, B_4, B_5$$
$$w_2: B_1, B_4, B_6 \qquad w_4: B_2, B_4, B_7 \qquad w_6: B_2, B_5, B_6 \qquad w_7: B_3, B_6, B_7$$

Proof of Axiom 3.   Check that every pair of lines (there are 21) has at least one point in common.

Proof of Axiom 5.   This follows easily since there are seven points and only three on any line.

Thus, we have produced a model of our system.

The importance of the existence of a model lies in the fact that had the axioms been inconsistent no model could possibly exist. We now conclude that Axioms 1-5 constitute a consistent system.

Similarly, it can be shown that had the wires been identified as points and the beads as lines while "on" is defined as before, the ornament remains a model for the system of axioms. Can this be anticipated?

When there are exactly three points on any line, there are (Theorem 7.5) exactly $3^2 - 3 + 1 = 7$ points in all. A geometry of seven points will be called a *7-point geometry*.

Example 7.2.

Using the number system consisting of exactly two elements — 0 and 1 — and having the following addition and multiplication tables

+	0	1
0	0	1
1	1	0

·	0	1
0	0	0
1	0	1

Fig. 7-2

consider the set of all possible triples (x_1, x_2, x_3) where each x is either 0 or 1. There are eight triples in all, but we shall find no use for the triple $(0,0,0)$. The remaining seven triples (identify them as points) are:

$$A = (0,0,1) \qquad D = (0,1,1)$$
$$B = (0,1,0) \qquad E = (1,0,1) \qquad G = (1,1,1)$$
$$C = (1,0,0) \qquad F = (1,1,0)$$

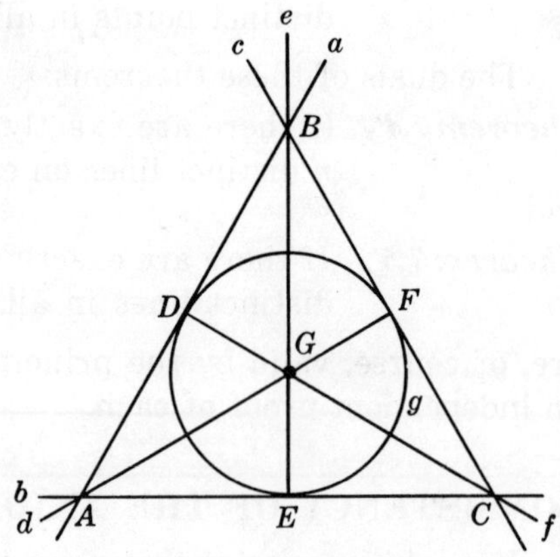

For lines take all possible equations of the form

$$a_1 x_1 + a_2 x_2 + a_3 x_3 = 0$$

obtained by replacing each a_i by 0 or 1, excepting only $0x_1 + 0x_2 + 0x_3 = 0$. They are

$$a: \ x_1 = 0 \qquad d: \ x_1 + x_2 = 0$$
$$b: \ x_2 = 0 \qquad e: \ x_1 + x_3 = 0 \qquad g: \ x_1 + x_2 + x_3 = 0$$
$$c: \ x_3 = 0 \qquad f: \ x_2 + x_3 = 0$$

Finally, let a point be on a line (a line be on a point) if and only if the triple satisfies the equation. As an aid in proving the axioms, locate the points on each line, as follows: (remember $1 + 1 = 0$ here)

$$a: A, B, D \qquad d: A, F, G$$
$$b: A, C, E \qquad e: B, E, G \qquad g: D, E, F$$
$$c: B, C, F \qquad f: C, D, G$$

Proof of Axioms 1 and 2.   Check that each pair of points $A, B; A, C; \ldots : F, G$ is on one and only one of the lines.

Proof of Axiom 3.   Check that every two lines have a point in common.

Proof of Axioms 4 and 5.   Clear.

For a diagram, see Fig. 7-2 where each line is taken as a straight line except for line g.

The models in these two examples differ in many respects: in one there is a set of beads and a set of wires, in the other there is a set of number triples and a set of linear equations; Fig. 7-1 and 7-2 differ in shape, aesthetic appeal, etc. However, by properly associating each point and line of the 7-point geometry of Example 7.1 with a point and line of Example 7.2, it is shown below that any statement derived from the one yields a valid statement in the other. As a consequence, we say that the two geometries are *isomorphic*, that is, their differences, whatever they be, have no geometric significance.

Let us associate B_1 with A B_4 with D

$\qquad\qquad\qquad\quad$ B_2 with E B_5 with G B_7 with F

$\qquad\qquad\qquad\quad$ B_3 with C B_6 with B

and $\qquad\qquad\quad$ w_1 with b w_4 with g

$\qquad\qquad\qquad\quad$ w_2 with a w_5 with f w_7 with c

$\qquad\qquad\qquad\quad$ w_3 with d w_6 with e

In Fig. 7-1, we find, for example, the lines w_1 and w_7 have B_3 in common while their correspondents b and c in Fig. 7-2 have C, the correspondent of B_3, in common. Also, defining perspectivities and projectivities as in Chapter 1, we have in Fig. 7-1,

$$\text{(i)}\qquad w_3(B_1, B_5, B_7) \stackrel{B_2}{\overline{\wedge}} w_2(B_1, B_6, B_4) \stackrel{B_3}{\overline{\wedge}} w_3(B_1, B_7, B_5)$$

and so $$(B_1, B_5, B_7) \,\overline{\wedge}\, (B_1, B_7, B_5)$$

When each point and line in (i) is replaced by its associate, we have

$$\text{(i')}\qquad d(A, G, F) \stackrel{E}{\overline{\wedge}} a(A, B, D) \stackrel{C}{\overline{\wedge}} d(A, F, G)$$

and so $$(A, G, F) \,\overline{\wedge}\, (A, F, G)$$

It is left for the reader to verify (i') in Fig. 7-2. The two 7-point geometries are evidently isomorphic. It is frequently said that there is just one 7-point geometry; what is meant is: Any two 7-point geometries are isomorphic.

In Problem 7.6, we give a model of a 13-point geometry in which there are exactly 4 points on every line. At first glance it might appear that the procedure here and in Example 7.2 could be generalized to provide a model for a geometry in which on any line there are exactly n points where n is any natural number. The finite sets $\{0,1\}$ and $\{0,1,2\}$ used in building number triples have, with respect to addition and multiplication, certain properties in common with the set of all rational numbers and the set of all real numbers. It follows from a theorem of algebra that the set $\{0,1,2,3,\ldots,m-1\}$ of remainders when the natural numbers are divided by m has these properties (i.e. is a field) when and only when m is a prime. The important property here is: If $a \cdot b = 0$, then $a = 0$ or $b = 0$ or both. When $m = 4$, the set $\{0,1,2,3\}$ of remainders lacks this property since when $a = 2 \neq 0$ and $b = 2 \neq 0$ we have $a \cdot b = 0$. It can be shown that Problem 7.6 may be generalized to provide a model for the case of exactly n points on a line when and only when $n - 1$ is a prime.

In Problem 7.10, we give a model of a geometry in which there are an infinite number of points on each line. Our axiom system is, thus far, consistent with there being a finite number of points as well as there being an infinite number of points on any line.

ADDITIONAL AXIOMS

It is our purpose in this chapter to provide a set of axioms from which the theorems of the preceding chapters may be deduced. With the projective line as defined in Chapter 1 in mind, we see that the points of a line must be in one-to-one correspondence with the set of all real numbers together with the number ∞.

It follows from the preceding chapters that the key theorem of plane projective geometry is, as its name implies, the Fundamental Theorem. We have seen, for instance, that the Fundamental Theorem implies the Theorem of Pappus which, in turn, implies the Two-Triangle Theorem of Desargues which, in turn, implies the special case of the Theorem of Pappus. One objection to using the Fundamental Theorem as an axiom is that it can hardly be characterized as simple. Of these theorems, the simplest is the Two-Triangle Theorem of Desargues. For this reason we now add

Axiom 6. If two triangles are perspective from a point, they are perspective from a line.

For a proof of the dual, see Problem 7.11.

We shall also need

Axiom 7. The diagonal points of a complete quadrangle are never collinear.

In 7-point geometry it is found that the diagonal points of any complete quadrangle *are* collinear; Axiom 7 then rules out any further consideration of this geometry. However, both Axioms 6 and 7 are provable theorems in 13- and 31-point geometry.

Axiom 7 allows us to introduce harmonic sets and, hence, a harmonic net on a line. Let $R(A,B,C)$ be a harmonic net on a line p, let p' be any other line of the plane and let $p \cdot p' = 0$. Let $P \neq 0$ be a point on p and $P' \neq 0$ be a point on p'; let R, distinct from P and P', be a point on PP'. Points A', B', C' on p' may now be defined by means of the perspectivity $p(A,B,C) \overset{R}{\barwedge} p'(A',B',C')$. Since this perspectivity carries any harmonic set on p into a harmonic set on p', it carries the harmonic net $R(A,B,C)$ on p into the harmonic net $R(A',B',C')$ on p'. Thus the existence of a harmonic net on one line of the plane assures the existence of a harmonic net on every line of the plane.

When the number of points on a line is finite, the number of elements of the harmonic net on the line is also finite. As a consequence $H(S,T;U,V)$ cannot imply the separation, as defined in Chapter 4, of the pair of point S, T by the pair U, V. (This is to be verified in Problems 17.17(b) and 17.20(a) for the 13- and 31-point geometries.) Now the relation "separation", together with its ramifications, plays an important role in the geometry which we are attempting to construct. Consequently, we introduce this relation by means of

Axiom 8. If $H(A,B;D,E)$, then the pairs of points A, B and D, E separate each other.

Axiom 9. If the pairs of points A, B and D, E separate each other, then A, B, D, E are distinct.

Axiom 10. If the pairs A, B and D, E_1 separate each other and if the pairs A, E_1 and B, E_2 separate each other, then the pairs A, B and D, E_2 separate each other.

The effect of these axioms is to eliminate further consideration of geometries with a finite number of points on a line, as shown in

Theorem 7.6. In any projective geometry satisfying Axioms 1-10 inclusive, there are infinitely many points on any line.

For a proof, see Problem 7.12.

The infinitude of points obtained in the proof of Theorem 7.6 is precisely the harmonic net, the elements of which (see Chapter 4) may be placed in one-to-one correspondence with the set of all rational numbers. When (by assumption) the points of this harmonic net constitute the totality of points on the line, we have what will be called *rational projective geometry in the plane*. In this geometry a projectivity which leaves any three distinct points A, B, C of a line fixed also leaves fixed the harmonic conjugate of any one of these points with respect to the other two and, hence, leaves fixed every point of $R(A,B,C)$, that is, every point of the line. We state this as the

Lemma: If a projectivity leaves each of three distinct points of a line of the rational projective plane invariant, it leaves every point of the line invariant.

In Problem 7.13 we prove a

Fundamental Theorem: Given in the rational projective plane three distinct collinear points A, B, C and another three distinct collinear points A', B', C', on the same line or on distinct lines, there is one and only one projectivity which carries the triple A, B, C respectively into the second triple A', B', C'.

In algebra, the set of rational numbers is obtained by means of the rational operations on the unit 1 and may be extended to the set of real numbers either by a construction (Dedekind cut) or by postulating the existence of non-repeating infinite decimals which obey the same laws of operations as the rationals. Here, the set of rational points on a line has been obtained from a triple of points on the line by means of harmonic sets and may be extended to the set of all real points on a line either by a construction or by

Axiom 11.   There exists a projective line the totality of whose points are in one-to-one correspondence with the extended real number system and a separation relation: A, B separates C, D if and only if $(A, B; C, D) < 0$ such that Axioms 1-10 are satisfied.

Solved Problems

7.1.   Prove: If p and q are distinct lines, then there is at most one point on both.

By Axiom 3 there is at least one point, say R, on both p and q. Suppose there is a second point $S \neq R$ on both p and q. Then on the distinct points R and S there are two distinct lines. Since this contradicts Axiom 2, the proof of the theorem is complete.

7.2.   Prove: Not all lines are on the same point.

Let P be a point and q be a line. Should P not be on q, the proof is complete. Suppose then that P is on q. Let $Q \neq P$ be another point on q (Axiom 4) and R be a point not on q (Axiom 5). Now $R \neq Q$ (why?); hence, by Axiom 1, there is a line $p = QR$ on both Q and R. Since $p \cdot q = Q$ and $P \neq Q$, then p is not on P (Theorem 7.1) and the proof is complete.

7.3.   Prove: There are at least three distinct lines on any point.

Let P be a point and p be a line not on P (Theorem 7.2). On p take three distinct points (Axiom 4); call them A, B, C. Since P is not on p, then $P \neq A$, $P \neq B$, $P \neq C$ and so $a = PA$, $b = PB$, $c = PC$ are three lines on P. Consider the lines a and b. Since $a \cdot p = A$, $b \cdot P = B$, and $B \neq A$, then B is not on a and A is not on b; hence, $a \neq b$. By repeating the argument, we have finally $a \neq b$, $b \neq c$, $c \neq a$ which completes the proof.

7.4.   Prove: If there are exactly $n \geqq 3$ distinct points on some line, there are exactly n distinct points on every line.

Denote by p the line on which there are exactly n distinct points $A_1, A_2, A_3, \ldots, A_n$. On any of these points, say A_2, take the distinct lines q and r, where $p \neq q$, $p \neq r$ (Theorem 7.3). On r take another point $B \neq A_2$ (Axiom 4) and consider the n joins of B and the points $A_1, A_2, A_3, \ldots, A_n$. As in the proof of Problem 7.3, these n lines are themselves distinct and each is distinct from p and q. Thus q sections these n lines in n distinct points one of which is $q \cdot p = A_2$. Thus, there are at least n points on q. Suppose there are $n + 1$ points on q. The reader will show there are then at least $n + 1$ points on p and, hence, exactly n points on q.

7.5. Prove: If there are exactly n distinct points on any line, there are exactly $n^2 - n + 1$ distinct points in all.

Let p be a line and P be a point not on p (Axiom 5). On p there are n distinct points (by hypothesis) which with P determine n distinct lines (proof of Problem 7.4). On each of these lines are n distinct points (Theorem 7.4) one of which is P. Thus, exclusive of P, there are $n(n-1)$ distinct points and, including P, there are $n(n-1)+1 = n^2 - n + 1$ distinct points in all.

7.6. Construct (if possible) a model for a geometry in which there are four points on any line.

By Theorem 7.5 there will be $4^2 - 4 + 1 = 13$ points in all. We shall proceed as in Example 2 using the set $\{0, 1, 2\}$ with addition and multiplication tables

+	0	1	2		·	0	1	2
0	0	1	2		0	0	0	0
1	1	2	0		1	0	1	2
2	2	0	1		2	0	2	1

Note. The elements of the set $\{0, 1, 2\}$ consist of the remainders when the natural numbers are divided by 3. The natural numbers are then separated into three classes: (1) $3, 6, 9, \ldots$ represented by 0, (2) $1, 4, 7, \ldots$ represented by 1, and (3) $2, 5, 8, \ldots$ represented by 2. The result of any operation with these numbers is again replaced by its representative. Thus $2 + 2 = 2 \cdot 2 = 4$ is replaced by 1 and $1 + 2 = 2 + 1 = 3$ is replaced by 0.

The number of triples $(x_1, x_2, x_3) \neq (0, 0, 0)$ is $3^3 - 1 = 26$. Having twice as many triples as points, let us agree that the triples (a, b, c) and $(2a, 2b, 2c)$ be the same point. The required points may now be given as follows:

$$A = (0,0,1) \qquad F = (1,1,0) \qquad J = (1,2,0)$$
$$B = (0,1,0) \qquad G = (1,1,1) \qquad K = (2,2,1)$$
$$C = (1,0,0) \qquad H = (0,1,2) \qquad L = (2,1,2)$$
$$D = (0,1,1) \qquad I = (1,0,2) \qquad M = (1,2,2)$$
$$E = (1,0,1)$$

(Note that $(2,1,1) = (2 \cdot 2, 2 \cdot 1, 2 \cdot 1) = (1,2,2) = M$, $(2,0,1) = (1,0,2) = I$, etc.)

These points lie by fours on the lines, as follows:

$$x_1 = 0: \quad A, B, D, H \qquad\qquad x_1 + x_2 = 0: \quad A, J, L, M$$
$$x_2 = 0: \quad A, C, E, I \qquad\qquad x_1 + x_3 = 0: \quad B, I, K, M$$
$$x_3 = 0: \quad B, C, F, J \qquad\qquad x_2 + x_3 = 0: \quad C, H, K, L$$

$$x_1 + 2x_2 = 0: \quad A, F, G, K \qquad\qquad x_1 + x_2 + x_3 = 0: \quad G, H, I, J$$
$$x_1 + 2x_3 = 0: \quad B, E, G, L \qquad\qquad x_1 + 2x_2 + x_3 = 0: \quad D, F, I, L$$
$$x_2 + 2x_3 = 0: \quad C, D, G, M \qquad\qquad x_1 + x_2 + 2x_3 = 0: \quad D, E, J, K$$

$$2x_1 + x_2 + x_3 = 0: \quad E, F, H, M$$

The reader may try his hand at constructing a figure. None is given since it is of no appreciable help in the study of the 13-point geometry.

7.7. Prove: The four points on any line in 13-point geometry when taken in any order form a harmonic set.

Consider the collinear points B, E, G, L. We first establish $H(B, E; G, L)$. The line $x_1 + x_2 + x_3 = 0$ on G meets the lines $x_1 = 0$ and $x_3 = 0$ on B in the points H and J respectively. The line EH meets $x_3 = 0$ in F and the line EJ meets $x_1 = 0$ in D. The line FD is on L and so $H(B, E; G, L)$. By construction we also have $H(B, E; L, G)$, $H(E, B; G, L)$ and $H(E, B; L, G)$.

The line $2x_1 + x_2 + x_3 = 0$ meets $x_1 = 0$ in H and $x_3 = 0$ in F. The line GH meets $x_3 = 0$ in J and the line GF meets $x_1 = 0$ in A. Since the line AJ is on L, we have $H(B, G; E, L)$ and also $H(G, B; E, L)$, $H(B, G; L, E)$, $H(G, B; L, E)$.

Consider the points D and J in which the line $x_1 + x_2 + 2x_3 = 0$ on E meets $x_1 = 0$ and $x_3 = 0$ respectively. The line LD meets $x_3 = 0$ in F and the line LJ meets $x_1 = 0$ in A. Since AF is on G, we have $H(B, L; E, G)$; then also, $H(L, B; E, G)$, $H(B, L; G, E)$, $H(L, B; G, E)$.

The remaining cases: $H(G, E; B, L)$, $H(E, L; B, E)$ and $H(E, L; B, G)$ are left for the reader.

7.8. For a model of the 31-point geometry (6 points on any line) take as points the triples $(x_1, x_2, x_3) \neq (0, 0, 0)$, where each of the x_i is from the set $\{0, 1, 2, 3, 4\}$ of remainders when the natural numbers are divided by 5, and as lines the equations of the form $a_1 x_1 + a_2 x_2 + a_3 x_3 = 0$ where each of the a_i is from the same set of remainders excepting only the choice $a_1 = a_2 = a_3 = 0$. We also agree that (jx_1, jx_2, jx_3), where $(j = 1, 2, 3, 4)$, are the same point and that $j(a_1 x_1 + a_2 x_2 + a_3 x_3) = 0$ are the same line.

(a) Show that $(1, 2, 3)$ and $(4, 3, 2)$ are the same point.

Here $(1, 2, 3) = (4 \cdot 1, 4 \cdot 2, 4 \cdot 3) = (4, 8, 12)$ which, when each entry is replaced by the remainder obtained by dividing by 5, becomes $(4, 3, 2)$.

(b) Show that $(2, 4, 3)$ and $(3, 1, 2)$ are the same point.

Consider the first entries 2 and 3 of the triples; 2 must be multiplied by some j such that the product $2j$, when divided by 5, has remainder 3. With $j = 4$, we have $(4 \cdot 2, 4 \cdot 4, 4 \cdot 3)$ or $(3, 1, 2)$ as required.

(c) Obtain the line on $H = (1, 2, 3)$ and $I = (3, 0, 2)$.

Let $a_1 x_1 + a_2 x_2 + a_3 x_3 = 0$ be the required line. Since H is on the line, we have (1) $a_1 + 2a_2 + 3a_3 = 0$; since I is on the line, we have (2) $3a_1 + 2a_3 = 0$. Solutions of this pair of simultaneous equations will be found by inspection. Clearly, (2) is satisfied when $a_1 = 1$ and $a_3 = 1$. Then, substituting in (1), we have $2a_2 + 4 = 0$ whence $a_2 = 3$. The required line is $x_1 + 3x_2 + x_3 = 0$.

(d) Obtain the line on $J = (2, 3, 2)$ and $K = (1, 1, 3)$.

We are to solve the simultaneous system: $2a_1 + 3a_2 + 2a_3 = 0$ and $a_1 + a_2 + 3a_3 = 0$. Adding the two equations, we find $3a_1 + 4a_2 = 0$ and take $a_1 = 2$, $a_2 = 1$. From the first equation we obtain $2a_3 + 1 = 0$; then $a_3 = 2$. The required line is $x_1 + 3x_2 + 2x_3 = 0$.

(e) Obtain the line on $J = (2, 3, 2)$ and $L = (1, 1, 2)$.

The equations are $2a_1 + 3a_2 + 2a_3 = 0$ and $a_1 + a_2 + 2a_3 = 0$. To the first equation add 3 times the second; we have $a_2 + 3a_3 = 0$ and take $a_2 = 2$, $a_3 = 1$. From the first equation we obtain $2a_1 + 3 = 0$. Then $a_1 = 1$ and the required line is $x_1 + 2x_2 + x_3 = 0$.

(f) Find the point common to the lines $2x_1 + 3x_2 + x_3 = 0$ and $x_1 + 2x_2 + 2x_3 = 0$.

To the second equation add twice the first; we have $3x_2 + 4x_3 = 0$ and take $x_2 = 2$, $x_3 = 1$. From the first equation $2x_1 + 2 = 0$ and $x_1 = 4$. The required point is $(4, 2, 1)$ or $(2, 1, 3)$ or $(3, 4, 2)$ or $(1, 3, 4)$.

(g) Find the six points on the line $x_1 + 3x_2 + x_3 = 0$.

We find easily the points $(0, 1, 2)$, $(2, 0, 3)$, $(2, 1, 0)$ and $(1, 1, 1)$. For the remaining points, we take $(1, 2, 3)$ and $(3, 2, 1)$.

(h) Obtain the equations of the six distinct lines on $J = (2, 3, 2)$ of (d) and (e).

Partial Ans. $2x_1 + 3x_2 + x_3 = 0$.

7.9. Consider on the line $x_1 = 0$ of Problem 7.8 the six points $A = (0, 0, 1)$, $B = (0, 1, 0)$, $D = (0, 1, 1)$, $E = (0, 1, 2)$, $F = (0, 2, 1)$ and $G = (0, 2, 3)$. Show that the selection of a pair of these points, say A and B, separates the remaining points into two pairs D, G and E, F so that $H(A, B; D, G)$ and $H(A, B; E, F)$.

On A take the lines $x_1 + x_2 = 0$ and $2x_1 + x_2 = 0$. On D take the line $3x_1 + 3x_2 + 2x_3 = 0$ meeting $x_1 + x_2 = 0$ in $P = (2, 3, 0)$ and $2x_1 + x_2 = 0$ in $R = (2, 1, 3)$. Let BP: $x_3 = 0$ meet $2x_1 + x_2 = 0$ in $Q = (2, 1, 0)$ and BR: $x_1 + x_3 = 0$ meet $x_1 + x_2 = 0$ in $S = (2, 3, 3)$. Then QS: $2x_1 + x_2 + x_3 = 0$ meets $x_1 = 0$ in G. Thus, $H(A, B; D, G)$.

On E take the line $x_1 + x_2 + 2x_3 = 0$ meeting $x_1 + x_2 = 0$ in P and $2x_1 + x_2 = 0$ in $T = (2, 1, 1)$. Let $BT: 2x_1 + x_3 = 0$ meet $x_1 + x_2 = 0$ in $U = (2, 3, 1)$ and $BP: x_3 = 0$ meet $2x_1 + x_2 = 0$ in $Q = (2, 1, 0)$. Then $QU: 2x_1 + x_2 + 3x_3 = 0$ meets $x_1 = 0$ in F and so $H(A, B; E, F)$.

7.10. Using known properties of Euclidean geometry, give a model for the projective plane in which there are infinitely many points on any line.

It is to be noted that we have used in the first six chapters one such model. As points of the projective plane we took the totality of points of the Euclidean plane together with certain additional elements called ideal points, and as lines the totality of lines of the Euclidean plane together with an ideal line. It is easily verified that this was all done in such a manner as to satisfy our five axioms. We give now another model.

Let O be a fixed point in ordinary space and consider the set of all lines on O and the set of all planes on O. Every pair of distinct lines determines a unique plane and every pair of distinct planes determines a unique line. Thus a model having the required properties is obtained by taking the lines on O as points of the projective plane, the planes on O as the lines of the projective plane, and defining the relation "point on line" to mean "a line lies in a plane".

7.11. Prove: If two triangles are perspective from a line, they are perspective from a point.

Consider in Fig. 7-3 the triangles ABC and $A'B'C'$ so situated that the points $P = AB \cdot A'B'$, $Q = BC \cdot B'C'$, $R = CA \cdot C'A'$ are on a line o. Now the triangles $AA'R$ and $BB'Q$ are perspective from P; hence, by Axiom 6, they are perspective from a line. On this line are the points $O = AA' \cdot BB'$, $C = AR \cdot BQ$ and $C' = A'R \cdot B'Q$. Thus the lines AA', BB', CC' are on O and the triangles ABC and $A'B'C'$ are perspective from this point.

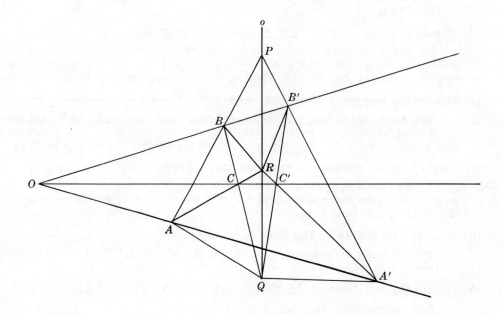

Fig. 7-3

7.12. Prove: In any projective geometry satisfying Axioms 1-10 inclusive, there are infinitely many points on any line.

Refer to Fig. 7-4 below in which A, B, D are distinct points on a line o. By Axiom 7, there exists a point F_1 on the line such that $H(A, B; D, F_1)$. By Axiom 8, the pairs A, B and D, F_1 separate each other and, by Axiom 9, the points A, B, D, F_1 are distinct.

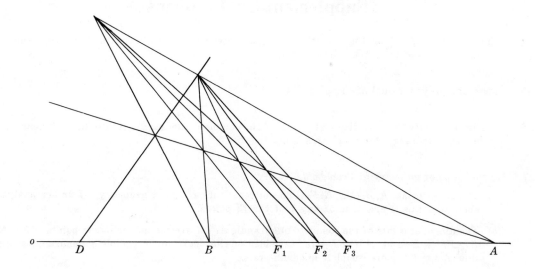

Fig. 7-4

By Axiom 7, there exists a point F_2 on the line o such that $H(A, F_1; B, F_2)$. By Axiom 8, the pairs A, F_1 and B, F_2 separate each other and, by Axiom 9, the points A, B, F_1, F_2 are distinct. Now A, B and D, F_1 also separate each other; hence, by Axiom 10, we have that the pairs A, B and D, F_2 separate each other. Then D and F_2 are distinct and so A, B, D, F_1, F_2 are distinct.

By Axiom 7, there exists a point F_3 on the line such that $H(A, F_2; F_1, F_3)$. Then A, F_2 and F_1, F_3 separate each other and A, F_1, F_2, F_3 are distinct. Since A, F_1 and B, F_2 also separate each other, we have, by Axiom 10, that A, F_1 and B, F_3 separate each other; hence, B and F_3 are distinct. Similarly, since A, B and D, F_1 separate each other, we have A, B and D, F_3 separate each other and so D and F_3 are distinct. Thus A, B, D, F_1, F_2, F_3 are distinct.

By Axiom 7, there exists a point F_4 on the line such that $H(A, F_3; F_2, F_4)$. By repeating the argument above, we find that $A, B, D, F_1, F_2, F_3, F_4$ are distinct; and so on, without end. We have proved (see the first paragraph of the concluding section of Chapter 4) that the points of a harmonic sequence on a line are infinite in number.

Now, in fact, Axiom 7 assures us of more: there is a harmonic net $R(A, B, D)$ on the line o. Since this net contains the harmonic sequence discussed above as a subset, the points of the harmonic net are also infinite in number.

7.13. Prove the Fundamental Theorem: Given in the rational projective plane three distinct collinear points A, B, C and another three distinct collinear points A', B', C' on the same line or on distinct lines, there is one and only one projectivity which carries the triple A, B, C respectively into the triple A', B', C'.

First, we establish Theorem 1.3, page 14, in the rational projective plane, validating the constructions by the axioms. Thus, one projectivity is assured. Now suppose there were two projectivities

$$(A, B, C, D) \ \overline{\wedge} \ (A', B', C', D') \text{ and } (A, B, C, D) \ \overline{\wedge} \ (A', B', C', D'')$$

where D is some point collinear with, but distinct from the points of the first triple. Then

$$(A', B', C', D') \ \overline{\wedge} \ (A', B', C', D'')$$

But this projectivity has three invariant points; hence, by the Lemma, $D' = D''$ and so there can be but one projectivity.

Supplementary Problems

7.14. Show, using Fig. 7-1 or Fig. 7-2, that in 7-point geometry there exist projectivities which interchange any two of the three points on a line.

7.15. State and prove the dual of Problem 7.14.

7.16. In Fig. 7-2 verify that the diagonal points of the complete quadrangle $ABCG$ are collinear. Is there a complete quadrangle for which this is not true?

7.17. In the 13-point geometry of Problem 7.6 verify:

(a) The seven points A, B, C, D, E, F, G do not constitute a 7-point geometry. (**Further** analysis will show that this is also true of any 7 of the 13 points.)

(b) The diagonal points of the complete quadrangle $AFIM$ are the non-collinear points $AF \cdot IM = G$, $AI \cdot FM = E$ and $AM \cdot FI = L$. Consider one or more other complete quadrangles; also the dual for one or more complete quadrilaterals.

(c) The triangles DEL and GIK, being perspective from the point C, are perspective from the line $x_3 = 0$. Check other pairs of triangles.

(d) The Theorem of Pappus holds, but always in the form of the special case (page 40). *Hint.* Take C, E, I and K, F, G and show that the line of Pappus is $x_1 + x_2 = 0$. Consider one or more other pairs of collinear triples.

(e) The Fundamental Theorem holds.

7.18. Given the points $K = (2, 3, 2)$, $L = (3, 1, 1)$, $R = (2, 1, 3)$, $P = (2, 3, 0)$, $Y = (1, 1, 1)$ of the 31-point geometry (Problem 7.8). Show: (a) The line KP is on $A = (0, 0, 1)$. (b) The line KR is on L. (c) The lines KL and PY are on $Z = (2, 0, 1)$.

7.19. For the points on $x_1 = 0$ of Problem 7.9, show that
$$(A, B, D, G, E, F) \; \overline{\wedge} \; (A, B, G, D, F, E)$$

Hint. Project from $P = (2, 3, 0)$ onto $x_1 + x_3 = 0$ and then from $V = (2, 3, 1)$ back onto $x_1 = 0$.

Thus, the selection of a pair of points (see Problem 7.9) on a line separates the remaining points into two pairs, each of which is a reciprocal pair in the hyperbolic involution on the line having the initial pair as double points.

7.20. (a) On the line $x_1 = 0$ of Problem 7.9, verify $H(A, D; B, E)$ and $H(B, D; E, G)$.

(b) Assume $H(A, B; D, G)$ in Problem 7.9 implies that the pair D, G separates the pair A, B, and likewise for $H(A, D; B, E)$ and $H(B, D; E, G)$ of (a). Obtain a contradiction.

(c) On the line $x_1 + 3x_2 + x_3 = 0$ of Problem 7.8(g) take any two points as diagonal points of a complete quadrangle. Show that the third diagonal point is not on the line.

(d) Select any three lines of Problem 7.8(h). On the first line select distinct points A, A' each distinct from J; similarly, select B, B' on the second line and C, C' on the third. The triangles ABC and $A'B'C'$ are perspective from the point J. Show that they are also perspective from a line.

7.21. On each of the lines $x_1 + x_2 = 0$ and $x_1 + 2x_1 = 0$ on the point $A = (0, 0, 1)$ of Problem 7.8, select a triple of distinct points — X, Y, Z on one and X', Y', Z' on the other — such that XX', YY', ZZ' are not concurrent. Find the equation of the line of Pappus. Is this line on A?

Chapter 8

Point Conics and Line Conics

INTRODUCTION

The study of conics or conic sections, that is, of the plane sections of a right circular cone, began quite early (perhaps, around 430 B.C.) in the history of mathematics. Among the conic sections are included the so-called *degenerate* or *singular conics* (for example, a single point, a pair of distinct lines, two coincident lines) as well as the *non-singular conics*. In turn, the non-singular conics are classified as parabolas, hyperbolas and ellipses (including circles).

In projective geometry the term point conic is applied to certain non-degenerate loci of points and dually, the term line conic is applied to certain non-degenerate envelopes of lines. That these point and line conics are indeed the familiar non-singular conics of analytic geometry will become apparent as we continue our study. In this chapter, for example, we show that they have in common the properties: (*a*) a line (point) cannot intersect (join) a point (line) conic in more than two points (lines) and (*b*) five points (lines), no three on the same line (point), determine a unique point (line) conic. Throughout this book it is to be understood that the terms point conic, line conic and conic refer to non-degenerate loci and envelopes unless explicitly stated otherwise.

THE POINT CONIC

Consider in Fig. 8-1 below the non-perspective projectivity

$$R(a, b, c, \ldots) \ \overline{\wedge} \ S(a', b', c', \ldots)$$

between pencils of lines on distinct centers R and S. The totality of points $A = a \cdot a'$, $B = b \cdot b'$, $C = c \cdot c'$, ... common to pairs of corresponding lines of the two pencils is called a *point conic*. In analytic geometry a conic was thought of as being traced out by a moving point; in projective geometry a point conic is constructed point by point. This consists essentially in taking any other line x on R and constructing its correspondent x' on S. To do this, we use the given pairs of lines a, a'; b, b'; c, c' to locate the point of Pappus (center of projectivity) P as the intersection of $(a \cdot c')(a' \cdot c)$ and $(b \cdot c')(b' \cdot c)$. For any other line (as d) on R, let the join of P and $a' \cdot d$ meet the line a in D'; then $SD' = d'$ is the correspondent of d in the projectivity and $d \cdot d' = D$ is another point on the point conic. Again, for any other line (as e) on R, join P and $b' \cdot e$ meeting b in E'; then $SE' = e'$ is the correspondent of e and $e \cdot e'$ is another point on the point conic; ....

Let the line of centers RS be called s when considered as a member of the pencil on R and be called r' when considered as a member of the pencil on S. Now $s \cdot c' = S$; let $PS \cdot c = S'$. Then $SS' = s'$ is the correspondent of s and $s \cdot s' = S$ is a point on the point conic. Similarly, R is also a point on the point conic.

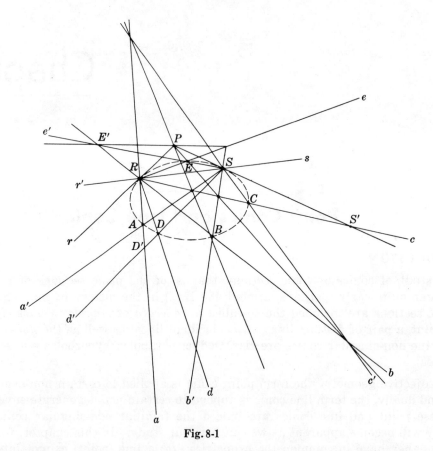

Fig. 8-1

Generally, on any line x of the pencil on R there are two distinct points of the point conic — R and $X = x \cdot x'$ where x' on S is the correspondent of x in the projectivity. The line $PR = r$ is an exception since its correspondent is $r' = RS$ and $r \cdot r' = R$. We define

   A *tangent* to a point conic is any line on which there is one and only one point of the conic.

Thus for the point conic of Fig. 8-1 both PR and PS are tangents. We have proved

Theorem 8.1.   The tangent to a point conic at the center R (S) of one of the pencils of lines generating it is the correspondent of the line of centers RS thought of as a member of the pencil on S (R).

To show that R and S play no particular roles as points of the point conic generated by projective pencils of which they are centers, we prove in Problem 8.1

Theorem 8.2.   Any two distinct points on a point conic may be used as the centers of two projective pencils by which it is generated.

There follow

Theorem 8.3.   A point conic is uniquely determined by any five of its points.

For a proof, see Problem 8.2.

Theorem 8.4.   Three distinct points of a point conic are never collinear.

Theorem 8.5.   Five distinct points, no three of which are collinear, determine a unique point conic.

Theorem 8.6.   On any point of a point conic there is one and only one tangent to the point conic.

THE LINE CONIC

The dual of a point conic, that is, the totality of lines joining corresponding points of two non-perspective projective pencils of points on distinct lines (*axes*) is called a *line conic*. Consider in Fig. 8-2 the non-perspective projectivity

$$r(A, B, C, \ldots) \;\overline{\wedge}\; s(A', B', C', \ldots)$$

on distinct axes r and s. Then AA', BB', CC', r, s are lines of the line conic generated by the projectivity. To construct other lines, we first locate the line of Pappus (axis of projectivity) p by joining $AC' \cdot A'C$ and $BC' \cdot B'C$. For any other point (as D) on r, let $A'D \cdot p = D_1$; then $AD_1 \cdot s = D'$ is the correspondent of D in the projectivity and DD' is another line of the line conic.

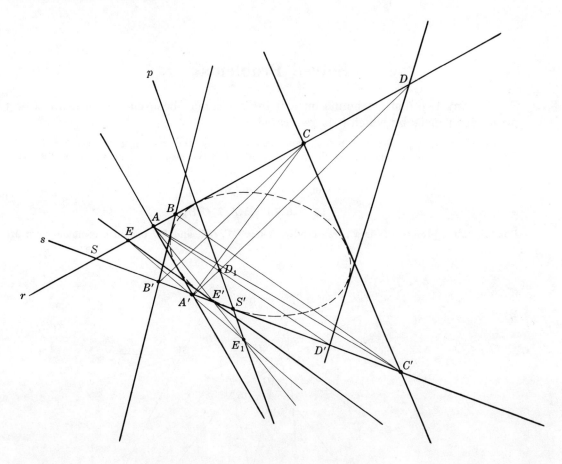

Fig. 8-2

Generally, on any point X of the pencil on r (also X' on s) there are two distinct lines of the line conic — r and $x = XX'$ where X' on s is the correspondent of X in the projectivity. The point $p \cdot s = S'$ is an exception since its correspondent is $S = r \cdot s$ and $SS' = s$. We define

> A *point of contact* of a line conic is any point on which there is one and only one line of the conic.

It is not difficult to prove

Theorem 8.1'.   The point of contact of a line conic on the axis r (s) of the pencils of points generating it is the correspondent of the point $r \cdot s$ thought of as a member of the pencil on s (r).

The duals of Theorems 8.2-8.6 follow readily.   In particular, there are

Theorem 8.5'.   Five distinct lines, no three of which are concurrent, determine a unique line conic.

and

Theorem 8.6'.   On any line of a line conic there is one and only one point of contact with the line conic.

Solved Problems

8.1.   Prove: Any two distinct points on a point conic may be used as the centers of two projective pencils by which it is generated.

Consider in Fig. 8-3 the point conic generated by the two pencils $R(a, b, c, d, \ldots)$ and $S(a', b', c', d', \ldots)$ of which at least the points A, B, C, D have been constructed as in Fig. 8-1. Let $AB = m$, $AC = n$; $m \cdot c = K$, $m \cdot d = L$; $n \cdot b' = T$, $n \cdot d' = U$; $b' \cdot d = V$. Then since

$$(a, b, c, d) \ \overline{\wedge}\ (a', b', c', d')$$

we have

$$(A, B, K, L) \ \overline{\wedge}\ (A, T, C, U)$$

But this latter projectivity is a perspectivity; hence BT, CK, and LU are on a common point M.

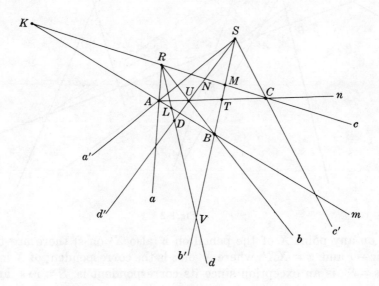

Fig. 8-3

Consider now

$$(BA, BR, BS, BD) \ \overline{\wedge}\ (L, R, V, D) \ \overset{M}{\overline{\wedge}}\ (U, N, S, D) \ \overline{\wedge}\ (CA, CR, CS, CD)$$

Then

$$(BA, BR, BS, BD) \ \overline{\wedge}\ (CA, CR, CS, CD)$$

and, moreover, the intersections of corresponding lines are on the point conic; hence this projectivity also generates the point conic.

8.2.　Prove: A point conic is uniquely determined by any five of its points.

Let A, B, C, R, S be any five points of a given point conic. By Theorem 8.2 any two of the points (say, R and S) may be taken as the centers of two pencils of lines which generate the point conic. By the Fundamental Theorem the remaining points establish a unique projectivity

$$(RA, RB, RC) \; \overline{\wedge} \; (SA, SB, SC)$$

which, in turn, generates the given point conic.

8.3.　For the point conic generated by $R(a, b, c, \dots) \; \overline{\wedge} \; S(a', b', c', \dots)$ of Fig. 8-1, construct the tangent at B.

Let the given point conic be considered (see Fig. 8-4) as generated by projective pencils with centers at B, the point at which the tangent is to be constructed, and any other point of the conic, say R; that is, consider the point conic as generated (see Theorem 8.2) by the projectivity

$$B(a'', c'', b', \dots) \; \overline{\wedge} \; R(a, c, s, \dots)$$

where $BA = a''$ and $BC = c''$. Locate the center of projectivity Q on $(a'' \cdot s)(a \cdot b')$ and $(c'' \cdot s)(c \cdot b')$. The required tangent is QB.

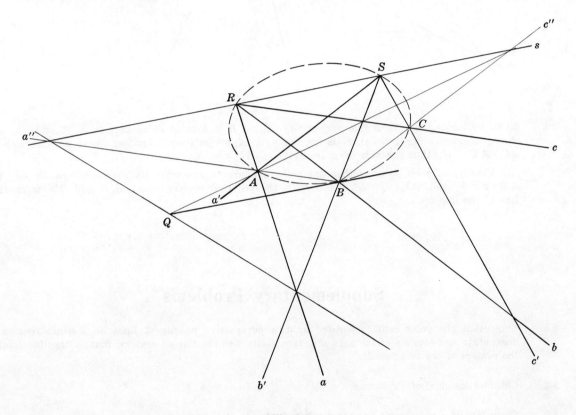

Fig. 8-4

8.4.　Construct the line conic determined by four lines, no three on the same point, and the point of contact on one of the lines.

In Fig. 8-5 below, let the given lines be a_1, a_2, a_3, a_4 and the given point of contact B on a_2. Take a_2 and any other line, say a_4, as the axes of the projective pencils of points which generate the line conic. We now establish the projectivity. Let $a_2 \cdot a_1 = A$, $a_4 \cdot a_1 = A'$; $a_2 \cdot a_3 = C$, $a_4 \cdot a_3 = C'$. By the dual of Theorem 8.1, the correspondent on a_4 of B on a_2 is the point $a_2 \cdot a_4 = B'$. Thus the generating projectivity is

$$a_2(A, B, C, \dots) \; \overline{\wedge} \; a_4(A', B', C', \dots)$$

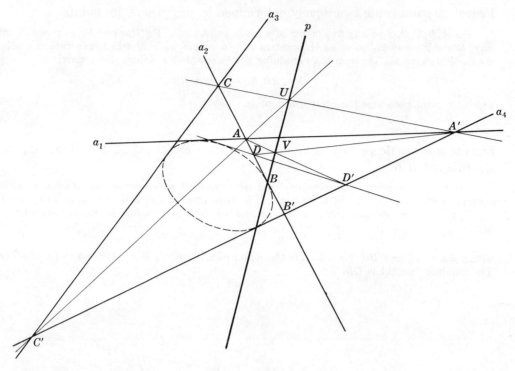

Fig. 8-5

Next, we locate the axis of this projectivity. Since B is a point of contact on a_2 (that is, by our choice of generating pencils), B is on the axis of projectivity. Another point on this axis is $AC' \cdot A'C = U$; thus the axis of projectivity is $BU = p$.

Finally, take on a_2 any other point D. We are to construct its correspondent on a_4. Let $A'D \cdot p = V$ and $AV \cdot a_4 = D'$. Then D' is the required correspondent of D and DD' is another line of the line conic.

Supplementary Problems

8.5. Show that the point conic generated by two perspective pencils of lines on distinct centers is degenerate and consists of the axis of perspectivity and the line of centers, that is, the line joining the centers of the two pencils.

8.6. Consider the dual of Problem 8.5.

8.7. Show that if a point conic has three collinear points, it is degenerate.

8.8. In Fig. 8-1, prove R is on the conic; in Fig. 8-2 prove $R = p \cdot r$ is a point of contact.

8.9. Prove: Theorem 8.4, Theorem 8.5, Theorem 8.6.

8.10. Take any five points, no three of which are collinear, and construct several additional points of the point conic on the given points.

Hint. It will be noted that an ellipse is used in the figures; this is done in order to keep the constructions on the page. It is suggested that for all constructions an ellipse be traced first and then the given points and lines be taken with respect to this ellipse. Why is there no loss in generality by so doing?

8.11.   Take any five lines, no three concurrent, and construct additional lines of the line conic on the given lines.

8.12.   Construct the dual of Problem 8.3.

8.13.   Given three lines a, b, c of a line conic C' and the points of contact A and B on a and b respectively, construct: (i) additional lines of C', (ii) the point of contact on c.
    Hint.   Take A and B as the centers of the projective pencils generating C'.

8.14.   Given three points A, B, C of a point conic C and the tangents a and b at A and B respectively, construct: (i) additional points of C, (ii) the tangent at C.

8.15.   If A, B, C are three distinct points on a line k and if P, Q are distinct points neither of which is on k, show that $R = PA \cdot QB$, $S = PA \cdot QC$, $T = PB \cdot QA$, $U = PB \cdot QC$, $V = PC \cdot QA$, $W = PC \cdot QB$ are on a conic.

8.16.   Prove: If XYZ is a variable triangle (see Fig. 8-6) whose sides $x = YZ$, $y = XZ$, $z = XY$ are on the fixed points U, E, D respectively while Y moves along the fixed line $b = CB$ and Z moves along the fixed line CA, the vertex X moves on a conic.

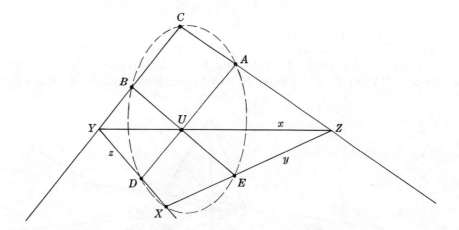

Fig. 8-6

8.17.   In Problem 8.16, show that C is on A, B, C, D, E.
    Hint.   When x coincides with BE so also does y and then X is at B.

8.18.   Use the results of Problems 8.16 and 8.17 to devise another procedure for constructing a conic on five given points A, B, C, D, E of which no three are collinear.

8.19.   Dualize Problems 8.16, 8.17, 8.18.

8.20.   Prove: If the vertices of two triangles are on a conic C, the sides of the triangles are tangent to another conic C'.

Chapter 9

Poles and Polar Lines

THE POLAR LINE

On the point conic C of Fig. 9-1, take any three pairs of distinct points $X, X_1; A, A_1;$ B, B_1 such that each pair is collinear with a given point P, not on C. Let $XA = a$, $XA_1 = a_1$, $XB = b$, $XB_1 = b_1$; $X_1A = a'$, $X_1A_1 = a_1'$, $X_1B = b'$, $X_1B_1 = b_1'$. Then (see Theorem 8.2, page 84)

$$X(a, b, a_1, b_1) \; \overline{\wedge} \; X_1(a', b', a_1', b_1')$$

The center T of this projectivity is on $(a \cdot b')(a' \cdot b)$ and $(a \cdot a_1')(a' \cdot a_1)$. Denote by p the line on $a \cdot a_1' = K$, $a' \cdot a_1 = L$, and T. Since K, L, P are the vertices of the diagonal triangle of the complete quadrangle XAA_1X_1, we have $H(a, a_1'; p, KP)$ by Theorem 4.12, page 48. Then easily $H(X, X_1; P_1, P)$, where $P_1 = XX_1 \cdot p$, and $H(A, A_1; A_1', P)$, where $A_1' = AA_1 \cdot p$.

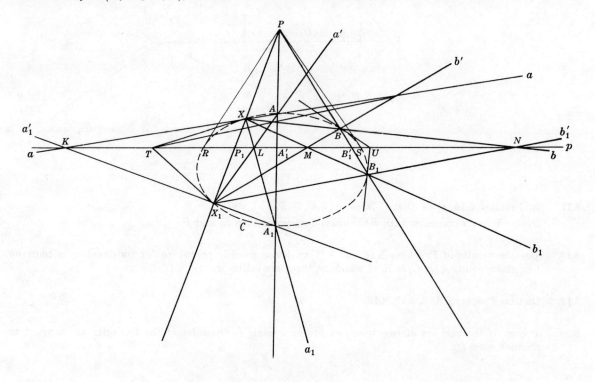

Fig. 9-1

Similarly, using the complete quadrangle XX_1B_1B, we have $b \cdot b_1' = N$, $b' \cdot b_1 = M$ and T on a line, say p', and $H(X, X_1; P_1', P)$, where $P_1' = XX_1 \cdot p'$. But $H(X, X_1; P_1, P)$ and $H(X, X_1; P_1', P)$ imply $P_1' = P_1$; hence $p' = p$ and $H(B, B_1; B_1', P)$, where $B_1' = BB_1 \cdot p$.

90

Now p is determined (see Theorem 8.1, page 84) by T, the common point of the tangents to C at X and X_1, and by P_1, the harmonic conjugate of P with respect to X and X_1. Thus for the given conic C, the line p depends at most on the collinear points P, X, X_1. We will show that p depends solely on the point P.

Let R and S be points of C on p. Then PR (also PS) is tangent to C. For, suppose that PR meets C again in $R_1 \neq R$ and let R_1' on PR be such that $H(R, R_1; R_1', P)$. By repeating the argument above for the complete quadrangle XX_1R_1R, we find the point R_1' to be on p. But now (see Problem 4.7, page 53) P and R coincide, contrary to the assumption that P is not on C. Thus since R and S depend solely upon P, so also does the line $p = RS$.

At this point we have proved a number of theorems regarding the points of the line p which will be listed below. First, we take one of them (our choice) as a definition:

> If P, any point not on a given point conic C, is a diagonal point of any complete quadrangle whose vertices are points of C, then the other two diagonal points determine a unique line p, called the *polar line (polar)* of P with respect to C. If P is a point of C, the polar line of P with respect to C is the tangent to C at P.

There follows

Theorem 9.1. On the polar line of P with respect to C are the harmonic conjugates of P with respect to every pair of points of C collinear with P.

Theorem 9.2. For any complete quadrangle whose vertices are points of C, the polar line with respect to C of each vertex of the diagonal triangle is its opposite side.

Theorem 9.3. On the polar line of P with respect to C are the points of intersection of the tangents to C at every two of its points collinear with P.

Theorem 9.4. On the polar line of P with respect to C are the points of contact of the tangents, if any, to C through P.

The insertion of "if any" in Theorem 9.4 requires an explanation. The reader is quite aware that, in Euclidean geometry, no tangents to an ellipse can be drawn through its center since the center is "inside" the ellipse. In projective geometry, a point conic separates the projective plane into two regions — a region *inside* the conic through each point of which every line meets the conic in two points and a region *outside* the conic through each point of which some lines meet the conic and others do not. Thus through no point inside the conic can a tangent to the conic be drawn while through every point outside the conic two tangents to the conic can be drawn.

The dual of Fig. 9-1 is illustrated by Fig. 9-1' below. Here x, x_1; a, a_1; b, b_1 are three pairs of distinct lines of the line conic C' which are concurrent with any selected line p not on C' (that is, not a line of C'). Let $x \cdot a = A$, $x \cdot a_1 = A_1$, $x \cdot b = B$, $x \cdot b_1 = B_1$; $x_1 \cdot a = A'$, $x_1 \cdot a_1 = A_1'$, $x_1 \cdot b = B'$, $x_1 \cdot b_1 = B_1'$. Then, by definition,

$$x(A, B, A_1, B_1) \ \overline{\wedge}\ x_1(A', B', A_1', B_1')$$

The axis of projectivity t is the line determined by $AB' \cdot BA'$ and $AA_1' \cdot A'A_1$; P is the intersection of $AA_1' = l$, $A'A_1 = k$ and t; and k, l, p are the sides of the diagonal triangle of the complete quadrilateral xaa_1x_1. We leave for the reader to complete the discussion and define:

> If p, any line not on a given line conic C', is a diagonal line of any complete quadrilateral whose sides are lines of C', then the other two diagonal lines determine a unique point P, called the *pole* of p with respect to C'. If p is a line of C', the pole of p with respect to C' is the point of contact on p.

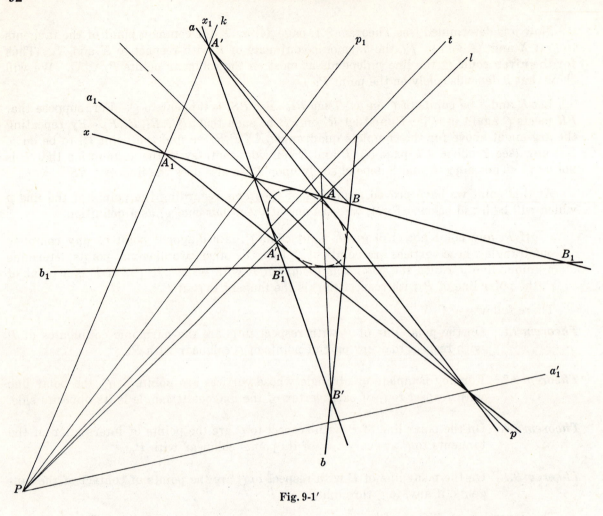

Fig. 9-1'

The duals of Theorems 9.1-9.4 follow readily:

Theorem 9.1'. On the pole of p with respect to C' are the harmonic conjugates of p with respect to every pair of lines of C' concurrent with p.

Theorem 9.2'. For any complete quadrilateral whose sides are lines of C', the pole with respect to C' of each side of the diagonal triangle is its opposite vertex.

Theorem 9.3'. On the pole of p with respect to C' are the joins of the points of contact of C' on every two of its lines concurrent with p.

Theorem 9.4'. On the pole of p with respect to C' are the tangents to C' (lines of C'), if any, at the points of contact of C' collinear with p.

RELATION BETWEEN POINT AND LINE CONIC

On the point conic C of Fig. 9-2 below, take four points X, X_1, B, B_1 and denote by x, x_1, b, b_1 respectively the tangents to C at these points. Let $BX_1 \cdot B_1X = M$, $BX \cdot B_1X_1 = N$, $BB_1 \cdot XX_1 = P$; $x \cdot x_1 = T$, $b \cdot b_1 = U$, $b \cdot x = V$, $b_1 \cdot x_1 = W$, $b \cdot x_1 = Y$, $b_1 \cdot x = Z$. Now

$$b(U, B, V, Y) \overset{P}{\doublebarwedge} b_1(U, B_1, W, Z) \doublebarwedge b_1(B_1, U, Z, W)$$

by Theorem 2.10, page 25; hence $b(U, B, V, Y) \doublebarwedge b_1(B_1, U, Z, W)$ with BB_1 as axis.

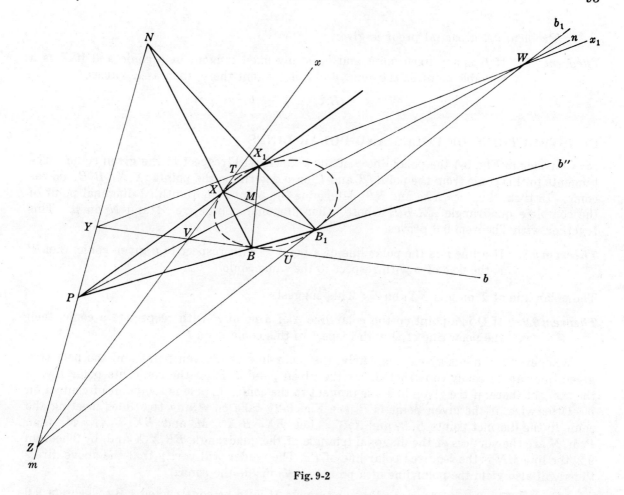

Fig. 9-2

Suppose PBB_1 is held fixed while PXX_1 varies; then Y and W move along b and b_1 respectively, generating C as a line conic. We have proved

Theorem 9.5.   The tangents to a point conic constitute a line conic.

Its dual is

Theorem 9.5′.   The points of contact of a line conic constitute a point conic.

As a consequence, we now define the term *conic* to apply equally to a point conic together with its tangents or to a line conic together with its points of contact. We may then speak of the complete quadrangle XAX_1A_1 of Fig. 9-1 as inscribed in the conic and of the complete quadrilateral xax_1a_1 of Fig. 9-1′ as circumscribed about the conic. Theorems 9.2 and 9.2′ may be restated as follows:

Theorem 9.2.   If a complete quadrangle is inscribed in a conic, each side of its diagonal triangle is the polar line of the opposite vertex.

Theorem 9.2′.   If a complete quadrilateral is circumscribed about a conic, each vertex of its diagonal triangle is the pole of the opposite side.

since, by definition, any conic is self-dual.

In Problem 9.1 we prove

Theorem 9.6.   The complete quadrangle whose vertices are any four points of a conic and the complete quadrilateral whose sides are the tangents to the conic at these points, have the same diagonal triangle.

In Problem 9.2 a partial proof is given of

Theorem 9.7. If B is any fixed point and b_1 is any fixed tangent of a conic and if X is a variable point on the conic and x is tangent there, then as X varies,

$$(BX, \ldots) \; \overline{\wedge} \; (b_1 \cdot x, \ldots)$$

CONSTRUCTIONS OF POLES AND POLAR LINES

Consider in Fig. 9-1 the polar line p of the point P with respect to the given conic. The tangents to this conic from the points T and U on p determine the points $X, X_1; B, B_1$ on the conic. In turn, the lines XX_1 and BB_1 determine the point P as the third diagonal point of the complete quadrangle XX_1B_1B whose other diagonal points are M and N on p. This together with Theorem 9.6 proves

Theorem 9.8. If a line p is the polar line of a point P with respect to a given conic, then P is the pole of p with respect to the same conic.

The polar line of T on p is XX_1 on P. This suggests

Theorem 9.9. If Q is a point on the polar line p of a point P with respect to a conic, then the polar line of Q with respect to this conic is on P.

Whenever a conic is given completely, the polar line of a given point and the pole of a given line can be easily constructed. If the given point P is on the conic, its polar line is the tangent there; if the given line p is tangent to the conic, its pole is the point of contact on p. Otherwise, if the given point is P (see Fig. 9-2), take on P any two lines meeting the conic in the distinct points B, B_1 and X, X_1. Let $BX_1 \cdot B_1X = M$ and $BX \cdot B_1X_1 = N$; then P, M, N are the vertices of the diagonal triangle of the quadrangle BB_1X_1X and, by Theorem 9.2, the line MN is the required polar line of P. The reader will verify that the above directions will also yield the polar line of a point (as M) inside the conic.

Let m of Fig. 9-2 be the given line whose pole M is to be constructed. By Theorem 9.9 the polar line of any point on m passes through M. Thus the intersection of the polar lines of any two points on m is the required pole. An alternate construction when the given line meets the conic in distinct points (as n in Fig. 9-2) is as follows: At each of the points of intersection, construct the tangent to the conic; the point of intersection of these tangents is the required pole.

In Problem 9.3 similar constructions are considered when the conic is not given completely.

CONJUGATE POINTS AND LINES

The pole-polar relation of Theorem 9.8 establishes a correspondence between the points and lines of the plane such that to each point there corresponds a unique line, its polar line with respect to the conic, and to each line there corresponds a unique point, its pole with respect to the conic.

In Problem 9.4 we prove

Theorem 9.10. As a point varies over a line (pencil of points) its polar line with respect to a given conic varies over a pencil of lines projectively related to the pencil of points.

Two points P and Q are called *conjugate* with respect to a conic if each lies on the polar line of the other with respect to this conic. Dually, two lines p and q are called *conjugate* with respect to a conic if each is on the pole of the other with respect to the conic.

In Problem 9.5 we prove

Theorem 9.11.   If two pairs of opposite vertices of a complete quadrilateral are conjugate points with respect to a conic, so also is the third pair of opposite vertices.

SELF-POLAR TRIANGLES

A triangle is said to be *self-polar* with respect to a conic if each vertex is the pole of the opposite side.   Clearly, each side is then the polar line of the opposite vertex.

From Theorem 9.2 follows

Theorem 9.12.   The diagonal points of the complete quadrangle whose vertices are any four distinct points on a conic are, in turn, the vertices of a self-polar triangle with respect to the conic.

Theorem 9.12 and its dual provide simple constructions of a self-polar triangle with respect to a conic having a given point (line), not on the conic, as a vertex (side).   For, if two lines (secants) are drawn through the given point A to meet the conic in four distinct points, the other two vertices B, C of the diagonal triangle of the complete quadrangle determined by these four points, together with A, are the vertices of a self-polar triangle with respect to the conic.

In Problem 9.6 we prove

Theorem 9.13.   If a triangle is inscribed in a conic, any line conjugate to one of the sides meets the other two sides in a pair of conjugate points.

and its converse

Theorem 9.14.   If a triangle is inscribed in a conic, any line which meets two sides of the triangle in a pair of conjugate points is conjugate to the other side.

Their duals are

Theorem 9.13′.   If a triangle is circumscribed about a conic, any point conjugate to one of the vertices joins the other vertices in a pair of conjugate lines.

and its converse

Theorem 9.14′.   If a triangle is circumscribed about a conic, any point which joins two vertices of the triangle in a pair of conjugate lines is conjugate to the other vertex.

Solved Problems

9.1.   Prove: The complete quadrangle whose vertices are any four points of a conic and the complete quadrilateral whose sides are the tangents to the conic at these points have the same diagonal triangle.

Consider in Fig. 9-2 the complete quadrangle XBB_1X_1 and the complete quadrilateral xbb_1x_1 meeting the conditions of the theorem. Let $x \cdot x_1 = T$, $b \cdot b_1 = U$, $x \cdot b = V$, $x_1 \cdot b_1 = W$, $x_1 \cdot b = Y$, $x \cdot b_1 = Z$. The sides of the diagonal triangle of xbb_1x_1 are TU, VW, YZ. By Theorem 9.3, $TU = MN$, $VW = PM$, and $YZ = PN$. But these are the sides of the diagonal triangle of XBB_1X_1; hence the theorem.

9.2. Prove: If B is any fixed point and b_1 is any fixed tangent of a conic and if X is a variable point on the conic and x is tangent there, then as X varies,

$$(BX, \ldots) \,\overline{\wedge}\, (b_1 \cdot x, \ldots)$$

Refer to Fig. 9-2 where now $B, b_1, B_1, b'' = BB_1, X_1, Y$ are fixed points and lines. As X varies on the conic, we have by definition

$$(BX, \ldots) \,\overline{\wedge}\, (X_1X, \ldots)$$

But
$$(X_1X, \ldots) \,\overset{b''}{\overline{\wedge}}\, (P, \ldots) \,\overset{Y}{\overline{\wedge}}\, (b_1 \cdot x, \ldots)$$

the point P varying over b''. Thus $(BX, \ldots) \,\overline{\wedge}\, (b_1 \cdot x, \ldots)$ as required.

This proof covers the case when b_1 is not the tangent at B. The case when b_1 is taken as the tangent b at B is left for the reader.

9.3. Let a conic C be given by five of its points A, B, C, D, E. Construct: (a) the polar line x with respect to C of any point X, (b) the pole X with respect to C of any line x.

Take B and C as centers of the projective pencils generating C.

(a) Refer to Fig. 9-3(a). Let $XB = b$ and $XC = c$. Suppose the polar line of X with respect to C meets b and c in the points B'' and C''. Then X and B'' separate harmonically the two points of the conic on b while X and C'' separate harmonically the two points of the conic on c. The construction then consists in (i) locating the second points B' and C' of C on b and c respectively, (ii) locating the harmonic conjugates B'' and C'' of X with respect to B, B' and C, C' respectively, (iii) joining B'' and C''.

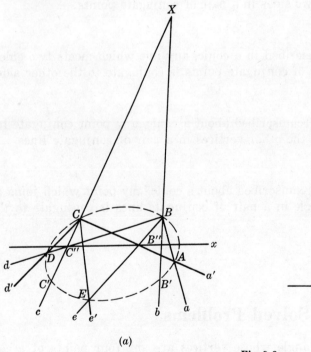
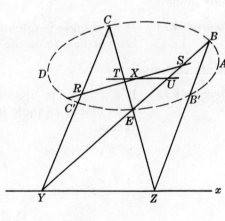

(a)   (b)

Fig. 9-3

(b) Refer to Fig. 9-3(b). Let $BE \cdot x = Y$ and $CE \cdot x = Z$. By Theorem 9.9, the pole of x is on the polar lines of both Y and Z with respect to C. The construction consists in (i) locating B' and C' the second points of C on BZ and YC respectively, (ii) locating the harmonic conjugates R and S of Y with respect to C, C' and E, B respectively and the harmonic conjugates T and U of Z with respect to E, C and B, B' respectively, (iii) intersecting RS and TU. (Or locate the point of intersection of the polar lines of any two points on x.)

9.4. Prove: As a point varies over a line (pencil of points) its polar line with respect to a given conic varies over a pencil of lines projectively related to the pencil of points.

Although the conclusion is the same, the proofs differ according as the line is or is not tangent to the conic. Suppose, first that the point P [see Fig. 9-4(a)] varies over a non-tangent m whose pole with respect to the conic is M. Take R as a fixed point on the conic. Let the lines PR and RM determine the points S and T respectively on the conic and the line PT determine the point U. Finally, let $RU \cdot ST = N$. By construction, PMN is the diagonal triangle of the complete quadrangle $RSTU$ and $MN = n$ is the polar line of P. Now in the figure, the line m and the points R, M, T are fixed. Clearly, as P varies on m, the line n varies on M. That

$$(P, \ldots) \ \overline{\wedge} \ (MN, \ldots)$$

follows from   (i) $(P, \ldots) \ \overline{\wedge} \ (RP, \ldots) \ \overline{\wedge} \ (TS, \ldots) \ \overline{\wedge} \ (MN, \ldots)$

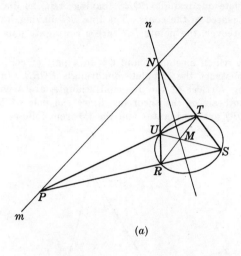

(a)   (b)

Fig. 9-4

Next, suppose that P [see Fig. 9-4(b)] varies over the line q which is tangent to the conic at Q. Let R be the point of intersection of two fixed tangents to the conic. Let $PR = s$ and construct its pole S. Since P is on s, its polar line is on S by Theorem 9.9 and since P is on q, its polar line is on Q, the point of tangency. Thus QS is the polar line of P. Now as P varies on q,

        (ii) $(P, \ldots) \ \overline{\wedge} \ (RP, \ldots) \ \overline{\wedge} \ (S, \ldots) \ \overline{\wedge} \ (QS, \ldots)$

and $(P, \ldots) \ \overline{\wedge} \ (QS, \ldots)$ as required.

9.5. Prove: If two pairs of opposite vertices of a complete quadrilateral are conjugate points with respect to a conic, so also is the third pair of opposite vertices.

In Fig. 9-5, let A and L, B and M be the two pairs of opposite vertices of the complete quadrilateral $pqrs$ which are conjugates with respect to a conic C not shown. We are to prove that D and N are also conjugates with respect to the same conic. Denote by A', B', D' the respective inter-

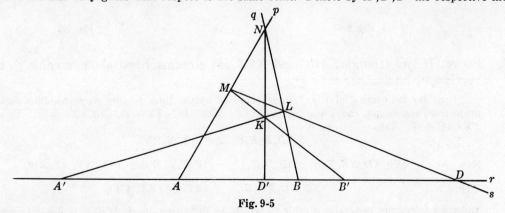

Fig. 9-5

sections of the polar lines of A, B, D with respect to C and the side r. Now the points A', B', D' are also the harmonic conjugates of the points A, B, D with respect to the two points of C on the line r; hence A, A'; B, B'; D, D' are three reciprocal pairs of an involution of the points on r.

By Theorem 6.4, page 66, the lines $A'L$, $B'M$, and $D'N$ are on a point K. Since both A' and L are on the polar line of A with respect to C, that polar line is $A'L$. Similarly, $B'M$ is the polar line of B with respect to C and, then, K is the pole of r. Also, D is on r and D' is on its polar line with respect to C; hence $D'K$ is the polar line of D. Finally, since $D'K$ is on N, the points D and N are conjugates with respect to C.

9.6. Prove: If a triangle PQR is inscribed in a conic, any line conjugate to one of the sides meets the other two sides in a pair of conjugate points and conversely.

Suppose A is the pole of the given line conjugate to PQ; then (see Fig. 9-6) A is on PQ. Let AR meet the conic again in S and construct the complete quadrangle $PQRS$, having ABC as diagonal triangle. Now ABC is a self-polar triangle with respect to the conic. The line BC, having its pole on PQ, satisfies the hypothesis of the theorem. Moreover, the points B, C are a conjugate pair with respect to the conic and the theorem follows.

Conversely, let A' be the pole of the given line which meets PR and QR in a pair of conjugate points. Let $A'R$ meet the conic again in S and construct the complete quadrangle $PQRS$. In this quadrangle the points B on PR and C on QR, being vertices of the diagonal triangle, are a pair of conjugates and thus the line BC satisfies the hypothesis of the theorem. Since the pole of BC is unique, A' is at A on PQ. Thus the lines BC and PQ are conjugates and the theorem follows.

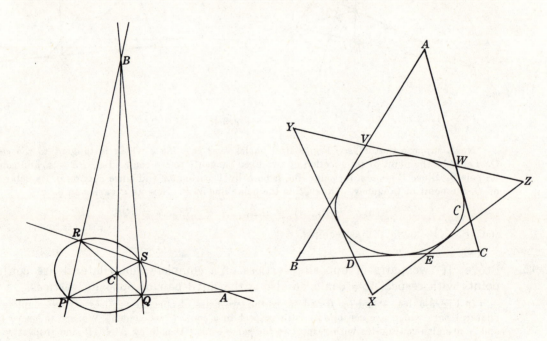

Fig. 9-6   Fig. 9-7

9.7. Prove: If two triangles ABC and XYZ are circumscribed about a conic C, their six vertices lie on another conic C'.

Consider the conic C of Fig. 9-7 as generated by the lines joining corresponding points of two projective pencils on the lines BC and YZ. Let $BC \cdot XY = D$, $BC \cdot XZ = E$; $YZ \cdot AB = V$, $YZ \cdot AC = W$. Then

$$(B, D, E, C) \;\overline{\wedge}\; (V, Y, Z, W)$$

Now $$(XB, XD, XE, XC) \;\overline{\overline{\wedge}}\; (B, D, E, C) \;\overline{\wedge}\; (V, Y, Z, W) \;\overline{\overline{\wedge}}\; (AV, AY, AZ, AW)$$

Then $$(XB, XD, XE, XC) \;\overline{\wedge}\; (AV, AY, AZ, AW)$$

and the projectivity generates a conic on which, by definition, lie A, X, B, Y, C, Z as required.

9.8. Prove: If two triangles are self-polar with respect to a conic, their six vertices lie on a conic.

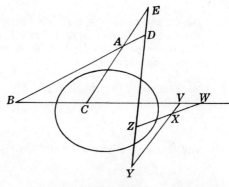

Consider in Fig. 9-8 the self-polar triangles ABC and XYZ with respect to a conic C. Let $BC \cdot XY = V$, $BC \cdot XZ = W$; $YZ \cdot AB = D$, $YZ \cdot AC = E$. Since V lies on BC, its polar line with respect to C is on A and, since V lies on XY, its polar line is on Z (Theorem 9.9); thus the polar line of V with respect to C is AZ. Similarly, the polar lines of W, D, E with respect to C are AY, XC, XB respectively. By Theorems 9.10 and 2.10, page 25,

Fig. 9-8

$$(B, W, V, C) \;\overline{\wedge}\; (AC, AY, AZ, AB) \;\overset{YZ}{\overline{\wedge}}\; (E, Y, Z, D) \;\overline{\wedge}\; (XB, XZ, XY, XC) \;\overline{\wedge}\; (XC, XY, XZ, XB)$$

Then
$$(AC, AY, AZ, AB) \;\overline{\wedge}\; (XC, XY, XZ, XB)$$

and A, B, C, X, Y, Z are on a conic.

9.9. Prove: If two tangents to a conic C vary in such a way that the joins of their points of contact are tangent to another conic C', then their points of intersection are on a third conic C''.

Let x be a variable tangent to C'. Then its pole X with respect to C is the point of intersection of the tangents to C at the two points of C on x. We are to prove that as x varies on C', X varies on a conic C''.

Denote by a and b two fixed tangents to C'; let $a \cdot x = X_1$ and $b \cdot x = X_1'$. Then C' may be considered as generated by the projectivity

$$a(X_1, \ldots) \;\overline{\wedge}\; b(X_1', \ldots)$$

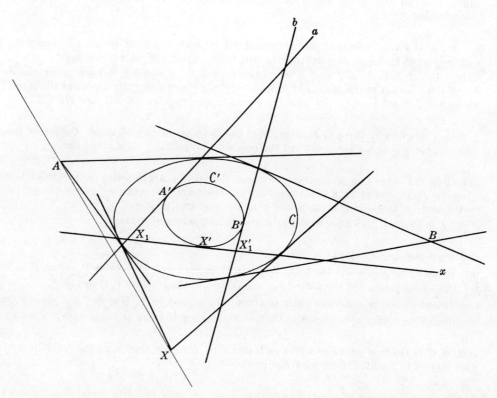

Fig. 9-9

Denote the poles of a, b, x with respect to C by A, B, X respectively. Since a and b are considered fixed while x varies, so also A and B are fixed while X varies. Now AX is the polar line of X_1 and BX is the polar line of X_1' with respect to C. As X_1 varies on a, its polar line with respect to C varies on A such that

$$a(X_1, \ldots) \; \overline{\wedge} \; (AX, \ldots)$$

by Theorem 9.10.   Similarly,

$$b(X_1', \ldots) \; \overline{\wedge} \; (BX, \ldots)$$

Thus

$$(AX, \ldots) \; \overline{\wedge} \; (BX, \ldots)$$

and this projectivity generates the conic C''.

Supplementary Problems

9.10. Complete the proof of Theorem 9.7, that is, prove: If B is any fixed point and X is a variable point on a conic and if b, x are their respective tangents to the conic, then $(BX, \ldots) \; \overline{\wedge} \; (b \cdot x, \ldots)$.

9.11. With respect to a conic completely drawn, construct the polar line of P when (a) P is inside, (b) P is on, (c) P is outside the conic.

9.12. With respect to the conic of Problem 9.11, construct the pole of the line p when (a) p has two distinct points in common with the conic, (b) p is tangent to the conic, (c) p has no real points in common with the conic.

9.13. Prove Theorem 9.9. There are three cases to be considered: P inside the conic, P on the conic, P outside the conic.

9.14. A conic C is given by four of its points and the tangent at one of them. Construct the polar line with respect to C of a given point and the pole with respect to C of a given line.
      Hint.   Let A, B, C, D on C and b the tangent at B be given; let X be a point not on C. Take A and B as centers of the pencils of lines generating C and construct the second points A', B', C' in which XA, XB, XC meet C. Consider the complete quadrangles $AA'BB'$ and $BB'CC'$.

9.15. A conic C is given by three of its points and the tangents at two of them. Construct the polar line with respect to C of a given point and the pole with respect to C of a given line.

9.16. Construct the polar line of a given point and the pole of a given line with respect to the conic C when (a) C is given by four of its tangents and the point of contact on one of them, (b) C is given by three of its tangents and the points of contact on two of them.
      Hint. Use Problem 9.15 after first locating sufficient points of contact.

9.17. Construct the conic C, given:
      (a) Three of its points and the polar line p with respect to C of a point P not on C.
      (b) One of its points and the polar lines p, q of two distinct points P, Q not on C.
      (c) Two of its points and a self-polar triangle with respect to it. Why is C not always unique here?
      (d) One of its points, the tangent at that point, and a self-polar triangle with respect to it.

9.18. If A, B, C, D are four points on a line such that $H(A, B; C, D)$, show that the polar lines of A, B, C, D with respect to a conic C form a harmonic set.

9.19. Prove: If two vertices of a triangle are the poles of their opposite sides with respect to a given conic, the third vertex is the pole of its opposite side.

9.20. Prove: If, with respect to a given conic, each of the points P, Q is conjugate to a third point R, the polar line of R with respect to the conic is PQ.

9.21. Show that if XYZ is a self-polar triangle with respect to a conic, one and only one of its vertices is inside the conic.

9.22. In Fig. 9-7, prove that A, D, E, X, V, W are on a conic.

9.23. In Fig. 9-8, prove that the six sides of the given triangles are on a conic.

9.24. Let XYZ be a self-polar triangle with respect to a conic C. (a) Take any point A on C and construct a complete quadrangle inscribed in C and having XYZ as diagonal triangle. Is this quadrangle unique? (b) Take any line a on C and construct a complete quadrilateral circumscribed about C and having XYZ as diagonal triangle. Is this quadrilateral unique?

9.25. Verify (i) and (ii) in Problem 9.4.
Hint. (i) requires only previous definitions; (ii) requires in addition the first part of the proof.

9.26. Prove: If Q, S are two points on a conic C and A, C are a pair of conjugate points on any line conjugate to QS with respect to C, then $AQ \cdot CS$ and $AS \cdot CQ$ are on C.
Hint. Let AQ, AS meet C again in P, R respectively; then A, C are diagonal points of the quadrangle $PQRS$.

9.27. Prove: Any line meeting two sides of a triangle in conjugate points with respect to a conic C passes through the pole of the third side.

9.28. State and prove the dual of Problem 9.9.

9.29. Prove: If A and A' are conjugate points with respect to a conic C and if AA' meets C in the points P and Q, then $H(P, Q; A, A')$.
Hint. Let A be outside C; consider any inscribed quadrangle having P, Q as vertices and A as diagonal point.

9.30. Prove: If the lines a and b on C are two fixed tangents to a conic C and if x, X are any other tangent to C and its point of contact, then $a \cdot x$ and $b \cdot x$ are separated harmonically by X and the intersection Y of x and the polar line of C.

9.31. Let p and q be two non-conjugate lines with respect to a conic C. Prove: (a) Every point P of p has a conjugate point Q on q. (b) With Q defined as in (a), the pencils of points $p(P, \dots)$ and $q(Q, \dots)$ are projective, the projectivity being a perspectivity only when $p \cdot q$ is on C.

9.32. Prove: As X varies over a given line, the polar lines of X with respect to two given conics intersect in points on a third conic.

9.33. Prove: If the tangents to a given conic meet another conic C in pairs of points, the tangents to C at these points meet on a third conic. *Hint.* Use Fig. 9-9.

9.34. Prove: Two conics C and C' having four points and four tangents in common have also a self-polar triangle in common.
Hint. Let P, Q, R, S be the four points and A, B, C be the vertices of the diagonal triangle of the quadrilateral whose sides are the four tangents. Let PA meet BC in X and the conics again in X_1 and X_2. Then $H(P, X_1; A, X)$, $H(P, X_2; A, X)$, and the points A, P, Q are collinear. Similarly, the points A, R, S are collinear and A is a vertex of the diagonal triangle of $PQRS$.

9.35. Prove: If two complete quadrangles have the same diagonal triangle, their eight vertices are either (a) by fours on two lines or (b) on a conic.
Hint. Let the quadrangles be $PQRS$ and $P'Q'R'S'$ having $A = PQ \cdot RS = P'Q' \cdot R'S'$, $B = PR \cdot QS = P'R' \cdot Q'S'$, $C = PS \cdot QR = P'S' \cdot Q'R'$ as diagonal points. For (a), suppose P, Q, P' collinear and show that P, Q, P', Q' must be collinear. For (b), suppose no three of the eight vertices collinear. Denote by C the conic on P, Q, R, S, P' and apply Problem 9.29 using the conjugate points A and $D = AP' \cdot BC$.

9.36. Prove: If two conics intersect in four points, their eight tangents at these points are either (a) by fours on two points or (b) envelope a conic.

Chapter 10

Theorems of Pascal and Brianchon

PASCAL'S THEOREM

In 1640 Pascal proved [see Fig. 10-1(a)]

Theorem 10.1 (Pascal's Theorem). If a simple hexagon $A_1A_2A_3A_4A_5A_6$ is inscribed in a conic, the intersections

$$R = A_1A_2 \cdot A_4A_5, \ \ S = A_2A_3 \cdot A_5A_6, \ \ T = A_3A_4 \cdot A_6A_1$$

of the three pairs of opposite sides are collinear.

For a proof, see Problem 10.1.

The line RST of Theorem 10.1 is called the *Pascal line* of the hexagon $A_1A_2A_3A_4A_5A_6$.

The converse of Pascal's Theorem:

Theorem 10.2. If the points of intersection of the three pairs of opposite sides of a simple hexagon are collinear, the vertices of the hexagon are on a conic.

is also valid and suggests an alternate construction of the conic determined by five given points, no three of which are collinear.

For a proof, see Problem 10.2.
For the construction, see Problem 10.3.

For a singular conic consisting of two distinct pencils of points, Pascal's Theorem becomes the Theorem of Pappus. For this reason the Theorem of Pappus is sometimes called the Theorem of Pappus-Pascal.

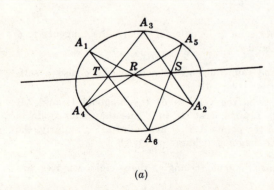

(a)

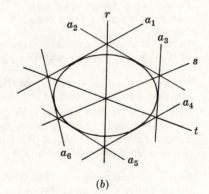

(b)

Fig. 10-1

BRIANCHON'S THEOREM

In 1806 Brianchon proved [see Fig. 10-1(b)] the dual of Pascal's Theorem

Theorem 10.1' (Brianchon's Theorem). If a simple hexagon $a_1a_2a_3a_4a_5a_6$ is circumscribed about a conic, the joins

$$r = (a_1 \cdot a_2)(a_4 \cdot a_5), \ \ s = (a_2 \cdot a_3)(a_5 \cdot a_6), \ \ t = (a_3 \cdot a_4)(a_6 \cdot a_1)$$

of the three pairs of opposite vertices are concurrent.

For a proof, see Problem 10.4.

The point $r \cdot s \cdot t$ of Theorem 10.1′ is called the *Brianchon point* of the hexagon $a_1 a_2 a_3 a_4 a_5 a_6$.

The converse of Brianchon's Theorem:

Theorem 10.2′. If the joins of the three pairs of opposite vertices of a simple hexagon are concurrent, the sides of the hexagon circumscribe a conic.

is also valid and suggests an alternate construction of the conic determined by five given lines, no three of which are concurrent.

For a singular conic consisting of two distinct pencils of lines, Brianchon's Theorem becomes the dual of the Theorem of Pappus.

SPECIAL CASES OF PASCAL'S THEOREM

In the proof of Pascal's Theorem use was made [see Fig. 10-1(a)] of the projectivity between two pencils of lines on the vertices (our choice) A_2 and A_4 of the hexagon:

$$\text{(i)} \qquad (A_2A_1, A_2A_3, A_2A_5, A_2A_6) \ \overline{\wedge} \ (A_4A_1, A_4A_3, A_4A_5, A_4A_6)$$

In this projectivity the correspondent of A_2A_4, considered as a line of the pencil on A_2, is the tangent t at A_4 to the conic. Thus (i) may also be given as

$$\text{(ii)} \qquad (A_2A_1, A_2A_3, A_2A_4, A_2A_6) \ \overline{\wedge} \ (A_4A_1, A_4A_3, t, A_4A_6)$$

which (see Fig. 10-2) is concerned with a simple pentagon $A_1A_2A_3A_4A_6$ inscribed in the conic and the tangent at one of the vertices. Then, as in Problem 10.1, we have

$$(U, S, A_4, A_6) \ \overline{\overline{\wedge}} \ (A_2A_1, A_2A_3, A_2A_4, A_2A_6) \ \overline{\wedge} \ (A_4A_1, A_4A_3, t, A_4A_6) \ \overline{\overline{\wedge}} \ (A_1, T, V, A_6)$$

so that
$$(U, S, A_4, A_6) \ \overline{\overline{\wedge}} \ (A_1, T, V, A_6)$$

and $R = UA_1 \cdot VA_4$ is on ST.   We have proved

Theorem 10.3. If a simple pentagon is inscribed in a conic, the tangent at one of its vertices meets the side opposite this vertex in a point collinear with the points of intersection of the other two pairs of non-consecutive sides.

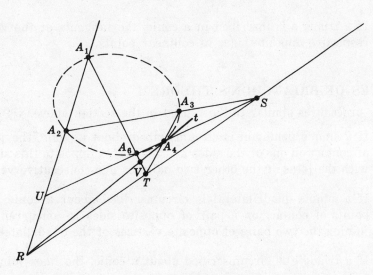

Fig. 10-2

For our immediate purpose it will be more convenient to consider the pentagon and its tangent as a degenerate hexagon obtained when two of the vertices (here, A_4 and A_5) coincide and to choose appropriate notation. This latter consists merely (see Fig. 10-3) in labeling the point of coincidence $A_{4,5}$ and the tangent there $t_{4,5}$. (Note that as A_4 and A_5 approach coincidence along the conic, the side A_4A_5 of the hexagon approaches the tangent $t_{4,5}$ as limiting position.) In accordance with the *principle of continuity*, first formulated by Poncelet, we assume that any property of the figure which continues to be a property as the figure is so varied, will be a property of the figure in its limiting position. Thus we have $A_1A_2 \cdot t_{4,5} = R$, $A_2A_3 \cdot A_6A_{4,5} = S$, $A_3A_{4,5} \cdot A_6A_1 = T$ with R, S, T collinear as required by the theorem.

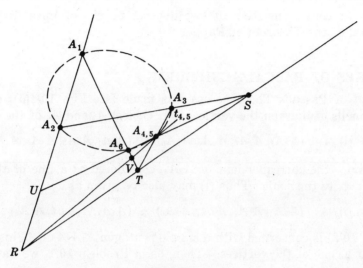

Fig. 10-3

Similarly, we may obtain

Theorem 10.4. If a simple quadrangle is inscribed in a conic the tangents at a pair of opposite vertices intersect in a point which is collinear with the points of intersection of the two pairs of opposite sides of the quadrangle.

and

Theorem 10.5. If a triangle is inscribed in a conic, the tangents at the vertices meet the respective opposite sides in collinear points.

SPECIAL CASES OF BRIANCHON'S THEOREM

Dually or by procedures similar to those used in the section above, we obtain

Theorem 10.3'. If a simple pentagon is circumscribed about a conic, the join of the point of contact on one of the sides and the vertex opposite this side is concurrent with the joins of the other two pairs of non-consecutive vertices.

Theorem 10.4'. If a simple quadrilateral is circumscribed about a conic, the join of the points of contact on a pair of opposite sides is concurrent with the lines joining the two pairs of opposite vertices of the quadrilateral.

Theorem 10.5'. If a triangle is circumscribed about a conic, the lines joining the vertices to the points of contact on the respective opposite sides are concurrent.

THE CONVERSE THEOREMS

The converses of Theorems 10.3-10.5:

Theorem 10.6.   If a line through a vertex of a simple pentagon meets the opposite side in a point collinear with the intersections of the non-consecutive sides remaining, the line is tangent to the conic which circumscribes the pentagon.

Theorem 10.7.   If through a pair of opposite vertices of a simple quadrangle lines are drawn which meet on the joins of the points of intersection of the opposite sides of the quadrangle, the two lines are tangent to a conic which circumscribes the quadrangle.

Theorem 10.8.   If through each vertex of a triangle a line is drawn to meet the opposite side of the triangle and if the three points of intersection so obtained are collinear, the three lines are tangent to a conic which circumscribes the triangle.

are valid as, also, are the converses of Theorems 10.3'-10.5' (the duals of Theorems 10.6-10.8). Each suggests an alternate construction to those given in Chapter 8. For certain of these, see Problems 10.5-10.7.

ASSOCIATED n-POINTS AND n-LINES

Call any simple n-point inscribed in a conic C (any complete n-point whose vertices are on C) and the simple n-line formed by the tangent to C at the vertices (the complete n-line whose sides are the tangents to C at the vertices of the n-point) *associates*.

Consider the simple hexagon $A_1A_2A_3A_4A_5A_6$ inscribed in the conic C (see Fig. 10-4) and its associate $a_1a_2a_3a_4a_5a_6$. The Pascal line of the inscribed hexagon is $p = RST$. Let $a_1 \cdot a_2 = P_1$, $a_2 \cdot a_3 = P_2$, $a_3 \cdot a_4 = P_3$, $a_4 \cdot a_5 = P_4$, $a_5 \cdot a_6 = P_5$, $a_6 \cdot a_1 = P_6$. Since P_1P_4 is the polar line of R, P_2P_5 is the polar line of S and P_6P_3 is the polar line of T with respect to C, these lines meet in a point B, the pole of p with respect to C. The point B is also the Brianchon point of the circumscribed hexagon. Thus,

Theorem 10.9.   The Pascal line of any simple hexagon inscribed in a conic C is the polar line of the Brianchon point of its associated hexagon with respect to C. Conversely, the Brianchon point of any simple hexagon circumscribed about a conic C is the pole of the Pascal line of its associated hexagon with respect to C.

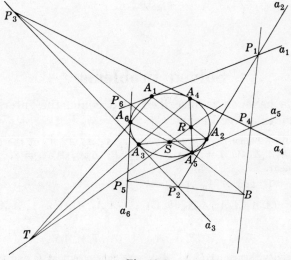

Fig. 10-4

Consider next the complete quadrangle (see Fig. 10-5) whose vertices P, Q, R, S are on a conic C and its associated complete quadrilateral. Applying Theorem 10.4 to the quadrangle

$PQRS$, we have $PQ \cdot RS = A$, $PS \cdot QR = B$, $p \cdot r$, $q \cdot s$ collinear;

$PRSQ$, we have $PR \cdot QS = C$, $PQ \cdot RS = A$, $p \cdot s$, $q \cdot r$ collinear;

$PSQR$, we have $PS \cdot QR = B$, $PR \cdot QS = C$, $p \cdot q$, $r \cdot s$ collinear.

The vertices of the diagonal triangle of $PQRS$ are A, B, C; the sides of the diagonal triangle of $pqrs$ are $(p \cdot q)(r \cdot s)$, $(p \cdot s)(q \cdot r)$, $(p \cdot r)(q \cdot s)$ which, in turn, are the lines BC, AC, AB. We have proved

Theorem 10.10. Any complete quadrangle whose vertices are on a conic and its associated complete quadrilateral with respect to that conic have the same diagonal triangle.

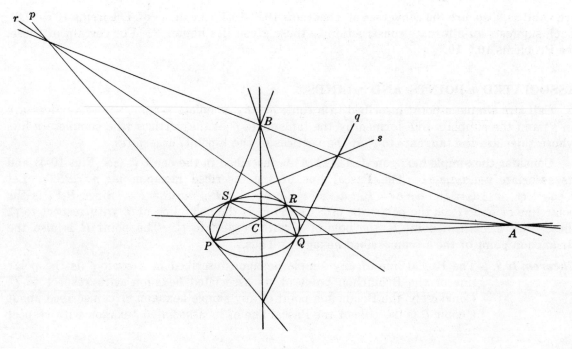

Fig. 10-5

Solved Problems

10.1. Prove: If a simple hexagon is inscribed in a conic, the intersections of the three pairs of opposite sides are collinear.

Consider in Fig. 10-6 below the six points $A_1, A_2, A_3, A_4, A_5, A_6$ of the conic C as vertices of the simple hexagon $A_1A_2A_3A_4A_5A_6$. Let $A_1A_2 \cdot A_4A_5 = R$, $A_2A_3 \cdot A_5A_6 = S$, $A_3A_4 \cdot A_6A_1 = T$; we are to prove R, S, T collinear.

Let $A_1A_2 \cdot A_5A_6 = U$ and $A_1A_6 \cdot A_4A_5 = V$. From

$$(U, S, A_5, A_6) \; \overline{\wedge} \; (A_2A_1, A_2A_3, A_2A_5, A_2A_6) \; \overline{\wedge} \; (A_4A_1, A_4A_3, A_4A_5, A_4A_6) \; \overline{\wedge} \; (A_1, T, V, A_6)$$

there follows

$$(U, S, A_5, A_6) \; \overline{\wedge} \; (A_1, T, V, A_6)$$

But this is a perspectivity; hence A_1U, ST, A_5V are concurrent at R and so R, S, T are collinear, as required.

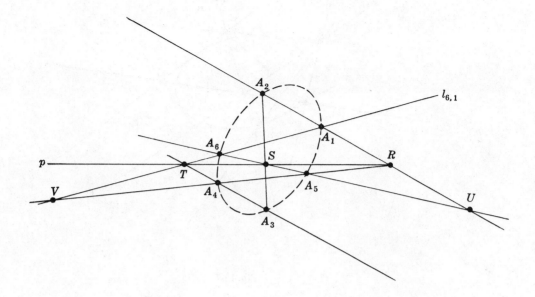

Fig. 10-6

10.2. Prove: If the points of intersection of the three pairs of opposite sides of a simple hexagon are collinear, the vertices of the hexagon are on a conic.

Consider in Fig. 10-6 the simple hexagon $A_1A_2A_3A_4A_5A_6$ with $R = A_1A_2 \cdot A_4A_5$, $S = A_2A_3 \cdot A_5A_6$, $T = A_3A_4 \cdot A_6A_1$ collinear. Let $U = A_1A_2 \cdot A_5A_6$ and $V = A_1A_6 \cdot A_4A_5$. Since

$$(U, S, A_5, A_6) \overset{R}{\barwedge} (A_1, T, V, A_6)$$

there follows the projectivity

$$(A_2A_1, A_2A_3, A_2A_5, A_2A_6) \; \barwedge \; (A_4A_1, A_4A_3, A_4A_5, A_4A_6)$$

This projectivity generates a conic on which are by definition the points $A_1, A_2, A_3, A_4, A_5, A_6$.

10.3. Given five distinct points on a conic C, use Theorem 10.2 to construct another point on C.

Let the given points be labeled A_1, A_2, A_3, A_4, A_5 as in Fig. 10-6. On A_1 take any line, distinct from the joins of A_1 and the other given points. Our problem can be made more specific, namely, locate on this line its second intersection A_6 with C. For this reason, label the line as $l_{6,1}$. Now $A_1A_2 \cdot A_4A_5 = R$; $A_3A_4 \cdot l_{6,1} = T$; $RT = p$, the Pascal line of the proposed hexagon; and $S = A_2A_3 \cdot p$. But $S = A_2A_3 \cdot A_5A_6$; hence A_6 is on A_5S. Thus $A_6 = A_5S \cdot l_{6,1}$.

Note 1. It is possible that the line selected is, in reality, the tangent to C at A_1. In this case, another line on A_1 is selected as $l_{6,1}$ and the above construction is repeated.

Note 2. The particular labels of the set of six attached to the given points are immaterial. It is suggested that the reader relabel the given points as A_1, A_2, A_3, A_5, A_6 and, choosing on A_3 some line as $l_{3,4}$, locate A_4 on it.

10.4. Prove: If a simple hexagon is circumscribed about a conic, the joins of the three pairs of opposite vertices are concurrent.

Consider in Fig. 10-7 below the six tangents $a_1, a_2, a_3, a_4, a_5, a_6$ of the conic C as sides of the simple hexagon $a_1a_2a_3a_4a_5a_6$. Let $(a_1 \cdot a_2)(a_4 \cdot a_5) = r$, $(a_2 \cdot a_3)(a_5 \cdot a_6) = s$, $(a_3 \cdot a_4)(a_6 \cdot a_1) = t$. We are to prove r, s, t concurrent.

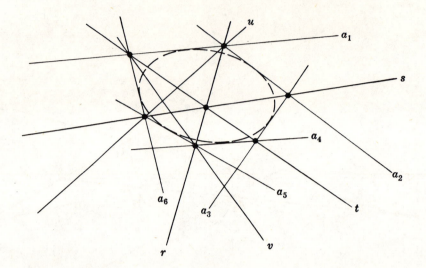

Fig. 10-7

Let $(a_1 \cdot a_2)(a_5 \cdot a_6) = u$ and $(a_1 \cdot a_6)(a_4 \cdot a_5) = v$. From

$$(u, s, a_5, a_6) \;\overline{\overline{\wedge}}\; (a_2a_1, a_2a_3, a_2a_5, a_2a_6) \;\overline{\wedge}\; (a_4a_1, a_4a_3, a_4a_5, a_4a_6) \;\overline{\overline{\wedge}}\; (a_1, t, v, a_6)$$

there follows $(u, s, a_5, a_6) \;\overline{\wedge}\; (a_1, t, v, a_6)$

But this is a perspectivity; hence $a_1 \cdot u,\ s \cdot t,\ a_5 \cdot v$ are collinear on r and so r, s, t are concurrent, as required.

   Note.   The problem — Given five distinct lines on a conic C, i.e. five distinct tangents of a conic C, construct another line of C. — is the dual of Problem 10.3. It is left for the reader to provide a construction.

10.5.   Given four points on a conic C and the tangent at one of them, locate another point on C.

   Let the given points be A_2, A_3, A_4, A_5 and the tangent be on A_2. For our purpose, relabel A_2 as $A_{1,2}$ and the tangent as $t_{1,2}$ (see Fig. 10-8). On A_5 take any line, distinct from the joins of A_5 and the other given points, and label it $l_{5,6}$. We are to construct the point A_6 in which $l_{5,6}$ again meets C. Now $t_{1,2} \cdot A_4A_5 = R$, $A_2A_3 \cdot l_{5,6} = S$, $p = RS$ is the Pascal line, and $T = A_3A_4 \cdot p = A_3A_4 \cdot A_6A_1$. Then $A_6 = TA_1 \cdot l_{5,6}$.

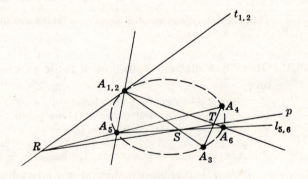

Fig. 10-8

10.6.   Given three lines (tangents) on a conic C and the points of contact of two of them, construct another line on C.

   Refer to Fig. 10-9 below in which the given parts are: the line $a_{1,2}$ with point of contact $T_{1,2}$, the line $a_{3,4}$ with point of contact $T_{3,4}$, and the line a_5. On a_5 take a point, distinct from any known point, and label it $P_{5,6}$. We are to construct on $P_{5,6}$ the line a_6 on C. Now $r = T_{1,2}(a_4 \cdot a_5)$, $s = (a_2 \cdot a_3)P_{5,6}$, $B = r \cdot s$ is the Brianchon point, and $t = BT_{3,4} = T_{3,4}(a_6 \cdot a_{1,2})$. Then $a_6 = P_{5,6}(a_{1,2} \cdot t)$.

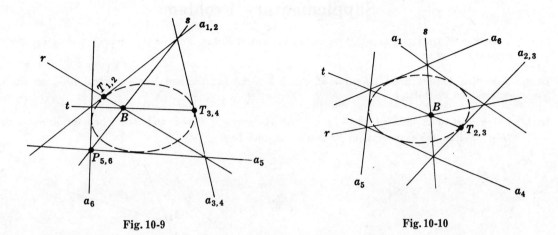

Fig. 10-9   Fig. 10-10

10.7. Given five lines on a conic C, locate the point of contact on one of them.

Refer to Fig. 10-10 in which the given lines are labeled $a_1, a_{2,3}, a_4, a_5, a_6$; we are then to construct the point of contact $T_{2,3}$ on $a_{2,3}$. Now $r = (a_1 \cdot a_{2,3})(a_4 \cdot a_5)$, $s = (a_{2,3} \cdot a_4)(a_6 \cdot a_1)$, $B = r \cdot s$ is the Brianchon point, and $t = B(a_5 \cdot a_6)$. Then $T_{2,3} = t \cdot a_{2,3}$.

10.8. Use the Theorem of Pappus to prove Desargues' Two-Triangle Theorem.

Consider in Fig. 10-11 the triangles $A_1 A_2 A_3$ and $B_1 B_2 B_3$ perspective from the point P. We are to prove $A_1 A_2 \cdot B_1 B_2 = L$, $A_2 A_3 \cdot B_2 B_3 = M$ and $A_3 A_1 \cdot B_3 B_1 = N$ collinear. In order to use the Theorem of Pappus, we must have additional triples of collinear points to work with. As a minimum, we locate $A_1 A_3 \cdot B_2 B_3 = S$, $A_1 B_2 \cdot A_3 B_3 = T$, $A_1 A_2 \cdot PS = U$, $B_1 B_2 \cdot PS = V$.

Using the triples of collinear points P, B_2, A_2 and A_1, A_3, S we have $PA_3 \cdot A_1 B_2 = T$, $PS \cdot A_1 A_2 = U$, $A_2 A_3 \cdot B_2 S = M$ collinear by the Theorem of Pappus.

Similarly, using the collinear triples P, A_1, B_1 and B_2, B_3, S we find T, V, N collinear. Finally, using the collinear triples A_1, B_2, T and V, U, S we have L, M, N collinear, as required.

The proof of the converse of the Desargues Theorem is left as an exercise.

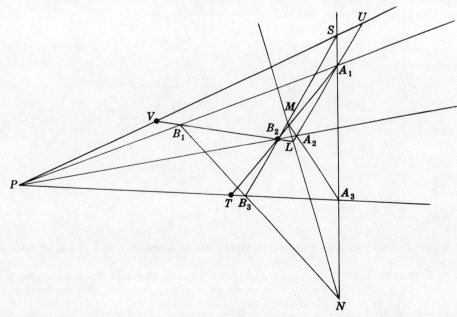

Fig. 10-11

Supplementary Problems

10.9. Take five distinct points, no three of which are collinear, and locate several points on the conic determined by them.

10.10. Take five distinct lines, no three of which are concurrent, and locate several lines on the conic determined by them.

10.11. Using Fig. 10-12 and Fig. 10-12′, prove the Theorems of Pascal and Brianchon for singular conics. *Hint.* Follow the arguments in Problems 10.1 and 10.4.

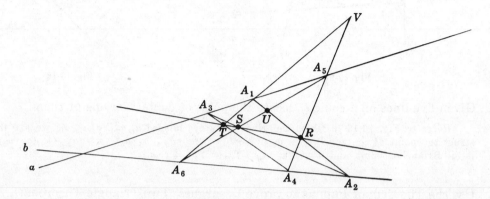

Fig. 10-12

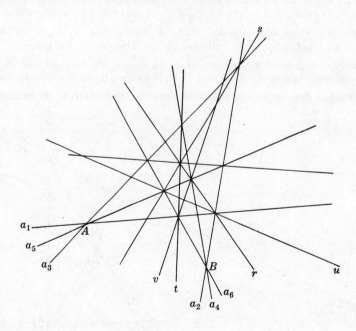

Fig. 10-12′

10.12. Given five distinct points on a conic C, construct the tangents to C at any two of these points.

10.13. Given four distinct points on a conic C and the tangent at one of them, construct the tangent to C at any one of the other points. Also, construct another point on C and the tangent at this point.

10.14. Given three distinct points on a conic C and the tangents at two of them, construct: (a) the tangent to C at the third given point, (b) another point of C, (c) the tangent at the point located in (b).

10.15. State and construct the dual of Problem 10.12.

10.16. State and construct the dual of Problem 10.13.

10.17. State and construct the dual of Problem 10.14.

10.18. Prove: If a triangle is inscribed in a conic C, the tangents to C at the vertices form a triangle perspective with the given one.

10.19. State and prove the dual of Problem 10.18.

10.20. Prove: If two triangles are perspective, the points of intersection of the sides of one with the non-corresponding sides of the other are on a conic.
Hint. In Fig. 10-4, let the sides of the given triangles be A_1A_2, A_3A_4, A_5A_6 and A_2A_3, A_4A_5, A_6A_1 with p the axis of perspectivity.

10.21. Prove: If a simple quadrangle $A_1A_2A_3A_4$ is inscribed in a conic, the tangent at A_1 and the side A_3A_4, the tangent at A_4 and the side A_1A_2, and the sides A_1A_4 and A_2A_3 intersect in collinear points.

10.22. State and prove the dual of Problem 10.21.

10.23. Prove: If a simple quadrangle is inscribed in a conic, the intersections of the tangents at the two pairs of opposite vertices and the intersections of the two pairs of opposite sides are four collinear points.

10.24. State and prove the dual of Problem 10.23.

10.25. Prove: If the vertices of a complete quadrangle are on a conic, the intersection of the tangents to the conic at a pair of vertices is collinear with the diagonal points of the quadrangle which are not on the join of the pair of vertices.

10.26. State and prove the dual of Problem 10.25.

10.27. If a simple hexagon $a_1a_2a_3a_4a_5a_6$ is circumscribed about a conic, prove that the points $a_1 \cdot a_3, a_3 \cdot a_5, a_5 \cdot a_1, a_2 \cdot a_4, a_4 \cdot a_6, a_6 \cdot a_2$ are on a conic.

10.28. Prove: If a triangle is inscribed in a conic, the sides on any vertex are separated harmonically by the tangent on that vertex and the join of that vertex and the intersection of tangents at the other two vertices.

10.29. State and prove the dual of Problem 10.28.

10.30. Complete the proof in Problem 10.8.

10.31. Given four tangents p, q, r, s to a conic and the point of contact P of p, construct the points of contact of the other tangents.

10.32. Two triangles ABC and $A'B'C'$ are perspective from P and p. If $p \cdot PA = A''$, $p \cdot PB = B''$, $p \cdot PC = C''$ and $H(A, A'; P, A'')$, prove: (a) $H(B, B'; P, B'')$ and $H(C, C'; P, C'')$, (b) the vertices of the two triangles are on a conic.

10.33. Prove: A simple quadrangle inscribed in a conic and its associated simple quadrilateral with respect to the conic have the properties: (i) the diagonals of the two form a harmonic set, (ii) the points of intersection of the pairs of opposite sides of the two form a harmonic set, (iii) the diagonals of the quadrilateral pass through the points of intersection of the pairs of opposite sides of the quadrangle.

10.34. Use Pascal's Theorem to show that each of the following sets of points in Fig. 4-7 is on a conic:

(a) P, S, D, E, F, G (c) P, R, F, G, I, J (e) P, Q, C, G, J, L

(b) Q, R, D, E, F, G (d) S, Q, F, G, I, J (f) R, S, C, F, J, N

Chapter 11

Desargues' Involution Theorem

INTRODUCTION

In Chapter 6 involutions on a given line were established by means of complete quadrangles having no vertex on the line. The involution was

(a) elliptic if no diagonal point of the quadrangle was on the line,

(b) hyperbolic, with double point A, if just one diagonal point A was on the line,

(c) hyperbolic, with double points A and B, if two diagonal points A and B were on the line.

The three reciprocal pairs of an involution determined by the complete quadrangle $PQRS$ on any line, not on a vertex, may be thought of (see Fig. 6-4) as determined by the three degenerate conics $PQ, RS; PR, QS; PS, QR$ determined by the vertices. In 1639 Desargues proved that the intersections, if any, of the line and a (non-degenerate) conic on the vertices P, Q, R, S are a reciprocal pair in the involution. In this chapter we consider involutions on a line or point established by one or more conics.

INVOLUTIONS DETERMINED BY A CONIC

Consider a conic C and a line k not tangent to C. Take any point X on k, construct its polar line x with respect to C, and let $x \cdot k = X'$. By definition, X and X' are conjugates with respect to the conic. Moreover, had we started with X', we would have obtained X. Thus the correspondence between X and X' is an involution and we have

Theorem 11.1. A given conic determines on every non-tangent line of the plane an involution whose reciprocal pairs consist of conjugate points with respect to the conic.

When the line intersects the conic, each point of intersection is on its own polar; hence these points are the double points of the involution of conjugate points (see also Problem 9.29, page 101) and the involution is hyperbolic. When the line does not meet the conic, the involution of conjugate points is elliptic.

Dually, we have

Theorem 11.1'. A given conic determines on every point, not on it, an involution whose reciprocal pairs are conjugate lines with respect to the conic.

The involution is elliptic if the point is an interior point of the conic and is hyperbolic, with the tangents on the point as double lines, if the point is an exterior point.

DESARGUES' THEOREM

Consider in Fig. 11-1(a) the simple quadrangle $PQRS$ and any line k which is not on a vertex of the quadrangle. By Theorems 6.3 and 6.2, page 65, the two pairs of opposite sides of the quadrangle determine on the line an involution of which the intersections of k and the diagonals of the quadrangle are a reciprocal pair. In Problem 11.1, we prove

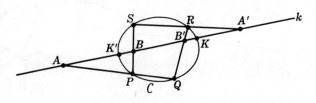

Fig. 11-1(a)

Theorem 11.2 (Desargues' Theorem). If a simple quadrangle is inscribed in a conic C and if a line k, not on a vertex of the quadrangle, is such that it meets C in two points, these points are a reciprocal pair in the involution on k determined by the pairs of opposite sides of the quadrangle.

This theorem provides an alternate construction of the conic determined by any five of its points. For, if P, Q, R, S, K are the given points and k is any line on K but not on any of the other given points, the opposite sides of the simple quadrangle $PQRS$ are met by k in two pairs of points which determine an involution. The mate K' of K in this involution, constructed as in Problem 6.2, page 68, is another point on the conic.

Dually, we have [see Fig. 11-1(b)]

Theorem 11.2'. If a simple quadrilateral is circumscribed about a conic C and if a point K, not on a side of the quadrilateral, is such that from it two tangents can be drawn to C, these tangents are a reciprocal pair in the involution on K determined by the pairs of opposite vertices of the quadrilateral.

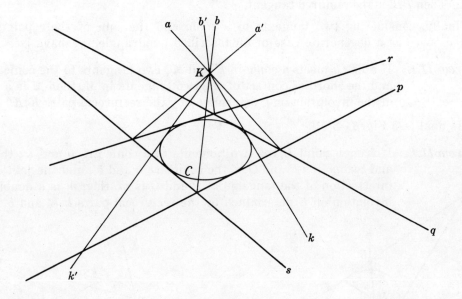

Fig. 11-1(b)

SPECIAL CASES OF DESARGUES' THEOREM

Either by considering a given triangle and a tangent on one vertex as a degenerate case of a simple quadrangle inscribed in a conic or by an independent proof, we have (see Fig. 11-2)

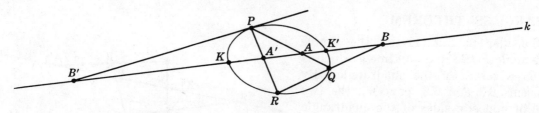

Fig. 11-2

Theorem 11.3.   If a triangle is inscribed in a conic C and if a line k, not on a vertex, is such that it meets C in two points, these points are a reciprocal pair in the involution on k determined by the pair of intersections with two sides of the triangle and the pair consisting of the intersection with the third side and with the tangent to C at the opposite vertex.

and its dual

Theorem 11.3′.   If a triangle is circumscribed about a conic C and if a point K, not on any side, is such that from it two tangents to C can be drawn, these tangents are a reciprocal pair in the involution on K determined by the pair of lines joining K to two of the vertices and the pair consisting of the joins of K to the third vertex and to the point of contact of the opposite side.

For an independent proof of Theorem 11.3′, see Problem 11.2.

Theorem 11.3 provides an alternate construction of the tangent to a conic, determined by any five of its points, at any one of them.   For, suppose K, K', P, Q, R are the given points and it is required to construct the tangent at P.   Let $KK' \cdot PQ = A$, $KK' \cdot PR = A'$, $KK' \cdot RQ = B$ and construct B' the mate of B in the involution determined by K, K' and A, A'.   Then PB' is the required tangent.

Finally, considering two tangents to a conic and the join of their points of contact counted twice as a degenerate case of an inscribed quadrangle, we have [see Fig. 11-3(a)]

Theorem 11.4.   If a line k meets a conic in K and K', two tangents to the conic in B and B', and the chord of contact of the two tangents in M, then M is a double point of the involution on k determined by the reciprocal pairs K, K' and B, B'.

and its dual [see Fig. 11-3(b)]

Theorem 11.4′.   If from a point K the two tangents to a conic are k and k', the joins of K and two points P and Q of the conic are b and b', and the join of K and the intersection of the tangents at P and Q is m, then m is a double line of the involution on K determined by the reciprocal pairs k, k' and b, b'.

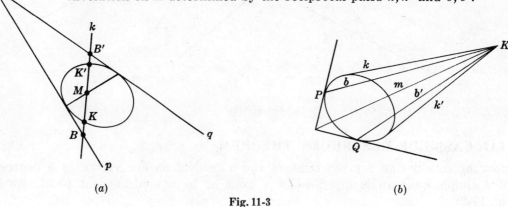

(a)   Fig. 11-3   (b)

QUADRANGULAR PENCILS OF CONICS

Five distinct points, no three of which are collinear, determine uniquely a conic. (Recall that, as used here, the term conic always implies a non-degenerate conic.) When a complete quadrangle is given, a conic on the vertices of this quadrangle is determined by selecting an additional point not on any side of the quadrangle. The totality of all such conics is called a *quadrangular pencil of conics*. By Theorem 11.2 we have

Theorem 11.5.   Those conics of a given quadrangular pencil which meet a given line k, not on a vertex of the defining quadrangle, do so in reciprocal pairs of the involution determined on k by the quadrangle.

It is clear from Fig. 11-1(a) that if K, K' is any reciprocal pair of the involution on k distinct from the pairs $A, A'; B, B'; C, C'$ then K and K' are on a conic of the quadrangular pencil. On the contrary, if K, K' is one of the excluded pairs, then P, Q, R, S, K, K' determine a pair of lines, that is, a degenerate conic. Thus the converse of Theorem 11.5, namely,

Theorem 11.6.   Any reciprocal pair of the involution on k determined by the quadrangle $PQRS$ is on a conic through the vertices of the quadrangle.

holds provided the degenerate conics $PQ, RS; PS, QR; PR, QS$ are included in the quadrangular pencil.

If the involution of Theorem 11.5 has a double point, it will be the point of contact of k and some conic of the quadrangular pencil. Since an involution has either no double points or two distinct double points, we have

Theorem 11.7.   If P, Q, R, S are distinct points, no three collinear, and if k is any line not on any of these points, there exists either two conics or no conic on the four points P, Q, R, S and tangent to the line k.

THE NINE-POINT CONIC

Let C_1 and C_2 be any distinct conics of the quadrangular pencil determined by the quadrangle $PQRS$ and let o be any line not on a vertex of the quadrangle or of its diagonal triangle ABC. Denote by O_1 the pole of o with respect to C_1 and by O_2 the pole of o with respect to C_2. As Z varies over o, (see Fig. 11-4), its polar line with respect to C_1 rotates about O_1 while its polar line with respect to C_2 rotates about O_2. By Theorem 9.10, page 94, each of the pencils of lines $O_1(z_1)$ and $O_2(z_2)$ are projectively related to the pencil of points $o(Z)$. Hence, $O_1(z_1) \barwedge O_2(z_2)$ and, as Z varies over o, $Z' = z_1 \cdot z_2$ describes a conic C on O_1 and O_2. We now show that this conic is independent of our choice of conics C_1 and C_2 by identifying nine well-defined points on it.

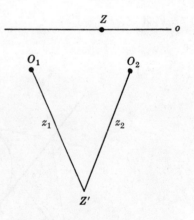

Fig. 11-4

First, let $AB \cdot o = C'$. With respect to any conic of the pencil, the pole of AB is C. When $Z = C'$ then $z_1 = O_1C$, $z_2 = O_2C$ and $Z' = C$ is on C. Similarly, A and B are on C, that is, C is on the vertices of the diagonal triangle. Now, let $PQ \cdot o = T$. When $Z = T$ then, by Theorem 9.1, $z_1 = O_1U$ and $z_2 = O_2U$ where $H(P, Q; T, U)$. We now have six additional points on C, namely, the harmonic conjugate, with respect to each pair of vertices of the quadrangle, of the point of intersection of the line on these vertices and o. The conic C is called the *nine-point conic* of the quadrangle with respect to the line o.

Solved Problems

11.1. Prove: If a simple quadrangle is inscribed in a conic C and if a line k, not on a vertex, is such that it meets C in two points, these points are a reciprocal pair in the involution on k determined by the pairs of opposite sides of the quadrangle.

Consider in Fig. 11-1(a) the simple quadrangle $PQRS$ inscribed in the conic C and the line k meeting C in K, K' and the opposite sides of the quadrangle in A, A' and B, B'. Think of C as generated by the projectivity

$$(PK, PK', PQ, PS) \; \overline{\wedge} \; (RK, RK', RQ, RS)$$

Then $(K, K', A, B) \; \overline{\overline{\wedge}} \; (PK, PK', PQ, PS) \; \overline{\wedge} \; (RK, RK', RQ, RS) \; \overline{\overline{\wedge}} \; (K, K', B', A')$

and $(K, K', A, B) \; \overline{\wedge} \; (K, K', B', A') \; \overline{\wedge} \; (K', K, A', B')$

by Theorem 2.10, page 25. From $(K, K', A, B) \overline{\wedge} (K', K, A', B')$ it follows that $K, K'; A, A'; B, B'$ are three reciprocal pairs in an involution on k. Since this involution is determined by the pairs $A, A'; B, B'$, we have the theorem.

11.2. Prove: If a triangle is circumscribed about a conic C and if a point K, not on any side, is such that from it two tangents to C can be drawn, these tangents are a reciprocal pair in the involution on K determined by the pair of lines joining K to two of the vertices and the pair consisting of the joins of K to the third vertex and to the point of contact of the opposite side.

Refer to Fig. 11-5 in which pqr is the circumscribed triangle and k, k' are the tangents to C drawn from K. Think of C as generated by a projectivity between pencils of points on p and r. Recalling that the correspondent of $p \cdot r$, thought of as a member of the pencil on p, is the point of contact R of r, we have

$$(p \cdot k, \; p \cdot k', \; p \cdot q, \; p \cdot r) \; \overline{\wedge} \; (r \cdot k, \; r \cdot k', \; r \cdot q, \; R)$$

Then $(k, k', a, b) \; \overline{\wedge} \; (k, k', b', a') \; \overline{\wedge} \; (k', k, a', b')$

and the theorem follows.

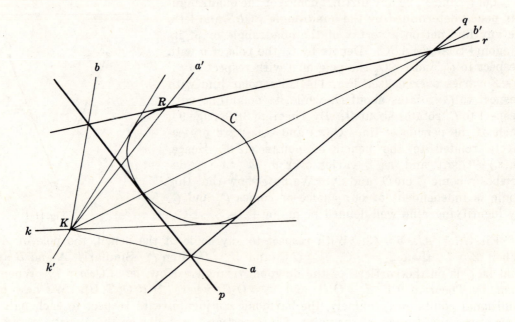

Fig. 11-5

Supplementary Problems

11.3. Prove: A simple quadrangle inscribed in a given conic determines, on any tangent to the conic not on a vertex, an involution having the point of tangency as a double point.

11.4. State and prove the dual of Problem 11.3.

11.5. Consider the case when the line k of Theorem 11.3 is tangent to the conic.

11.6. Given five distinct points, no three of which are collinear, construct (a) another point on the conic C determined by the given points, (b) the tangent to C at any one of the points.

11.7. Construct the dual of Problem 11.6.

11.8. Given four distinct points, no three of which are collinear, and a line on one of the points, construct at any other of the points the tangent to the conic on the four given points and having the given line as a tangent.

11.9. Construct the dual of Problem 11.8.

11.10. Prove: If a triangle PQR is inscribed in a conic C, then any two sides of the triangle are separated harmonically by the tangent to C at the common vertex and the line joining this vertex to the point of intersection of the tangents to C at the other two vertices.

11.11. State and prove the dual of Problem 11.10.

11.12. Given four distinct points, no three of which are collinear, and a line on one of them, construct the conic on the given points and tangent to the given line.
Hint. Consider Fig. 11-2 in which P, K, Q, R and the tangent at P are given.

11.13. State and prove the dual of Problem 11.12.

11.14. Prove Theorem 11.4.

11.15. Construct the conic given two of its tangents, their points of contact, and one other of its points.
Hint. Consider the involution on any line on the latter point.

11.16. State and prove the dual of Problem 11.15.

11.17. Prove Theorem 11.6 assuming the quadrangular pencil enlarged to include the degenerate conics.

11.18. Given a triangle ABC, a line x not on a vertex, and a point X on x but not on a side. Construct the conic C tangent to x at X and having ABC as a self-polar triangle.

11.19. Let C be the nine-point conic of a given quadrangular pencil of conics with respect to a given line o. Show that, if o is tangent to two conics of the pencil (see Theorem 11.7), the points of contact are on C.

Chapter 12

Pencils of Points and Lines on a Conic

DEFINITIONS

Consider a pencil of points $x(A_1, B_1, C_1, \ldots)$ and a point O not on x. Project the points of x from O by $OA_1 = a$, $OB_1 = b$, $OC_1 = c$, $\ldots$ and cut the pencil $O(a, b, c, \ldots)$ by any conic C on O. The set of points $A, B, C, \ldots$ thus determined on C by $O(a, b, c, \ldots)$ is called a *pencil of points* on C and will be denoted by $C(A, B, C, \ldots)$. In an earlier chapter, the correspondence between the pencil of points on x and the pencil of lines on O was called an elementary perspectivity. Similarly, the correspondence between the pencil of points $C(A, B, C, \ldots)$ and the pencil of lines obtained by projecting these points from *any point* O of C will be called an *elementary perspectivity* and denoted by $C(A, B, C, \ldots) \overline{\wedge} O(a, b, c, \ldots)$ while the correspondence between $C(A, B, C, \ldots)$ and the pencil $x(A_1, B_1, C_1, \ldots)$ will be called a *perspectivity* and denoted by

$$\text{(i)} \qquad C(A, B, C, \ldots) \overset{o}{\overline{\wedge}} x(A_1, B_1, C_1, \ldots)$$

Thus,

A pencil $C(A, B, C, \ldots)$ on a conic and a pencil $x(A_1, B_1, C_1, \ldots)$ on a line are *perspective* provided there is a point O on the conic, but not on the line, such that $AA_1, BB_1, CC_1, \ldots$ are on O.

It is clear from Fig. 12-1(a) that the perspectivity (i) is completely determined when two pairs of corresponding points are known as was the case for a perspectivity between two pencils of points on distinct lines. On the other hand, the perspectivity (i) will have $0, 1, 2$ self-corresponding points according as x and C have no points in common [Fig. 12-1(a)], x is tangent to C, x and C have two distinct points in common.

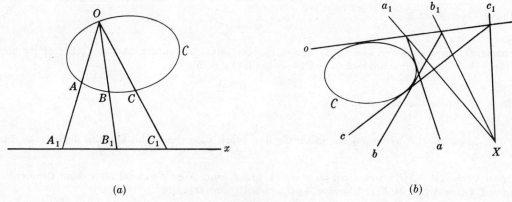

(a)   (b)

Fig. 12-1

Dually, see Fig. 12-1(b), the set of all lines $a, b, c, \ldots$ on a conic C is called a *pencil of lines* on C and denoted by $C(a, b, c, \ldots)$. The pencil $C(a, b, c, \ldots)$ and a pencil of lines $X(a_1, b_1, c_1, \ldots)$ are called *perspective* provided there is a line o on C, but not on X, such that $a \cdot a_1, \ b \cdot b_1, \ c \cdot c_1, \ldots$ are on o.

Consider next two pencils of points $x(A_1, B_1, C_1, \ldots)$ and $x'(A_2, B_2, C_2, \ldots)$ on distinct lines x and x' which are projectively, but not perspectively, related. Project the points of x' from any point O' on C, but not on x', and obtain the pencil $C(A, B, C, \ldots)$. Now

$$C(A, B, C, \ldots) \overset{O'}{\barwedge} x'(A_2, B_2, C_2, \ldots) \barwedge x(A_1, B_1, C_1, \ldots)$$

We define

A pencil $C(A, B, C, \ldots)$ and a pencil $x(A_1, B_1, C_1, \ldots)$ are projective, $C(A, B, C, \ldots) \barwedge x(A_1, B_1, C_1, \ldots)$, provided there exists a pencil $x'(A_2, B_2, C_2, \ldots)$ such that

$$x(A_1, B_1, C_1, \ldots) \barwedge x'(A_2, B_2, C_2, \ldots)$$

and
$$C(A, B, C, \ldots) \overset{O'}{\barwedge} x'(A_2, B_2, C_2, \ldots)$$

with O' on C but not on x'.

The projectivity $C(A, B, C, \ldots) \barwedge x(A_1, B_1, C_1, \ldots)$ is completely determined when three pairs of corresponding points are known.

Consider next (see Fig. 12-2) two pencils of points $x'(A_2, B_2, C_2, \ldots)$ and $x''(A_3, B_3, C_3, \ldots)$ each projectively, but not perspectively, related to the pencil $x(A_1, B_1, C_1, \ldots)$ and obtain as in the paragraph immediately above

(ii) $C(A, B, C, \ldots) \overset{O'}{\barwedge} x'(A_2, B_2, C_2, \ldots) \barwedge x(A_1, B_1, C_1, \ldots)$

and   (iii) $C(A', B', C', \ldots) \overset{O''}{\barwedge} x''(A_3, B_3, C_3, \ldots) \barwedge x(A_1, B_1, C_1, \ldots)$

with O' on C but not on x' and O'' on C but not on x''. Combining (ii) and (iii), we have

$$C(A, B, C, \ldots) \overset{O'}{\barwedge} x'(\ldots) \barwedge x(\ldots) \barwedge x''(\ldots) \overset{O''}{\barwedge} C(A', B', C', \ldots)$$

and define

Two pencils $C(A, B, C, \ldots)$ and $C(A', B', C', \ldots)$ on the same conic are projective provided there exists a pencil $x(A_1, B_1, C_1, \ldots)$ to which each is projective.

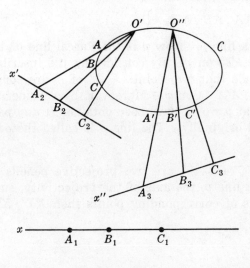

Fig. 12-2

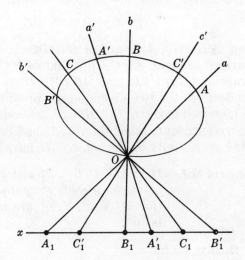

Fig. 12-3

Finally, consider in Fig. 12-3 two projectively related superposed pencils of points $x(A_1, B_1, C_1, \ldots)$ and $x(A_1', B_1', C_1', \ldots)$. Join the points of each pencil to any point O, not on x, to obtain the pencils of lines $O(a, b, c, \ldots)$ and $O(a', b', c', \ldots)$. Clearly, $O(a, b, c, \ldots) \barwedge O(a', b', c', \ldots)$. If now these pencils of lines are cut by any conic C on O, two pencils of points $C(A, B, C, \ldots)$ and $C(A', B', C', \ldots)$ respectively are obtained. In turn, these pencils are projectively related, i.e. $C(A, B, C, \ldots) \barparallel C(A', B', C', \ldots)$, since

$$C(A, B, C, \ldots) \;\underline{\overline{\wedge}}\; O(a, b, c, \ldots) \;\barparallel\; O(a', b', c', \ldots) \;\underline{\overline{\wedge}}\; C(A', B', C', \ldots)$$

For the dual, consider two projectively related superposed pencils of lines $X(a_1, b_1, c_1, \ldots)$ and $X(a'_1, b'_1, c'_1, \ldots)$. Section the two pencils by any line o not on X to obtain

$$o(A_1, B_1, C_1, \ldots) \;\barparallel\; o(A'_1, B'_1, C'_1, \ldots)$$

Let C be any conic tangent to o and from each of the points $A_1, B_1, C_1, \ldots; A'_1, B'_1, C'_1, \ldots$ draw the second tangent $a, b, c, \ldots; a', b', c', \ldots$ respectively to C. Then

$$C(a, b, c, \ldots) \;\underline{\overline{\wedge}}\; o(A_1, B_1, C_1, \ldots) \;\barparallel\; o(A'_1, B'_1, C'_1, \ldots) \;\underline{\overline{\wedge}}\; C(a', b', c', \ldots)$$

and
$$C(a, b, c, \ldots) \;\barparallel\; C(a', b', c', \ldots)$$

TWO PROJECTIVE PENCILS ON A CONIC

Let
$$C(A, B, C, D, \ldots)$$

and
$$C(A', B', C', D', \ldots)$$

be two projective pencils of points on a conic C. When these pencils are projected respectively from two points O and O' on C, we have

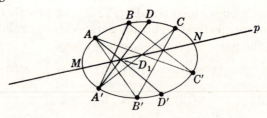

Fig. 12-4

$$(OA, OB, OC, OD, \ldots) \;\underline{\overline{\wedge}}\; C(A, B, C, D, \ldots)$$
$$\barparallel C(A', B', C', D', \ldots) \;\underline{\overline{\wedge}}\; (O'A', O'B', O'C', O'D', \ldots)$$

and so
$$(OA, OB, OC, OD, \ldots) \;\barparallel\; (O'A', O'B', O'C', O'D', \ldots)$$

In particular, when $O = A'$ and $O' = A$, we have the perspectivity

$$(A'A, A'B, A'C, A'D, \ldots) \;\underline{\overline{\wedge}}\; (AA', AB', AC', AD', \ldots)$$

Then $A'B \cdot AB'$, $A'C \cdot AC'$, $A'D \cdot AD'$, $\ldots$ are on a line p. Now p is the Pascal line of the inscribed hexagon $AB'CA'BC'$ and so $B'C \cdot BC'$ is also on p. By considering the inscribed hexagons $AB'DA'BD'$ and $AC'DA'CD'$, we conclude that two points X, X' of C are correspondents in the two given pencils provided $AX' \cdot A'X$ is on p. Moreover, by considering the inscribed hexagon $BC'DB'CD'$, we conclude that p would have been obtained if *any* pair of correspondents other than A, A' had been used originally. The line p is called the *axis* of the projectivity on the conic. We have proved

Theorem 12.1.   If $C(A, B, C, D, \ldots)$ and $C(A', B', C', D', \ldots)$ are two projective pencils of points on a conic C, there exists a line p, the axis of the projectivity, such that if $X, X'; Y, Y'$ are two pairs of corresponding points then $XY' \cdot X'Y$ is on p.

Dually, we have

Theorem 12.1'.   If $C'(a, b, c, d, \ldots)$ and $C'(a', b', c', d', \ldots)$ are two projective pencils of lines on a conic C', there exists a point P, the center of the projectivity, such that if $x, x'; y, y'$ are two pairs of corresponding lines then $(x \cdot y')(x' \cdot y)$ is on P.

If, in Theorems 12.1 and 12.1', the conics are assumed to be the same while the lines $a, b, c, \ldots; a', b', c', \ldots$ are the tangents to this conic at the respective points $A, B, C, \ldots; A', B', C', \ldots$, then the center P of the projectivity

$$C(a, b, c, \ldots) \; \overline{\wedge} \; C(a', b', c', \ldots)$$

is the pole of the axis p of the projectivity $C(A, B, C, \ldots) \; \overline{\wedge} \; C(A', B', C', \ldots)$.

A projectivity on a conic C is called *hyperbolic, parabolic,* or *elliptic* according as it has two, one, or no double elements. We leave for the reader to show: (a) a projectivity between two pencils of points on C is hyperbolic, parabolic, or elliptic according as the axis p meets C in distinct points, is tangent to C, or does not meet C; (b) a projectivity between two pencils of lines on C is hyperbolic, parabolic, or elliptic according as the center P is outside, on, or inside C.

By definition a projectivity

$$C(A, B, C, D, \ldots) \; \overline{\wedge} \; x(A_1, B_1, C_1, D_1, \ldots)$$

implies $\qquad C(A, B, C, D, \ldots) \; \overset{O'}{\overline{\wedge}} \; x'(A_2, B_2, C_2, D_2, \ldots) \; \overline{\wedge} \; x(A_1, B_1, C_1, D_1, \ldots)$

where O' is on C but not on x'. By Theorem 2.10, page 25, there always exists a projectivity

$$x(A_1, B_1, C_1, D_1) \; \overline{\wedge} \; x(B_1, A_1, D_1, C_1)$$

There follows easily

Theorem 12.2. $C(A, B, C, D) \; \overline{\wedge} \; C(B, A, D, C)$ for any four distinct points A, B, C, D on a conic C.

STEINER'S CONSTRUCTION

Theorem 12.1 provides a method, given first by Steiner, for constructing the double points, if any, of a superposed projectivity $x(A_1, B_1, C_1, \ldots) \; \overline{\wedge} \; x(A_1', B_1', C_1', \ldots)$. In Fig. 12-5, O is any point not on x and C is any conic on O. From O project the points $A_1, B_1, C_1; A_1', B_1', C_1'$ into $A, B, C; A', B', C'$ respectively on C. Construct the Pascal line p for the inscribed hexagon $AB'CA'BC'$ and mark its intersections M, N with C. Then $M_1 = OM \cdot x$ and $N_1 = ON \cdot x$ are the required double points. To show this, consider

$$x(A_1, B_1, C_1, M_1, N_1) \; \overset{O}{\overline{\wedge}} \; C(A, B, C, M, N) \; \overset{B'}{\overline{\wedge}} \; p(T, U, V, M, N)$$
$$\overset{B}{\overline{\wedge}} \; C(A', B', C', M, N) \; \overset{O}{\overline{\wedge}} \; x(A_1', B_1', C_1', M_1, N_1)$$

which exhibits M_1, N_1 as the double points of the given superposed projectivity on x.

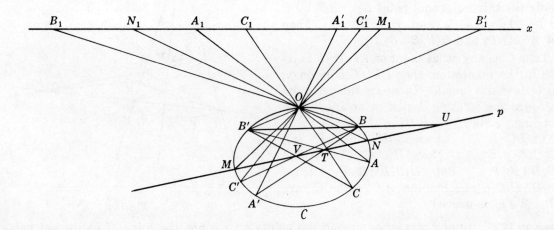

Fig. 12-5

The above construction is for a superposed projectivity having two double points. For any given projectivity the number of double points is not known until p has been constructed. Should the reader care for an example of a projectivity with one or with no double point, it is best to begin with a line x, a point O not on x, and a conic C on O. For one double point, take for p any tangent to C except that at O, and for no double point, take for p any line, not x, which has no point in common with C. Of course, we have now turned the problem around; we are constructing the projectivity having prescribed double points. Choose points $A, B, C; A', B', C'$ on C such that the Pascal line of $AB'CA'BC'$ is p and project these points from O onto x to determine the projectivity

$$x(A_1, B_1, C_1, \ldots) \; \overline{\wedge} \; x(A_1', B_1', C_1', \ldots)$$

The construction for the double points, if any, of an involution on a line is, of course, quite similar to that given above. Let $A_1, A_1'; B_1, B_1'$ be two reciprocal pairs of the given involution on a line x and let their projections on the conic from any point O of the conic be $A, A'; B, B'$ respectively. Since $x(A_1, A_1', B_1, B_1') \; \overline{\wedge} \; x(A_1', A_1, B_1', B_1)$, it follows that $C(A, A', B, B') \; \overline{\wedge} \; C(A', A, B', B)$ and we have only to construct the axis of involution p determined by the points $AB' \cdot BA'$ and $AB \cdot A'B'$. Note that Steiner's construction consists in transferring a given projectivity on a line onto a conic to take advantage of the simplicity of certain constructions when projectivities on a conic are involved. This, in turn, has led to a thorough investigation of just what constructions are possible when, in addition to a straight edge, a fixed conic is provided.

INVOLUTIONS ON A CONIC

A projectivity between two pencils of points (lines) on a conic, having a pair of corresponding elements which correspond reciprocally, is called an *involution on the conic*. In Problem 12.1 we prove

Theorem 12.3.   In an involution on a conic, every pair of corresponding elements correspond reciprocally.

Consider in Fig. 12-6 two reciprocal pairs $A, A'; B, B'$ in an involution of points on the conic C. From $(A, A', B) \; \overline{\wedge} \; (A', A, B')$, the axis of projectivity (called the *axis of involution*) p is on the points $AB' \cdot A'B$ and $AB \cdot A'B'$. Note that p is a side of the diagonal triangle of the complete quadrangle $AA'BB'$ inscribed in C. Denote the third diagonal point $AA' \cdot BB'$ by P and let $A_1 = AA' \cdot p$ and $B_1 = BB' \cdot p$. Then $H(A, A'; A_1, P)$ and $H(B, B'; B_1, P)$.

Take C as any other point on C. If C' is its mate in the involution, then $CB' \cdot C'B$ is on p; thus C' is easily found. (Compare this with the procedure for locating mates in an involution on a line.) Since $(B, B', C) \; \overline{\wedge} \; (B', B, C')$, p is also on $BC \cdot B'C'$. Suppose $BB' \cdot CC' = P'$ and let $CC' \cdot p = C_1$. Now $H(C, C'; C_1, P')$ and $H(B, B'; B_1, P')$. But $H(B, B'; B_1, P)$; hence $P' = P$, $H(C, C'; C_1, P)$, and AA', BB', CC' are on P. We have proved

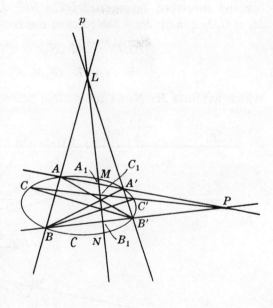

Fig. 12-6

Theorem 12.4.   In any involution among the points on a conic, the joins of reciprocal pairs are concurrent.

The point of concurrency P of Theorem 12.4 is called the *center of the involution*. Note that it is the pole of the axis of involution.

An involution on a conic C, being a special type of projectivity on C, is called *hyperbolic*, *parabolic*, or *elliptic* according as there are two, one, or no double elements.

Fig. 12-6 illustrates the case of a hyperbolic involution on C. Here the axis of involution p meets C in two distinct points M and N, the double points of the involution. The center of involution P is then outside the conic.

In a parabolic involution, the axis of involution is tangent to the conic; hence the center of involution P is on the axis and, in particular, is on the conic. As a consequence P is a member of each reciprocal pair in the involution. Parabolic involutions on a conic are of no particular importance and will not be considered here.

In an elliptic involution on a conic the axis does not meet the conic and the center of involution is therefore inside the conic.

In Problem 12.2 we prove

Theorem 12.5.   If the involution of conjugate points determined by a conic C on a line x is projected onto C from any of its points, the resulting involution on C has x as its axis.

HARMONIC SETS ON A CONIC

On any point O take four lines a, b, c, d such that $H(a, b; c, d)$. Let these lines be cut by any conic C on O in the respective points A, B, C, D. Now project these points from any other point O' on C by the lines a', b', c', d' respectively. By the definition of a conic

$$O(a, b, c, d) \; \barwedge \; O'(a', b', c', d')$$

and hence $H(a', b'; c', d')$. We define

Four distinct points on a conic form a *harmonic set* (or, are *harmonic*) provided they are projected from any point of the conic by a harmonic set of lines.

Dually,

Four distinct lines on a conic form a *harmonic set* (or, are *harmonic*) provided they are sectioned by any line of the conic in a harmonic set of points.

Note that (Fig. 12-7) when A, B, C, D on a conic C are projected from some one of them, say D, the projectors are DA, DB, DC, and the tangent to C at D. Dually, if a, b, c, d on a conic are sectioned by some one of them, say d, the intersections are $d \cdot a, d \cdot b, d \cdot c$, and the point of contact of d.

There follows from Theorem 9.7, page 94,

Theorem 12.6.   If four points on a conic C are harmonic, the tangents to C at these points are harmonic.

For a proof, see Problem 12.3.

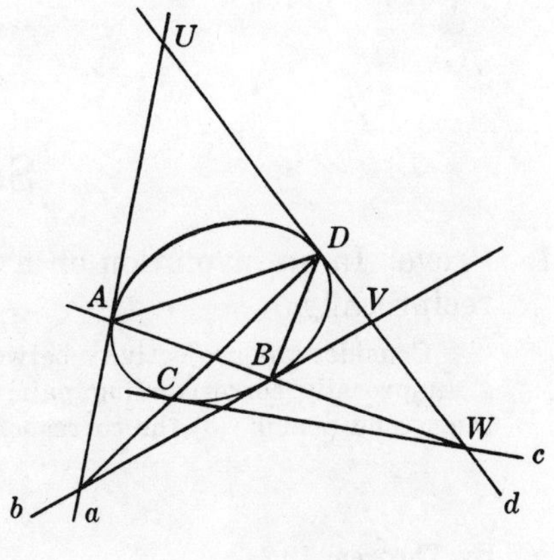

The dual of Theorem 12.6,

Fig. 12-7

Theorem 12.6'.   If four lines of a conic are harmonic, their points of contact are harmonic.

is also its converse.

In Problem 12.4 we prove

Theorem 12.7.   If A, B, C, D are four harmonic points on a conic C, that is, if $H(A, B; C, D)$, then the lines AB and CD are conjugates with respect to C.

The converse

Theorem 12.8.   If two conjugate lines with respect to a conic have in common with it the pairs of points A, B and C, D respectively, then $H(A, B; C, D)$.

is also valid.

In Problem 12.5 we prove the dual of Theorem 12.8,

Theorem 12.8'.   If two conjugate points with respect to a conic have in common with it the pairs of lines a, b and c, d respectively, then $H(a, b; c, d)$.

As a consequence of Theorem 12.8, we have

Theorem 12.9.   In a hyperbolic involution on a conic, the double elements form with each reciprocal pair a harmonic set.

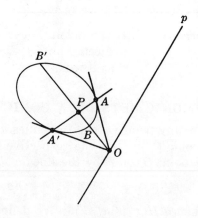

Finally, using Fig. 12-8, it is seen that for a given reciprocal pair A, A' of an elliptic involution on a conic, only the line PO on the center of involution P and the pole O of the line AA' is conjugate to AA'.   Hence,

Theorem 12.10.   In an elliptic involution on a conic there is one and only one reciprocal pair which forms with a given reciprocal pair a harmonic set.

Fig. 12-8

Solved Problems

12.1.   Prove: In an involution on a conic, every pair of corresponding elements correspond reciprocally.

Consider the projectivity between two pencils of points on a conic C in which: (1) A and B are a reciprocally corresponding pair; (2) the correspondent of $C \neq A, B$ in the first pencil is D in the second pencil; (3) the correspondent of D in the first pencil is E in the second pencil.   Then

$$C(A, B, C, D) \ \overline{\wedge}\ C(B, A, D, E)$$

By Theorem 12.2, $$C(A, B, C, D) \ \overline{\wedge}\ C(B, A, D, C)$$

Since $$C(B, A, D, C) \ \overline{\wedge}\ C(A, B, C, D) \ \overline{\wedge}\ C(B, A, D, E)$$

requires $E = C$, the pair of correspondents C, D and, thus, every pair of correspondents, correspond reciprocally.

12.2.   Prove: If the involution of conjugate points determined by a conic C on a line x is projected onto C from any of its points, the resulting involution on C has x as axis.

In Fig. 12-9, let $A_1, A_1'; B_1, B_1'$ be two reciprocal pairs of the involution

$$x(A_1, A_1', B_1, \ldots) \;\overline{\wedge}\; x(A_1', A_1, B_1', \ldots)$$

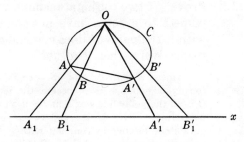

of conjugate points with respect to C. Let the projection of this involution onto C from the point O on it be the involution

$$C(A, A', B, \ldots) \;\overline{\wedge}\; C(A', A, B', \ldots)$$

The sides OA, OA' of the triangle OAA' are met by x in the pair of conjugates A_1, A_1'. By Theorem 9.14, page 95, x and AA' are conjugate. Then AA' is on the pole of x and x is the axis of the involution on C as required.

Fig. 12-9

12.3. Prove: If four points on a conic C are harmonic, the tangents to C at these points are also harmonic.

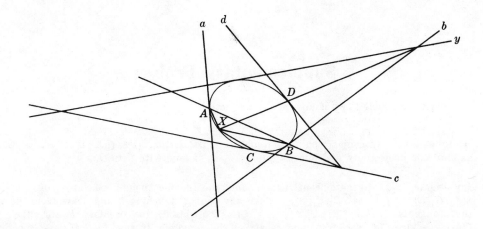

Fig. 12-10

Let the four points be A, B, C, D and let X be any fifth point on C; then, by hypothesis, (XA, XB, XC, XD) is a harmonic set. Denote the tangents to C at A, B, C, D as a, b, c, d respectively. By Theorem 9.7, page 94, with y any tangent to C,

$$(XA, XB, XC, XD) \;\overline{\wedge}\; (y \cdot a, y \cdot b, y \cdot c, y \cdot d)$$

Hence $(y \cdot a, y \cdot b, y \cdot c, y \cdot d)$ is a harmonic set and, by definition, (a, b, c, d) is also.

12.4. Prove: If A, B, C, D are four harmonic points on a conic C, that is, if $H(A, B; C, D)$, then the lines AB and CD are conjugates with respect to C.

By definition, the lines projecting A, B, C, D from any point of C are harmonic; in particular, we have

$$H(DA, DB; DC, DV)$$

where DV is the tangent to C at D. Let $AB \cdot DC = W$ and $AB \cdot DV = V$; then $H(A, B; V, W)$ and the polar line of V is $DW = CD$. Thus AB is on the pole of CD and the two lines are conjugates with respect to C.

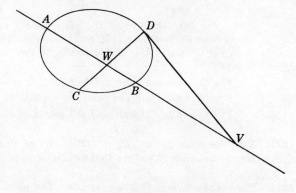

Fig. 12-11

12.5. Prove: If two conjugate points with respect to a conic have in common with it the pairs of lines a, b and c, d respectively, then $H(a, b; c, d)$.

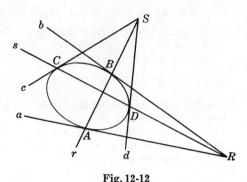

Since $R = a \cdot b$ and $S = c \cdot d$ are conjugate points with respect to the conic, R is on the polar line s of S and S is on the polar line r of R. Now r and s are conjugate lines having in common with C the pairs of points A, B and C, D. By Theorem 12.8, page 124, $H(A, B; C, D)$ and by Theorem 12.6, $H(a, b; c, d)$.

Fig. 12-12

Supplementary Problems

12.6. Construct the double lines, if any, of the projectivity

$$O(a, b, c, \ldots) \ \overline{\wedge}\ O(a', b', c', \ldots)$$

Hint. Cut by any conic C on O and obtain $C(A, B, C, \ldots) \ \overline{\wedge}\ C(A', B', C', \ldots)$. Construct the axis of this projectivity and join its intersections, if any, with C to O.

12.7. Investigate the alternate construction for the double points, if any, of the projectivity $x(A_1, B_1, C_1, \ldots) \ \overline{\wedge}\ x(A_1', B_1', C_1', \ldots)$. Take any conic C touching x and draw from the given points the tangents $a, b, c, \ldots; a', b', c', \ldots$ to C. Obtain the center B of the projectivity $C(a, b, c, \ldots) \ \overline{\wedge}\ C(a', b', c', \ldots)$ and construct the tangents, if any, to C from B meeting x in the required double points.

12.8. On a line take two pairs of points which do not separate each other and construct the pair which separate harmonically each of the given pairs.

12.9. Prove Theorem 12.4 by considering the triangles ABC and $A'B'C'$ where $A, A'; B, B'; C, C'$ are three reciprocal pairs of the involution.

12.10. Let a conic C be given by (a) five of its points, or (b) four of its points and the tangent at one of them, or (c) three of its points and the tangents at two of them, or (a') five of its tangents, or (b') four of its tangents and the point of contact on one of them, or (c') three of its tangents and the points of contact on two of them; and let p be a line neither on any of the given points of $(a), (b), (c)$ nor coincident with any of the given tangents of $(a'), (b'), (c')$. Construct the intersections, if any, of p and C.

Hint. For (a), take two of the given points as centers of projection and project the remaining points from each of them onto p. In turn, project these points on p onto any conic C' from any point O on it. Construct the double points, if any, of the projectivity on C' and project them through O onto p.

For (b), take the given point of contact as one of the centers of projection.

For (c), take the given points of contact as the centers of projection.

For $(a'), (b'), (c')$, first find the points of contact of all tangents and proceed as before.

12.11. Let C be given by $(a'), (b'), (c'), (a), (b), (c)$ of Problem 12.10 and let P be a point not on any of the given lines. Construct the tangents, if any, to C from P.

12.12. Prove the converse of Theorem 12.4: The pairs of points in which a conic is cut by the lines of a pencil whose center is not on the conic form an involution. What are its double points?

12.13. Prove: In an involution among the points on a conic, the tangents to the conic at reciprocal pairs intersect on the axis of the involution.

12.14. Given five points A, B, C, D, E on a conic C and a point R, construct the tangents, if any, to C from R.

    Hint. Use the results of Problems 10.3 and 12.10. First, find the second points of intersection A', B' of RA, RB and C; then construct p the axis of the involution $C(A, A', B) \barwedge C(A', A, B')$ and the points of intersection, if any, of p and C.

12.15. Given five tangents of a conic C and a line r, construct the points of intersection, if any, of r and C.

12.16. Given four points P, Q, R, S, no three of which are collinear, and a line x not on any of the given points, construct a conic on the four points and tangent to the line.

    Hint. Use Theorem 11.1. Obtain on x the involution determined by the simple quadrangle $PQRS$ and construct its double points, if any. Each double point, together with the given points, determines a conic satisfying the conditions. Discuss the number of conics possible.

12.17. Given four lines p, q, r, s, no three of which are concurrent, and a point X, not on any of the given lines, construct a conic on X and having the four lines as tangents.

12.18. Given three non-collinear points P, Q, R and two lines x and y, neither of which is on any of the given points, construct a conic on the given points and tangent to the given lines.

    Hint. Use Theorem 11.5. Let $PQ \cdot x = B$, $PQ \cdot y = B'$ and construct the double points, if any, of the involution $(P, Q, B) \barwedge (Q, P, B')$. Let $PR \cdot x = C$, $PR \cdot y = C'$ and repeat. Discuss the number of conics possible.

12.19. Given three non-concurrent lines p, q, r and two points X and Y, neither of which is on any of the given lines, construct a conic on the given points and tangent to the given lines.

12.20. Let P, Q be any distinct points on a conic C. Locate on C another pair of points A, B such that $H(A, B; P, Q)$. Are the points A, B unique?

12.21. Let A, P, Q be any three distinct points on a conic C. Locate on C a point B such that $H(A, B; P, Q)$. Is the point B unique?

12.22. Let P, Q, R, S be any distinct points on a conic C. Under what conditions will there exist points A, B on C such that $H(A, B; P, Q)$ and $H(A, B; R, S)$?

12.23. Take P, Q, R, S on C meeting the conditions of Problem 12.22 and construct the pair A, B.

12.24. State and solve the duals of Problems 12.20-12.23.

12.25. Prove: Reciprocal pairs of a hyperbolic involution on a line are conjugate pairs with respect to every conic which meets the line in the double points of the involution.

12.26. State and prove the converse of Theorem 12.8'.

12.27. Five points on a conic C are given. Construct on any line x an involution of conjugate points with respect to C.

Chapter 13

Plane Affine Geometry

INTRODUCTION

The preceding chapters contain the more basic propositions of plane projective geometry. In the earlier of these chapters the elementary geometry of high school was taken as the fundamental geometry. The projective plane was then obtained by modifying the ordinary plane as follows: (a) to the ordinary plane was adjoined an ideal line of ideal points, (b) the identity of the ideal points and the ideal line, as such, was then erased, that is, the ideal points were treated thereafter as ordinary points and the ideal line as an ordinary line of the projective plane.

In this chapter we reverse the procedure, that is, we take plane projective geometry as the fundamental geometry and, by properly modifying the projective plane, obtain other geometries. We begin by selecting a (any) line of the projective plane and call it the *ideal line* or *line at infinity* l_∞. When this line is removed from the plane, the points and lines remaining are said to constitute the *affine plane*. The geometry of this plane is called *plane affine geometry*. In developing this geometry, we propose to work in the projective plane with one of its lines (always on the page) labeled l_∞. A number of familiar theorems of elementary geometry will be "rediscovered" but the accompanying figures will for a time appear strange. (Should the reader insist on working with familiar figures, he has only to redraw them with l_∞ off the page and "at an infinite distance".)

Two distinct points A and B of the projective plane separate the unique line $p = AB$ into two segments (see Chapter 1) $\widehat{AB}$ and $\underset{\frown}{AB}$. Suppose we pass to the affine plane by choosing for l_∞ any line which is neither on A nor B. In the process, the line p loses one of its points, namely the point P_∞ in which it meets l_∞, and so becomes an open line. On this, the affine line p, the points A and B determine one and only one segment, namely that segment $\widehat{AB}$ or $\underset{\frown}{AB}$ of the projective line which does not contain P_∞. This segment will be labeled simply as AB.

No changes in labels have been made in the above paragraph to indicate a change from the projective plane to the affine plane. We could of course label a projective point, not on l_∞, as A^* and the corresponding affine point as A with similar changes in labels for a projective line, distinct from l_∞, and the corresponding affine line. This is not done here since all theorems of this chapter will be stated, of course, in terms of points and lines of the affine plane and, elsewhere, it will always be tolerably clear from the context whether the line p, for example, is being thought of as a projective line or as an affine line.

PARALLEL LINES

In the projective plane any two distinct lines p and q intersect. When this point of intersection is not on l_∞, the affine lines p and q intersect; when this point of intersection is a point of l_∞, the affine lines p and q do not intersect. We define:

Two lines of the affine plane which do not intersect are called *parallel*.

128

Here we have borrowed a familiar term of elementary geometry to cover the familiar situation of two lines which never meet. Now projective geometry is non-metric; hence the reader must be careful not to give to the term parallel (or to any of the other familiar terms to be introduced as we proceed) a metric interpretation. Thus it must *not* be assumed that parallel lines in the affine plane are everywhere equidistant.

From the above definition of parallel lines, there follow:

Theorem 13.1.   All lines parallel to a given line are parallel to one another.

and

Theorem 13.2.   Through a given point, not on a given line, one and only one line can be drawn parallel to the given line.

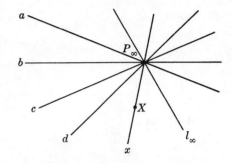

Refer to Fig. 13-1 in which each of the lines b, c, d is, by definition, parallel to the line a. Clearly, b is parallel to c and, in fact, the lines a, b, c, d are parallel to one another. Again, the line x, on X and parallel to the line a, is unique since, by definition, the projective line x must also be on P_∞, the ideal point of a.

We leave for the reader the proof of

Theorem 13.3.   If a line intersects one of two parallel lines, it intersects the other also.

Fig. 13-1

CONGRUENT SEGMENTS

Consider in Fig. 13-2 the complete quadrangle $PQRS$ having l_∞ as a side of its diagonal triangle. Since each pair of affine lines PQ, RS and PS, QR consists of distinct parallel lines, the simple quadrangle $PQRS$ is called a *parallelogram*. By definition, the *sides* of this parallelogram are the segments PQ, QR, RS, SP. On the remaining sides of the quadrangle are the diagonals (the segments PR and QS) of the parallelogram.

Two segments AB and CD are said to be *congruent by translation* provided $ABCD$ is a parallelogram. For example, the segments PQ and RS (also, PS and QR) of Fig. 13-2 are congruent. We now have a means of comparing segments on distinct parallel lines in that, of two such segments, we can determine whether they are congruent or not congruent. There is no difficulty in extending this to the case of two segments on the same line. Consider in Fig. 13-3 the parallelograms $PP'R'R$ and $QQ'R'R$ whose sides PP' and QQ' are segments on the same line o. The segment PP'

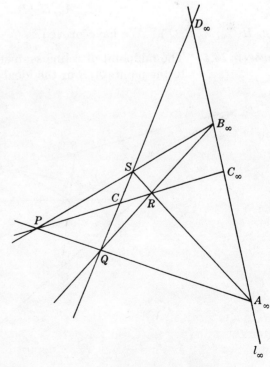

Fig. 13-2

is congruent to RR' which, in turn, is congruent to QQ'. It would seem natural then to define the segments PP' and QQ' to be congruent, i.e.,

Two segments AA' and BB' of the same line are said to be *congruent by translation* provided there exists a segment CC' such that $AA'C'C$ and $BB'C'C$ are parallelograms.

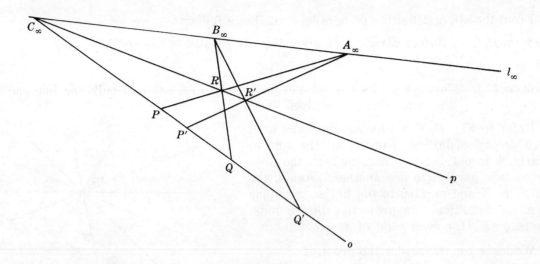

Fig. 13-3

MIDPOINT OF A SEGMENT

Suppose that the points P' and Q of Fig. 13-3 coincide at M as in Fig. 13-4. Then the segments PM and MQ' are congruent and we call M the *midpoint* of the segment PQ'. Let $PR' \cdot l_\infty = D_\infty$. The complete quadrangle $PMR'R$ defines on l_∞ the harmonic set $H(C_\infty, A_\infty; D_\infty, B_\infty)$ and, since

$$(C_\infty, A_\infty, D_\infty, B_\infty) \overset{R'}{\barwedge} (C_\infty, M, P, Q')$$

then $H(C_\infty, M; P, Q')$. We have proved

Theorem 13.4.   The midpoint of a line segment AB is the harmonic conjugate with respect to the points A, B of the ideal point on the line AB.

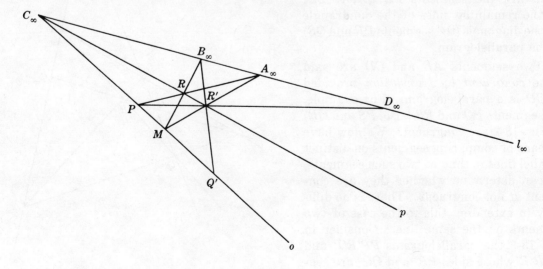

Fig. 13-4

Returning now to Fig. 13-2, let $PR \cdot QS = C$. Since $H(A_\infty, B_\infty; C_\infty, D_\infty)$, it follows from

$$(R, P, C_\infty, C) \stackrel{S}{\overline{\wedge}} (A_\infty, B_\infty, C_\infty, D_\infty) \stackrel{R}{\overline{\wedge}} (S, Q, C, D_\infty)$$

that $H(R, P; C_\infty, C)$ and $H(S, Q; D_\infty, C)$. By Theorem 13.4, C is the midpoint of each of the segments PR and QS. We have proved

Theorem 13.5. The point of intersection of the diagonals of a parallelogram is the midpoint of each.

Note 1. The above proof consists of two parts. In the first part, the results $H(R, P; C_\infty, C)$ and $H(S, Q; D_\infty, C)$ are obtained in the projective plane, that is, without giving any special significance to l_∞ and its points; in the second part, an affine interpretation of these results in the light of certain definitions previously introduced is made.

Note 2. An alternate proof is as follows: In the projective plane Fig. 13-2 exhibits a complete quadrangle $PQRS$ with C, A_∞, B_∞ its diagonal points and with C_∞, D_∞ two of its associated harmonic points. Then, by Theorem 4.11, page 48, $H(R, P; C_\infty, C)$ and $H(S, Q; D_\infty, C)$ and the proof continues as before.

Two familiar theorems concerning triangles now follow. In Problem 13.1, we prove

Theorem 13.6. The line on the midpoints of two sides of a triangle is parallel to the third side.

It seems natural to call QC, joining one vertex of the triangle PQR to the midpoint of the segment determined by the other two vertices, a *median* of the triangle. Denote by PB and RA the other medians of this triangle and let $PB \cdot QC = G$. We leave for the reader to prove

Theorem 13.7. The medians of a triangle are concurrent. The point G of concurrency is called the *centroid* or *median point* of the triangle.

Consider now in Fig. 13-5 the simple parallelograms $PP'R'R$ and $QQ'R'R$ of Fig. 13-3. Let $PR \cdot Q'R' = S$, $P'R' \cdot QR = T$, $m = ST$, $p = RR'$, $m \cdot p = C$, $m \cdot o = M$, and $m \cdot l_\infty = M_\infty$. Since C is the midpoint of the diagonal RR' of the simple parallelogram $RSR'T$, we have $H(R, R'; C_\infty, C)$. By projecting these four points first from S and then from T, we have $H(P, Q'; C_\infty, M)$ and $H(Q, P'; C_\infty, M)$. Then M is the midpoint of each of the segments PQ' and $P'Q$. We have proved

Theorem 13.8. If two segments PP' and QQ' on the same line are congruent, then the segments PQ' and $P'Q$ have a common midpoint.

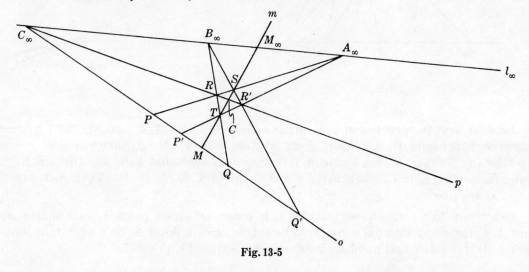

Fig. 13-5

Referring to the same figure, let it be given that M is the common midpoint of the segments PQ' and $P'Q$ on the line o and let $o \cdot l_\infty = C_\infty$. On M take any line $m \neq o$ and let $m \cdot l_\infty = M_\infty$. On l_∞ take another point A_∞ and construct its harmonic conjugate B_∞ with respect to C_∞, M_∞. Let $PA_\infty \cdot QB_\infty = R$, $PA_\infty \cdot Q'B_\infty = S$, $P'A_\infty \cdot QB_\infty = T$, $P'A_\infty \cdot Q'B_\infty = R'$. Now $RSR'T$ is a complete parallelogram having $C = RR' \cdot ST$ as a diagonal point. Suppose the perspectivity $(P, Q', C_\infty) \overset{S}{\underset{\wedge}{=}} (A_\infty, B_\infty, C_\infty)$ carries M into M'_∞. Since $H(P, Q'; C_\infty, M)$, then $H(A_\infty, B_\infty; C_\infty, M'_\infty)$. But by construction, $H(A_\infty, B_\infty; C_\infty, M_\infty)$; then $M'_\infty = M_\infty$ and S is on m. Similarly, we prove T on m and so C is on m. Now let $p \cdot l_\infty = C'_\infty$. Since $(R', R, C) \overset{T}{\underset{\wedge}{=}} (A_\infty, B_\infty, M_\infty)$ carries C'_∞ into C_∞ (why?), $C'_\infty = C_\infty$ and p is on C_∞. Then p is parallel to o; $PP'R'R$ and $QQ'R'R$ are parallelograms; and thus PP' and QQ' are congruent. We have proved

Theorem 13.9.   If the segments PQ' and $P'Q$ on the same line have a common midpoint, then the segments PP' and QQ' are congruent.

LENGTH

A construction of the parabolic projectivity

$$(A, B, C, F_1, F_2, \ldots) \; \overline{\underset{\wedge}{}} \; (A, C, F_1, F_2, F_3, \ldots)$$

is illustrated in Fig. 4-4, page 49. We now study this diagram in the affine plane (see Fig. 13-6) taking the line q as l_∞ and making certain helpful changes in notation, particularly for the points on o. By construction, A_0A_1 and A_1A_2 are congruent, A_1A_2 and A_2A_3 are congruent, $\ldots$ A_iA_{i+1} and $A_{i+1}A_{i+2}$ are congruent, $\ldots$. Then for any positive integer n, the segment A_0A_n on o consists of n non-overlapping congruent segments. We say

The *distance* A_0A_n is n times that of A_0A_1.

To construct A_0A_n when A_0A_1 is given, we have only to note that $H(M_\infty, A_1; A_0, A_2)$, $H(M_\infty, A_2; A_1, A_3), \ldots, H(M_\infty, A_{n-1}; A_{n-2}, A_n)$.

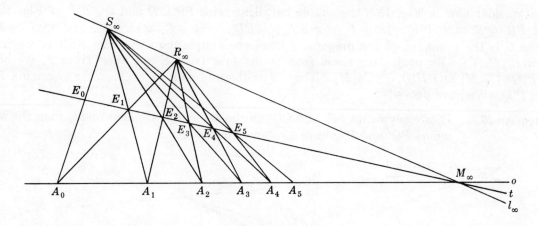

Fig. 13-6

Consider next the problem of separating a given segment A_0A_1 (see Fig. 13-7 below) into n congruent segments when n is any given positive integer. On A_0 take any line $p \neq o$ and on p take any point B_1. Then locate on p the point B_n such that B_0B_n is n times B_0B_1. Join A_1 and B_n meeting l_∞ in P_∞. Let $B_iP_\infty \cdot o = A_{i/n}$, $(i = 1, 2, 3, \ldots, n-1)$. Then A_0A_1 is n times $A_0A_{1/n}$, as required.

We are now in a position (see Chapter 7) to construct on o a point A_q such that A_0A_q is q times A_0A_1 for q any rational number. The existence of a point A_r on o such that A_0A_r is r times A_0A_1, for r any real number, is assured by Axiom 11, page 77.

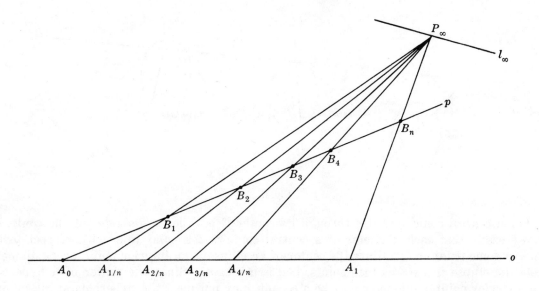

Fig. 13-7

The *length* of a segment A_0A_r may now be defined as $\left|\dfrac{A_0A_r}{A_0A_1}\right| = |r|$, where $|r|$ indicates the numerical value of r. Consider a segment A_0X on line o and a segment A_0Y on another line p meeting o at A_0. The length of A_0X may be found by choosing arbitrarily a point A_1 on the segment. Similarly, the length of A_0Y may be found by choosing arbitrarily a point B_1 on the segment. However, these lengths are of little significance since it is impossible to determine whether the unit of measure A_0A_1 on o is or is not congruent to the unit of measure A_0B_1 on p. In affine geometry, then, *segments on non-parallel lines cannot be compared.*

CONICS

In the projective plane all conics are equivalent; hence all conics have been diagrammed in Chapters 8-12 as curves which in analytic geometry are called ellipses. Let C be a conic in the projective plane. When we pass to the affine plane, either

  (a) C does not meet l_∞ and is called an *ellipse,*

  (b) C is tangent to l_∞ and is called a *parabola,*

or

  (c) C meets l_∞ in two distinct points and is called a *hyperbola.*

CENTER AND DIAMETERS

The pole, with respect to a given conic C, of l_∞ is called the *center* of C. The center C of an ellipse (see Fig. 13-8 below) is an interior point; the center C of a hyperbola is an exterior point; the pole, with respect to a parabola, of l_∞ is its point of contact C_∞ with the curve. Strictly speaking, a parabola has no center but we often find it convenient to speak of C_∞ as if it were the center. Ellipses and hyperbolas are called *central conics.*

When the conic is given completely, the center is constructed as the intersection of the polar lines of any two points on l_∞ (see Chapter 9); when the conic is not given completely, the simplest construction [see Problem 9.3(b), page 96] requires that three of its points be known.

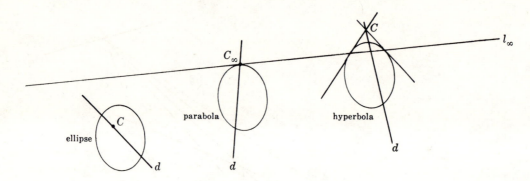

Fig. 13-8

For a central conic, any line through its center C is called a *diameter* of the conic. It follows easily that each diameter of a central conic is the polar line, with respect to the conic, of some point on l_∞. Since the center of an ellipse is an interior point, every diameter meets the ellipse in two distinct points, i.e., is a secant. Since the center of a hyperbola is an exterior point, a diameter may be a secant, may not meet the hyperbola at all, or may be tangent to the conic at one of its intersections with l_∞. These two tangents are called *asymptotes* of the hyperbola. Although the asymptotes are diameters, they have certain unique properties. As an example, the asymptotes separate those diameters which do not meet the hyperbola from those which are secants. For a parabola, the polar line of any point on l_∞ is a line on C_∞. Thus we may say that every line on the center of a conic is a diameter. Only in the case of a parabola are the diameters parallel to one another.

CONSTRUCTIONS

All constructions of Chapters 8-10 remain valid in affine geometry since it is possible to choose l_∞ so as neither to be on any given point nor to fall along any given tangent. On the other hand, from a projective construction a variety of affine constructions follow by choosing l_∞ properly.

Example 13.1.

Consider Problem 10.5, page 108: Given the points $A_{1,2}, A_3, A_4, A_5$ and the tangent $t_{1,2}$, construct another point on the conic. By choosing l_∞ as indicated, the following affine constructions result:

(1) Given the center and three points of a parabola, construct another point of the conic.

   Take l_∞ along $t_{1,2}$.

(2) Given three points on a hyperbola, the tangent at one of these points, and the ideal point on one asymptote, construct another point on the conic.

   Take l_∞ on A_3.

(3) Given two points on a hyperbola, the tangent at one of these points and the ideal points on both asymptotes, construct another point on the conic.

   Take l_∞ on A_3 and A_4.

(4) Given three points and an asymptote of a hyperbola, construct another point on the conic.

   Take $l_\infty \neq t_{1,2}$ on $A_{1,2}$.

(5) Given two points, an asymptote, and the ideal point on the other asymptote of a hyperbola, construct another point on the conic.

   Take l_∞ on $A_{1,2}$ and A_3.

(6) Given three points on a hyperbola, the tangent at one of these points, and the ideal point on one asymptote, construct the ideal point on the other asymptote.

   Take l_∞ on A_5 along $l_{5,6}$.

CONJUGATE DIAMETERS

Let P and Q be distinct points on a conic and call the segment PQ a *chord* of the conic. Consider now a system of parallel chords of a given conic. The projective lines on which these chords lie are on a common point, say A_∞, and the polar line of A_∞ with respect to the conic is on the midpoint of each chord. Thus, we have

Theorem 13.10. The midpoints of a system of parallel chords of a conic C are on a diameter of C.

As a consequence, the construction of a diameter of a given conic reduces to joining the midpoints of any two distinct parallel chords of the conic while the construction of the center reduces to intersecting any two distinct diameters of the conic.

Consider in Fig. 13-9 the central conic C and any one of its diameters d. Let $d \cdot l_\infty = D_\infty$. Now the polar line of D_∞ with respect to C is generally another diameter d' of C. (The reader should verify that this is true except when d is an asymptote of a hyperbola. To avoid the listing of such exceptions, we will work hereafter under the following convention: Whenever an arbitrary tangent or diameter of a central conic is given, the tangent or diameter is *not* to be taken as an asymptote of a hyperbola.) Let $d' \cdot l_\infty = D'_\infty$. Then, with respect to C, the line d is conjugate to d'. (Why?) As a consequence, d and d' are called a pair of *conjugate diameters* of C. Continuing in this manner, the diameters of C may be separated into conjugate pairs. Suppose now that we begin with the diameter d'. It is left for the reader to show that its conjugate is d and, thus, to establish

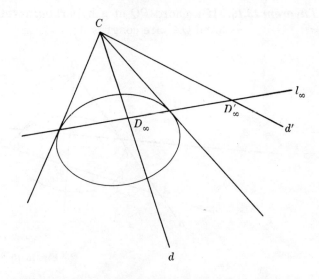

Fig. 13-9

Theorem 13.11. The pairs of conjugate diameters of a central conic C are reciprocal pairs of an involution on C, the center of C.

When C is an ellipse, the involution of Theorem 13.11 is without double elements and is called *elliptic*. When C is a hyperbola, the involution has two distinct double elements (the asymptotes) and is called *hyperbolic*.

We leave for the reader the proof of

Theorem 13.12. Any parallelogram inscribed in a conic has two diameters of the conic as diagonals and lines parallel to a pair of conjugate diameters as sides.

HYPERBOLAS

In this section we consider theorems concerned with the asymptotes of a hyperbola. Many of these arise simply as affine interpretations of projective theorems. For example, Theorem 9.13′ yields

Theorem 13.13. If a tangent to a hyperbola meets the asymptotes in A and A', then any two parallel lines on A and A' constitute a conjugate pair.

since the center of the hyperbola and any point on l_∞ constitute a conjugate pair.

Again, the theorem of Problem 9.30, page 101, yields

Theorem 13.14. Any pair of conjugate diameters of a hyperbola separate harmonically the asymptotes.

In Problem 13.8, we prove

Theorem 13.15. If the tangent to a hyperbola at M meets the asymptotes a and a' in A and A' respectively and if X is any point of the line on M and parallel to a', then the polar line x of X is on A' and parallel to AX.

Consider now in Fig. 13-10 the hyperbola C with asymptotes a and a'. Let d and d' be a pair of conjugate diameters meeting l_∞ in D_∞ and D'_∞ respectively. On D'_∞ take any line o meeting C in P, Q and the asymptotes in A, A'. Then $M = o \cdot d$ is the midpoint of the chord PQ and, by Theorem 13.14, is also the midpoint of the segment AA'. By Theorem 13.9 the segments AP and QA' are congruent. We have proved

Theorem 13.16. If a chord PQ of a hyperbola meets the asymptotes in A and A', then AP and QA' are congruent.

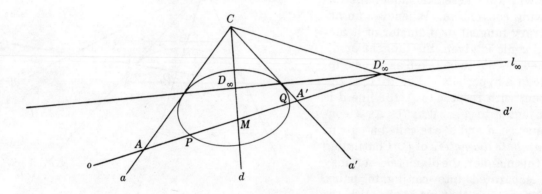

Fig. 13-10

As a special case of Theorem 13.16 or by an independent proof, we have

Theorem 13.17. The point of contact is the midpoint of the segment of a tangent to a hyperbola intercepted by the asymptotes.

Solved Problems

13.1. Prove: The line on the midpoints of two sides of a triangle is parallel to the third side.

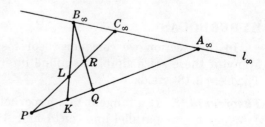

Fig. 13-11

Consider in Fig. 13-11 the triangle PQR and let $PQ \cdot l_\infty = A_\infty$, $QR \cdot l_\infty = B_\infty$, $RP \cdot l_\infty = C_\infty$. Denote by K the midpoint of PQ and by L the midpoint of RP. Since $H(P, Q; A_\infty, K)$ and $H(P, R; C_\infty, L)$, it follows that K and L are corresponding points in the perspectivity

$$(P, Q, A_\infty, \ldots) \stackrel{B_\infty}{\overline{\wedge}} (P, R, C_\infty, \ldots)$$

Then K, L, B_∞ are collinear and, hence, KL is parallel to QR.

This (affine) theorem may also be obtained from the projective theorem stated in the note to Problem 4.2, page 51, by taking as l_∞ the line cutting the sides of the triangle. For, then, the harmonic conjugates of the intersections are midpoints of the sides and the join of any two midpoints, being collinear with the ideal point on the third side of the triangle, is parallel to that side.

13.2. Given a triangle PQR and a line o not on any vertex. Let $PQ \cdot o = A$, $QR \cdot o = B$, $RP \cdot o = C$, $o \cdot l_\infty = O_\infty$ and let T be any other point on o. If A', B', C' are points on o such that each of the segments AA', BB', CC' has T as midpoint, show that the lines PB', QC', RA' are concurrent.

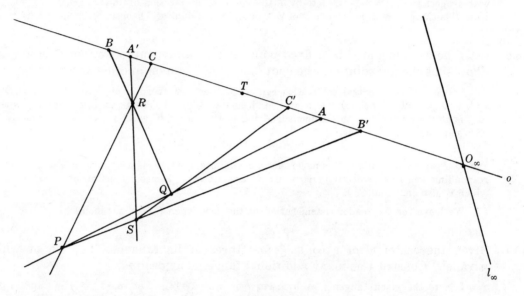

Fig. 13-12

Refer to Fig. 13-12. Let $PB' \cdot RA' = S$ and consider the complete quadrangle $PQRS$. Its pairs of opposite sides meet o in three reciprocal pairs of an involution on o of which A, A' and B, B' are two of the pairs. Since this involution has T, O_∞ as double points, the third pair is C, C'. Then $QS = C'$ and QC' is on S as required.

This affine theorem is a byproduct of the problem of constructing in the projective plane the companions of A, B, C in the hyperbolic involution on o having T and U as double points, the theorem resulting when U is taken as the ideal point on o.

13.3. If the midpoints of the sides of a triangle PQR are P', Q', R' respectively, show that

$$H(P'P, P'R; P'Q', P'R')$$

Refer to Fig. 13-13 in which $PR \cdot l_\infty = C_\infty$. By Theorem 13.6, $P'R'$ is parallel to PR and, hence, is also on C_∞. Now

$$(P'P, P'R, P'Q', P'R') \; \overline{\overline{\lambda}} \; (P, R, Q', C_\infty)$$

and, by construction, Q' is the midpoint of the segment PR. Then $H(P, R; Q', C_\infty)$ and so $H(P'P, P'R; P'Q', P'R')$ as required.

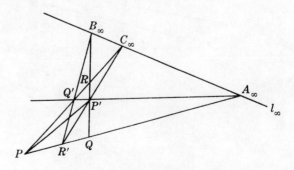

Fig. 13-13

13.4. Prove: If a diameter d of a conic C meets it in distinct points, the tangents drawn to C at these points are parallel.

Let d meet C in the points P and R. Since $d = PR$ is on C, the center of C, its pole with respect to C is some point, say A_∞, on l_∞. By Theorem 9.9, page 94, the tangents to C at P and Q are on A_∞ and, hence, are parallel.

When C is a parabola, one of its tangents on any point of l_∞ is l_∞ itself; hence two tangents to a parabola are never parallel.

13.5. Prove: If AA' is any chord of a conic C whose midpoint is on a diameter d of C, the tangents to C at A and A' intersect on d.

Let $AA' \cdot d = A''$ and $AA' \cdot l_\infty = A_\infty$; then $H(A, A'; A_\infty, A'')$ and d is the polar line of A_∞ with respect to C. (Why?) Denote by T the point of intersection of the tangents to C at A and A'. Now T is the pole with respect to C of AA' and, by the dual of Theorem 9.9, page 94, is on d.

13.6. Let C be a conic and P be a fixed point, not on an asymptote when C is a hyperbola. Show that the midpoints of the chords of C on P lie on another conic.

Suppose C is a central conic with center C. Let any chord x on P meet C in Q, Q' and let $x \cdot l_\infty = X_\infty$. (We do not exclude P on C and, hence, $Q = P$.) The polar line x' of X_∞ with respect to C is on the midpoint of the segment QQ' and is also on C. Now let x vary on P; then by Theorem 9.10, page 94,

$$P(x) \ \overline{\overline{\wedge}} \ l_\infty(X_\infty) \ \overline{\wedge} \ C(x')$$

and so $P(x) \overline{\wedge} C(x')$. Since P is never on an asymptote in case C is a hyperbola, PC is never self-corresponding and this projectivity is never a perspectivity. Hence by definition, $x \cdot x'$ is on a conic which is also on P and C.

We leave for the reader consideration of the case when C is a parabola.

13.7. Given the center C of a conic C and three of its tangents, no two of which are parallel. Construct as many additional tangents as desired.

Denote the given tangents as a_1, a_3, a_5 (see Fig. 13-14). Let $a_i \cdot l_\infty = A_{\infty,i}$ and $CA_{\infty,i} = c_i$, $(i = 1, 3, 5)$. On $A_{\infty,1}$ there is a second tangent to C; call it a_2. Now the polar line of $A_{\infty,1}$ is on C and also on the points of contact A_1, A_2 of the tangents a_1, a_2. Let $A_1 A_2 \cdot l_\infty = B_\infty$. Then since $H(C, B_\infty; A_1, A_2)$ also $H(c_1, l_\infty; a_1, a_2)$, and a_2 can be constructed. Similarly, the tangent a_4 parallel to a_3 can be constructed and then additional tangents may be obtained using Brianchon's theorem.

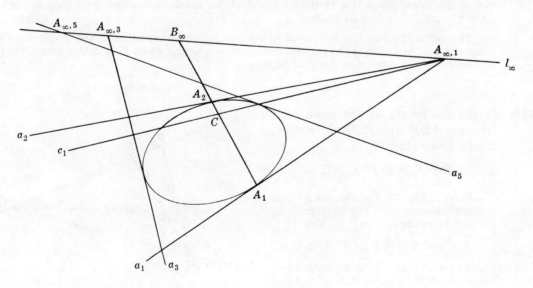

Fig. 13-14

13.8. Prove: If the tangent to a hyperbola at M meets the asymptotes a and a' at A and A' respectively and if X is any point of the line on M and parallel to a', then the polar line x of X is on A' and parallel to AX.

Refer to Fig. 13-15 in which m is the tangent at M and X is any point on MA'_∞. Let $AX \cdot l_\infty = B_\infty$. We are to prove that $x = A'B_\infty$.

With respect to the hyperbola the lines AB_∞ and $A'B_\infty$ are a conjugate pair by Theorem 13.13, while the lines $A'B_\infty$ and MA'_∞ are a conjugate pair since A' is the pole of MA'_∞. Then the pole of $A'B_\infty$ is $AB_\infty \cdot MA'_\infty = X$ or, what is the same, the polar line x of X is $A'B_\infty$.

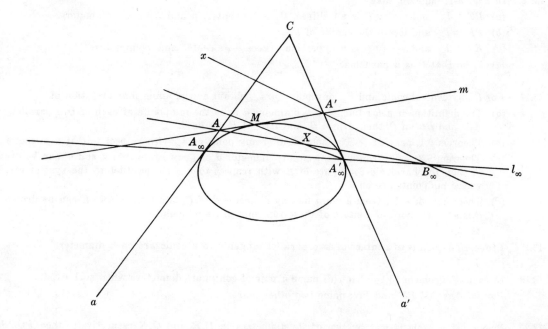

Fig. 13-15

Supplementary Problems

13.9. In the projective plane having one of its lines labeled l_∞, let there be given a line p meeting l_∞ in P_∞, four distinct points A, B, C, D on p, and a point E on neither line.

(a) Construct the line q on E and parallel to p.

(b) Construct the point F such that $ABEF$ is a parallelogram.

(c) Construct the point G such that $AEGB$ is a parallelogram.

(d) Give two proofs that E is the midpoint of the segment FG.

(e) Give a procedure for proving the segments AB and CD either congruent or not congruent.

13.10. Prove: If $PQRS$ is a parallelogram, then the complete quadrangle $PQRS$ has two of its diagonal points on l_∞ and conversely.

13.11. Prove: The harmonic conjugate of the median on a vertex of a triangle, with respect to the two sides on that vertex, is parallel to the third side.

13.12. Prove: The midpoints of the segments intercepted on two given parallel lines r and s by all non-parallels to these lines are on a line parallel to the given lines.

Hint. On any non-parallel p let $r \cdot p = R$, $s \cdot p = S$, $RS \cdot l_\infty = L_\infty$, and let T be the midpoint of the segment RS. Examine the projections from $r \cdot s$ of these points on other non-parallels.

13.13. In Fig. 13-11 let $LA_\infty \cdot QR = M$ and prove: The length of the segment joining the midpoints of two sides of a triangle is one-half the length of the third side.

13.14. Prove: If three distinct points A_1, A_2, A_3 on line x and three distinct points A_1', A_2', A_3' on line y are such that $A_1 A_2', A_1' A_2$ are parallel and $A_2 A_3', A_2' A_3$ are parallel, so also are $A_1 A_3', A_1' A_3$.

13.15. Prove: Theorem 13.7, page 131.

13.16. In Fig. 9-1, page 90, take
(a) $PT = l_\infty$ and verify C is an ellipse, P_1 is its center, p and XX_1 are diameters.
(b) $PK = l_\infty$ and locate the center of C.
(c) $KT = l_\infty$ and verify C is a hyperbola. Locate its center and asymptotes.
(d) l_∞ so that C is a parabola.

13.17. For C any central conic and P_∞ any point on l_∞, obtain as the affine interpretation of
(a) The definition of polar line: — If a diameter d is on the midpoints of each of two parallel chords AA' and BB' of C, then $AB \cdot A'B'$ and $AB' \cdot A'B$ are on d.
(b) Theorem 9.1, page 91: — The polar line of any point on l_∞ with respect to C is a diameter of C.
(c) Theorem 9.3: — If, from a point on any diameter d of C, two tangents a and a' can be drawn to C, then the harmonic conjugate of d with respect to a, a' is parallel to the system of chords whose midpoints are on d.
(d) Theorem 9.4: — If, from a point on any diameter d of C, two tangents to C can be drawn, then d meets the chord of contact of these tangents in its midpoint.

13.18. Prove: The points of contact of two parallel tangents to a conic are on a diameter.

13.19. In each of Problem 13.16(a), (c), (d) name a pair of conjugate diameters when such exist.
Partial Ans. (c) PM and PN; name two other pairs.

13.20. Prove: If a conic is met by two of its diameters in P, R and Q, S respectively, then $PQRS$ is a parallelogram.

13.21. Prove: Theorem 13.12, page 135.

13.22. Prove: If, from each of two points on a diameter of a conic, two tangents (a and a', b and b') can be drawn, then the lines $(a \cdot b)(a' \cdot b')$ and $(a \cdot b')(a' \cdot b)$ are each parallel to the system of chords whose midpoints are on d.

In Problems 13.23-13.28, C is a central conic.

13.23. Prove: If P is any point on a diameter d of C, then the polar line of P is parallel to the conjugate diameter d'.
Hint. Let $d \cdot l_\infty = D_\infty$ and $d' \cdot l_\infty = D_\infty'$; then D_∞' is the pole of d.

13.24. Prove: If d and d' are a pair of conjugate diameters of C, then d' (d) is parallel to the system of parallel chords whose midpoints are on d (d').

13.25. Let d, any diameter of C, meet C in D and D' and let e and e' be any pair of conjugate diameters of C. Show that lines on D and D' parallel to each of e and e' meet in F and G on C and that FG is a diameter of C.

13.26. Let the points of contact of parallel tangents a and a' to C be A and A' respectively and let B be the point of contact of any third tangent b to C. Show that $a \cdot b$ is on the diameter parallel to $A'B$ and $a' \cdot b$ is on the diameter parallel to AB.

13.27. Let the points of contact of parallel tangents a and a' to C be A and A' respectively, let b be any other tangent to C, and let d be the diameter of C on $a \cdot b$. Locate the point of contact B of b.

13.28. If the conjugate diameters d and d' of C meet C in A, B and D, E respectively, if P is any other point on C and q is any line parallel to d', show that $R = AP \cdot q$ and $S = BP \cdot q$ are conjugate points with respect to C.

13.29. If M is the midpoint of a chord PQ of a parabola and if O is the pole of PQ, show that the parabola meets the segment OM in its midpoint.

13.30. Given four tangents of a parabola C, construct the diameter which is conjugate to one of the given tangents.
Hint. A fifth tangent is l_∞.

13.31. Given a diameter d of a parabola and (a) three of its points or (b) three of its tangents, construct the parabola.

13.32. Prove: If d and d' are a pair of conjugate diameters of a hyperbola, one intersects the hyperbola and the other does not.

13.33. A parallelogram has its sides parallel to the asymptotes of a hyperbola C. If one of its diagonals is a chord of C, show that the other diagonal is a diameter of C.
Hint. See Problem 10.25, page 111.

13.34. Prove: The point of contact is the midpoint of the segment of a tangent to a hyperbola intercepted by the asymptotes.

13.35. The line p on any given point A and parallel to an asymptote meets the hyperbola in P and the polar line of A in B. Show that P is the midpoint of the segment AB.

13.36. Given two tangents, the point of contact of one of them, and an asymptote of a hyperbola, construct:
(a) the point of contact of the second tangent,
(b) the tangent from any point on the asymptote,
(c) the ideal point on the other asymptote,
(d) the center of the hyperbola,
(e) the other asymptote.

13.37. Prove: The nine-point conic of a quadrangle with respect to l_∞ is on the vertices of the diagonal triangle and the midpoints of the six segments determined by the vertices of the quadrangle, i.e., the midpoints of the sides of the complete quadrangle.

13.38. Show that the joins of the midpoints of the opposite sides of the complete quadrangle of Problem 13.37 are diameters of the nine-point conic.

Chapter 14

Plane Euclidean Geometry

INTRODUCTION

Affine geometry has been derived from real projective geometry by (1) selecting in the projective plane an arbitrary line, denoted thereafter as l_∞, and (2) defining as parallel any two affine lines x and y provided the projective lines x and y intersect on l_∞. In this geometry, two segments may be compared if and only if they are on the same line or on parallel lines.

Affine geometry becomes Euclidean geometry as soon as we have a means for comparing two segments on non-parallel lines. In turn, this will be provided as soon as we state when two lines are perpendicular. For, then, ellipses can be separated into two classes — circles and non-circles — and two non-parallel segments OA and OB can be defined as congruent by rotation provided A and B are on a circle whose center is O.

PERPENDICULAR LINES

The following properties of perpendicular lines in elementary metric geometry indicate how they may be defined:

If line a is perpendicular to line a', then a' is perpendicular to a.

No line is perpendicular to itself.

The first property suggests a connection between a pair of perpendicular lines and a reciprocal pair of an involution; the second property requires that this involution be elliptic.

We begin once more with the projective plane, having one of its lines labeled l_∞, and parallel lines defined as in Chapter 13. On l_∞ select an arbitrary elliptic involution to be denoted hereafter as $\mathcal{A}$. Call this involution the *absolute* or *orthogonal involution* and denote its reciprocal pairs as $\bar{A}_\infty, \bar{A}'_\infty; \bar{B}_\infty, \bar{B}'_\infty; \bar{C}_\infty, \bar{C}'_\infty; \ldots$. (In every diagram of Chapter 13, the line l_∞ appeared; in this chapter, the line l_∞ appears with at least two reciprocal pairs of the absolute involution located on it.)

Two affine lines x and x' are said to be perpendicular provided the projective lines x and x' are on a reciprocal pair of the absolute involution. In Fig. 14-1, the line a on $\bar{A}_\infty$ and the line a' on $\bar{A}'_\infty$ are perpendicular; similarly, b_1 and b' (also, b_2 and b') are perpendicular.

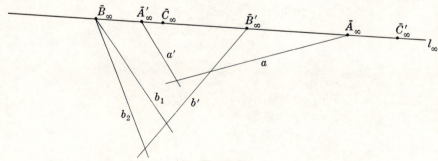

Fig. 14-1

142

There follow easily

Theorem 14.1. On any point there is one and only one line perpendicular to a given line.

Theorem 14.2. Two lines perpendicular to the same line are parallel.

Theorem 14.3. Any line perpendicular to one of two parallel lines is perpendicular to the other.

Theorem 14.4. In any pencil of lines, with center P, the pairs of perpendicular lines are reciprocal pairs of an involution on P.

The involution defined in Theorem 14.4, whose reciprocal pairs join P with reciprocal pairs of the absolute involution $\mathcal{A}$, is called the *circular involution* on P.

We now list the following definitions:

Any angle formed by a pair of perpendicular lines is called a *right angle*.

If PQ is a segment of a line o with midpoint M, the line on M and perpendicular to o is called the *perpendicular bisector* (*right bisector*) of the segment.

If two sides of a triangle are segments of perpendicular lines, the triangle is called a *right triangle*; otherwise, the triangle is called *oblique*.

The line on a vertex of a triangle and perpendicular to the opposite side is called an *altitude* of the triangle.

The intersection of an altitude, on any vertex, and the opposite side is called the *foot* of the altitude.

The feet of the altitudes of any oblique triangle are the vertices of the *pedal triangle* of the oblique triangle.

A parallelogram whose adjacent sides are perpendicular is called a *rectangle*.

In Problem 14.1, we prove

Theorem 14.5. The altitudes of any triangle are concurrent.

The point of concurrency in Theorem 14.5 is called the *orthocenter* of the triangle.

CIRCLES

Let P and Q be any two distinct ordinary points (see Fig. 14-2), let $PQ \cdot l_\infty = \bar{P}_\infty$, and denote as $\bar{P}'_\infty$ the mate of $\bar{P}_\infty$ in $\mathcal{A}$. Since

$$(\bar{A}_\infty, \bar{A}'_\infty, \bar{B}_\infty, \bar{B}'_\infty, \ldots, \bar{P}_\infty, \bar{P}'_\infty, \ldots) \; \bar{\wedge} \; (\bar{A}'_\infty, \bar{A}_\infty, \bar{B}'_\infty, \bar{B}_\infty, \ldots, \bar{P}'_\infty, \bar{P}_\infty, \ldots)$$

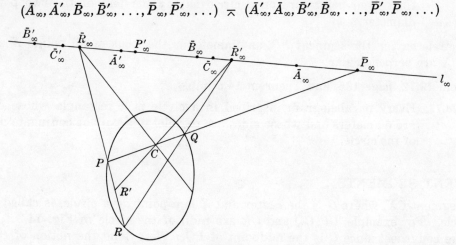

Fig. 14-2

we have

$$(P\bar{A}_\infty, P\bar{A}'_\infty, P\bar{B}_\infty, P\bar{B}'_\infty, \ldots, P\bar{P}_\infty, P\bar{P}'_\infty, \ldots) \; \overline{\overline{\wedge}} \; (\bar{A}_\infty, \bar{A}'_\infty, \bar{B}_\infty, \bar{B}'_\infty, \ldots, \bar{P}_\infty, \bar{P}'_\infty, \ldots)$$

$$\overline{\wedge} \; (\bar{A}'_\infty, \bar{A}_\infty, \bar{B}'_\infty, \bar{B}_\infty, \ldots, \bar{P}'_\infty, \bar{P}_\infty, \ldots) \; \overline{\overline{\wedge}} \; (Q\bar{A}'_\infty, Q\bar{A}_\infty, Q\bar{B}'_\infty, Q\bar{B}_\infty, \ldots, Q\bar{P}'_\infty, Q\bar{P}_\infty, \ldots)$$

and, hence,
$$(P\bar{A}_\infty, P\bar{A}'_\infty, P\bar{B}_\infty, P\bar{B}'_\infty, \ldots, P\bar{P}_\infty, P\bar{P}'_\infty, \ldots)$$
$$\overline{\wedge} \; (Q\bar{A}'_\infty, Q\bar{A}_\infty, Q\bar{B}'_\infty, Q\bar{B}_\infty, \ldots, Q\bar{P}'_\infty, Q\bar{P}_\infty, \ldots)$$

The conic defined by this projectivity is an ellipse (Why?) on the points P and Q. In the projectivity, the correspondent of PQ, as a member of the pencil on P, is the tangent $Q\bar{P}'_\infty$ while the correspondent, as a member of the pencil on Q, is the tangent $P\bar{P}'_\infty$. Since $P\bar{P}'_\infty$ and $Q\bar{P}'_\infty$ are a pair of parallel tangents, PQ is a diameter of the ellipse and C, the midpoint of PQ, is its center.

Let $\bar{R}_\infty, \bar{R}'_\infty$ be any other reciprocal pair of $\mathcal{A}$. Let $P\bar{R}_\infty \cdot Q\bar{R}'_\infty = R$ and $P\bar{R}_\infty \cdot C\bar{R}'_\infty = R'$. From
$$(P, Q, C, \bar{P}_\infty) \; \overset{\bar{R}'_\infty}{\overline{\wedge}} \; (P, R, R', \bar{R}_\infty)$$
we have $H(P, R; R', \bar{R}_\infty)$ so that R' is the midpoint of the chord PR. Then the polar line of $\bar{R}_\infty$ with respect to the ellipse is $CR' = C\bar{R}'_\infty$ and so the polar line of $\bar{R}'_\infty$ is $C\bar{R}_\infty$. Thus $C\bar{R}_\infty$ and $C\bar{R}'_\infty$ are a pair of conjugate diameters. Now $\bar{R}_\infty, \bar{R}'_\infty$ was any reciprocal pair of $\mathcal{A}$; hence the joins to C of every reciprocal pair of $\mathcal{A}$ constitute the involution of conjugate diameters of this ellipse. Hereafter, an ellipse whose pairs of conjugate diameters cut out on l_∞ the absolute involution or, what is the same, whose involution of conjugate diameters is a circular involution, will be called a *circle*. The term ellipse will now be restricted to any central conic which does not meet l_∞ and whose involution of conjugate diameters is not a circular involution.

From the above discussion, we obtain easily

Theorem 14.6. Any angle inscribed in a semicircle is a right angle.

Theorem 14.7. The perpendicular bisector of any chord of a circle is on its center.

Theorem 14.8. The tangent to a circle at an extremity of a diameter is perpendicular to the diameter.

Theorem 14.9. The diameter conjugate to a given chord of a circle is its perpendicular bisector.

Theorem 14.10. The points of intersection of every pair of perpendicular lines on two distinct fixed points are on a circle having the join of the two points as a diameter.

Thus the circle having the segment PP' as diameter is the locus of all points X such that PX and $P'X$ are perpendicular.

Theorem 13.12, page 135, and Theorem 14.6 imply

Theorem 14.11. Every parallelogram inscribed in a circle is a rectangle whose diagonals are diameters and whose sides are parallel to a pair of conjugate diameters of the circle.

CONGRUENT SEGMENTS

Any segment CX, where C is the center and X is a point on a circle, is called a *radius* of the circle. For example, CP, CQ and CR are radii of the circle in Fig. 14-2. Now CP and CQ are congruent since C is the midpoint of PQ. We extend the notion of congruent segments to (a) all radii of a given circle by defining

Two segments CX and CY are said to be *congruent by rotation* provided they are radii of the same circle.

and (b) two segments on non-parallel lines by defining

Two segments PQ and RS are said to be *congruent* provided there exist segments CX, congruent to PQ by translation, and CY, congruent to RS by translation, such that CX and CY are congruent by rotation.

A unit of measure selected for a given line may now be transferred to any other line of the plane. Thus any two segments may be compared since their lengths can be expressed in terms of a common unit of measure.

Let A, B, C be any three non-collinear points and let O be the point of intersection of the perpendicular bisectors of the segments AB and BC. Denote by C the circle having O as center and AB as a chord and denote by C' the circle having O as center and BC as a chord. (See Problem 14.2.) Since OB is a radius of both C and C', it follows that they are one and the same circle. We have proved

Theorem 14.12. Any three non-collinear points determine a unique circle.

ANGLES

In elementary geometry the angle at C in Fig. 14-3(a) is variously designated as $\angle C, \angle ACB, \angle BCA$. We shall use instead the directed angle, defined as follows:

The *directed angle* from AC to CB is that angle through which the line $a = AC$ must be rotated about C in the counterclockwise direction or sense to bring it into coincidence with $b = CB$.

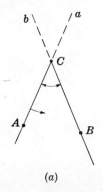 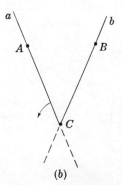

(a)   (b)

Fig. 14-3

This directed angle, always positive and less than 180°, will be denoted by (ab) or (ACB). It is important to note that the directed angle (ACB) is independent of the location of $A \neq C$ on a and of $B \neq C$ on b. For example, $(ab) = (ACB) = \angle ACB$ in Fig. 14-3(a) but $(ab) = (ACB)$ is the supplement of $\angle ACB$ in Fig. 14-3(b). Also, in Fig. 14-4, $\angle ACB$ and $\angle ADB$ are equal when C and D are on the same side of AB while $\angle ACB$ and $\angle ADB$ are supplementary when C and D are on opposite sides of AB. In either case, however, *the directed angles (ACB) and (ADB) are equal*. Thus, we have

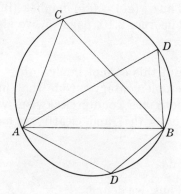

If four distinct points A, B, C, D are on a circle, then (ACB) and (ADB) are equal.

Fig. 14-4

We wish now to examine these angles in a projective setting. To motivate the definition which will be made later, the following problem is suggested: Given on the origin O

four lines a, b, a', b' with a' perpendicular
to a and b' perpendicular to b. Let the
equation of a be $y = mx$, $m = \tan \phi$, and
denote by θ the directed angle (ab). Obtain
the equations of b, a', b' and verify that
$(a, b; a', b')$ is a function of θ alone.

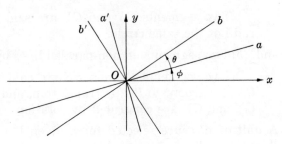

Fig. 14-5

Fig. 14-6 consists of a circle with four
distinct points A, B, C, D marked on it.
Consider on C the lines: $a = AC$, $b = CB$,
a' perpendicular to a and meeting the circle again in A', and b' perpendicular to b and meeting the circle again in B'. (Note that AA' and BB' are diameters of the circle.) Then join
D to A, B, A', B' to obtain the respective lines $\bar{a}, \bar{b}, \bar{a}', \bar{b}'$ of which $\bar{a}$ and $\bar{a}'$, also $\bar{b}$ and $\bar{b}'$, are
perpendicular. By Theorem 8.2, page 84, we have

$$\text{(i)} \qquad (a, b, a', b') \;\barwedge\; (\bar{a}, \bar{b}, \bar{a}', \bar{b}')$$

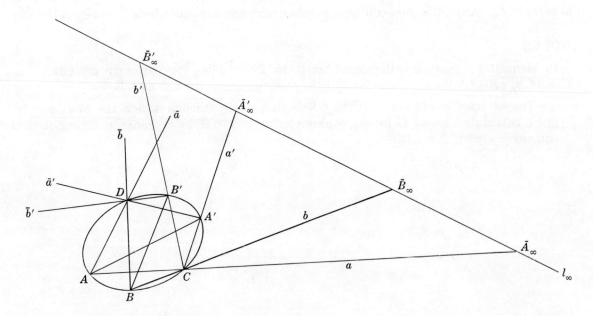

Fig. 14-6

Thus, $(ab) = (\bar{a}\bar{b})$ implies (i). The converse, however, is generally not true. For, using
Theorem 2.10, page 25, we also have

$$\text{(ii)} \qquad (a, b, a', b') \;\barwedge\; (\bar{b}, \bar{a}, \bar{b}', \bar{a}')$$

But this cannot imply $(ab) = (\bar{b}\bar{a})$ since then $(\bar{a}\bar{b})$ would be equal to $(\bar{b}\bar{a})$ whereas they are
supplementary. Note that in (i) we have the sense $\mathcal{S}(a, b, a', b') = \mathcal{S}(\bar{a}, \bar{b}, \bar{a}', \bar{b}')$, that is,
both are counterclockwise, whereas in (ii) we have $\mathcal{S}(a, b, a', b') \neq \mathcal{S}(\bar{b}, \bar{a}, \bar{b}', \bar{a}')$. Now
$(ba) = (\bar{b}\bar{a})$ and, clearly, we have

$$(b, a', b', a) \;\barwedge\; (\bar{b}, \bar{a}', \bar{b}', \bar{a})$$

and

$$\mathcal{S}(b, a', b', a) = \mathcal{S}(\bar{b}, \bar{a}', \bar{b}', \bar{a})$$

Thus, we define:

If two lines a, b and their respective perpendiculars a', b' are on point P and two
lines c, d and their respective perpendiculars c', d' are on point Q, then $(ab) = (cd)$
provided

$$(a, b, a', b') \; \overline{\wedge} \; (c, d, c', d')$$

and
$$\mathcal{S}(a, b, a', b') \; = \; \mathcal{S}(c, d, c', d')$$

obtains.

Example 14.1.

In Fig. 14-7 the lines a, b, a', b' on O are, in order, two lines and their respective perpendiculars. Since these lines meet l_∞ in two reciprocal pairs of the absolute involution $\mathcal{A}$, we have

$$(a, b, a', b') \; \overset{l_\infty}{\overline{\wedge}} \; (\bar{A}_\infty, \bar{B}_\infty, \bar{A}'_\infty, \bar{B}'_\infty) \; \overline{\wedge} \; (\bar{A}'_\infty, \bar{B}'_\infty, \bar{A}_\infty, \bar{B}_\infty) \; \overset{l_\infty}{\overline{\wedge}} \; (a', b', a, b)$$

Then
$$(a, b, a', b') \; \overline{\wedge} \; (a', b', a, b)$$

$$\mathcal{S}(a, b, a', b') \; = \; \mathcal{S}(a', b', a, b)$$

and so angles (ab) and $(a'b')$ are equal.

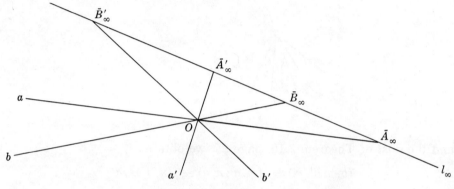

Fig. 14-7

Example 14.2.

Suppose a, c and b, d with a, b on a point O and c, d on another point O' are pairs of parallel lines. Then since

$$(a, b, a', b') \; \overset{l_\infty}{\overline{\wedge}} \; (\bar{A}_\infty, \bar{B}_\infty, \bar{A}'_\infty, \bar{B}'_\infty) \; \overset{l_\infty}{\overline{\wedge}} \; (c, d, c', d')$$

$$(a, b, a', b') \; \overline{\wedge} \; (c, d, c', d')$$

Also
$$\mathcal{S}(a, b, a', b') \; = \; \mathcal{S}(c, d, c', d')$$

since each agrees with $\mathcal{S}(\bar{A}_\infty, \bar{B}_\infty, \bar{A}'_\infty, \bar{B}'_\infty)$ on l_∞. Thus (ab) and (cd) are equal.

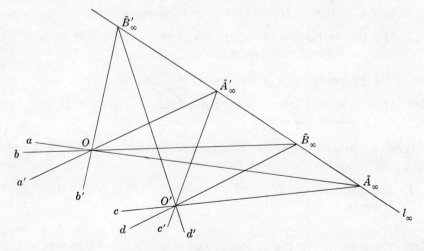

Fig. 14-8

BISECTORS OF ANGLES

If three concurrent lines a, b, m are such that (am) and (mb) are equal, the line m is said to *bisect* the angle (ab). Suppose m bisects (ab). Then (see Fig. 14-9) we are given

$$(a) \qquad (a, m, a', n) \; \overline{\wedge} \; (m, b, n, b')$$

and

$$(b) \qquad \mathscr{S}(a, m, a', n) \, = \, \mathscr{S}(m, b, n, b')$$

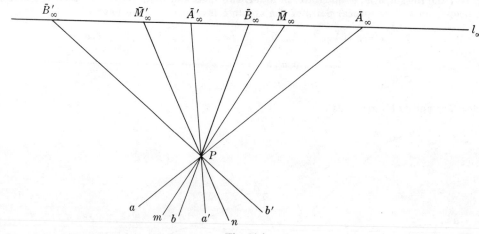

Fig. 14-9

From (a) and the dual of Theorem 2.10, page 25, we obtain

$$(b, n, b', m) \; \overline{\wedge} \; (m, a', n, a) \; \overline{\wedge} \; (n, a, m, a')$$

or

$$(b, n, b', m) \; \overline{\wedge} \; (n, a, m, a')$$

Since

$$\mathscr{S}(b, n, b', m) \, = \, \mathscr{S}(n, a; m, a')$$

it follows that (bn) and (na) are equal.

Again, by the dual of Theorem 2.10,

$$(a, m, a', n) \; \overline{\wedge} \; (m, b, n, b') \; \overline{\wedge} \; (b', n, b, m)$$

which exhibits m, n as a reciprocal pair of the involution $(a, a', b, b') \; \overline{\wedge} \; (b', b, a', a)$. But m, n is also a reciprocal pair in the circular involution $(a, a', b, b') \; \overline{\wedge} \; (a', a, b', b)$. By the dual of Theorem 6.13, page 67, the lines m, n are the double lines of the involution $(a, b, a', b') \; \overline{\wedge} \; (b, a, b', a')$; hence, $H(m, n; a, b)$. We have proved

Theorem 14.13. If a line m bisects the angle (ab), then (1) the line n, perpendicular to m, bisects (ba) and (2) $H(m, n; a, b)$.

The converse of Theorem 14.13 is also valid, that is,

Theorem 14.14. If $H(m, n; a, b)$ and m is perpendicular to n, then m bisects one of $(ab), (ba)$ and n bisects the other.

When m bisects (ab) and n bisects (ba), we call m and n the *internal* and *external* bisectors of (ab). We leave for the reader to prove

Theorem 14.15. The altitudes and sides of an oblique triangle bisect the angles of its pedal triangle.

and

Theorem 14.16. The bisectors of the angles of any triangle concur in sets of three to form a quadrangle.

THE AXES OF A CONIC

Consider a conic C with center C. On C are two involutions of lines of particular interest — the involution of conjugate diameters of C and the circular involution. Since the latter is always elliptic, these two involutions (see Theorems 6.11 and 6.12, page 67) will have at least one reciprocal pair in common. (Should they have two reciprocal pairs in common, the involutions are identical. This is the case when the conic is a circle.) Every ellipse (not a circle) and every hyperbola has, then, just one pair of perpendicular conjugate diameters. We define

> The *axes* of a central conic are the perpendicular conjugate diameters of the conic or, what is the same, an axis is a diameter which is perpendicular to the system of parallel chords which it bisects.

Each axis of an ellipse meets it in two points, called *vertices*. The *length* of an axis is the length of the segment joining these vertices. The lengths of the axes of an ellipse are unequal since, otherwise, we have a circle; the axis of greater length is called the *transverse (major) axis* while the other is called the *conjugate (minor) axis*. Only one axis of a hyperbola meets the curve (see Problem 13.32, page 141). This axis is called the *transverse axis*; the other is called the *conjugate axis*. The axis of a parabola is that diameter which is perpendicular to the system of chords which it bisects. (Why is there only one?) The *vertex* of a parabola is the ordinary point of intersection of the parabola and its axis.

To construct the axes of a central conic C, take any diameter d meeting C in the points A and B. The circle on AB as diameter either (a) has common tangents with C at A and B or (b) meets C in two additional points, say D and E. When (a) obtains, the line AB is perpendicular and conjugate to these tangents and, hence, to any chords parallel to them. Thus AB is an axis of C and its perpendicular bisector is the other. When (b) obtains, the points A, B, D, E are the vertices of a rectangle. By Theorem 14.10, the sides of the rectangle are parallel to a pair of conjugate diameters of the conic. These conjugate diameters, being perpendicular, are the required axes.

When the conic is a parabola meeting l_∞ in C_∞, take any diameter d and any chord AB perpendicular to d. Let M be the midpoint of AB; then the diameter on M is the required axis.

THE FOCI AND DIRECTRICES OF A CONIC

At any point P the involution of conjugate lines with respect to a given conic C contains one reciprocal pair of perpendicular lines. We define:

> A *focus* of a conic is a point at which the involution of conjugate lines with respect to the conic coincides with the circular involution.

The center of a circle is a focus but it is not immediately clear that such a point exists for any other type of conic.

Assume for the moment that the point F *is* a focus of a conic C which is not a circle. Any tangent to C on F would be self-conjugate and, hence, self-perpendicular. Thus,

Theorem 14.17. A focus of a conic is always an interior point.

Let C be the center of C, where $C = C_\infty$ in case C is a parabola. The chord on F, perpendicular to CF would be conjugate to CF and, hence, would be bisected by CF. Thus CF would be an axis of C and we have proved

Theorem 14.18. Any focus of a conic is on an axis.

Finally, suppose there were two foci F and F'. Then the perpendiculars to FF' at F and F' are conjugate to FF'. Since they are parallel, their intersection is, say F_∞, on l_∞. Thus FF' is a diameter and, hence, is an axis. We have proved

Theorem 14.19. If a conic has two foci F and F', then FF' is an axis.

Since no conic, except a circle, has more than two axes while a circle has but one focus, (see Problem 14.14), we conclude

Theorem 14.20. A conic can have no more than two foci.

There remains the problem of proving the existence of foci for conics other than circles. For this purpose, we make use of Theorem 9.13', page 95, in connection with Fig. 14-10.

(a) *Parabola.* Take l_∞ as u. Then C is a parabola with vertex V, v is the tangent at V, and t is any other tangent to C. The pole of the axis UV is $K = v \cdot l_\infty$; hence any point X on the axis together with K are a pair of conjugate points. Then, by Theorem 9.13', the lines $XL = X(t \cdot v)$ and $XM = X(t \cdot l_\infty)$ are a conjugate pair. When XL is perpendicular to t, XM is perpendicular to XL and X is a focus. In addition to establishing the existence of a focus, we have proved

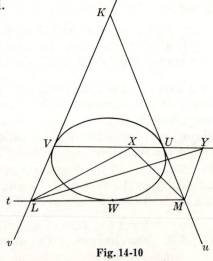

Fig. 14-10

Theorem 14.21. If t, any tangent to a parabola, meets the tangent at the vertex in L, then the join of L and the focus F is perpendicular to t.

Suppose there were a second focus X'. By Theorem 14.19 X' would be on the axis and then LX' would be perpendicular to t, contrary to Theorem 14.1. Thus the parabola has but one focus.

(b) *Hyperbola.* Take t as an asymptote and u, v tangents at the vertices U, V respectively. (Note that $KW = l_\infty$.) Now any point Y on the axis UV and K are a conjugate pair and Y will be a focus provided LY and MY are perpendicular. Reverting to the familiar sketch of a hyperbola in Plane Analytic Geometry, it is clear that the circle having LM as diameter meets UV in two points, say, Y_1 and Y_2. By Theorem 14.6, LY_1 and MY_1, also LY_2 and MY_2, are perpendicular; thus Y_1 and Y_2 are foci.

(c) *Ellipse.* Take u and v as tangents at the extremities U and V of the major axis and t the tangent at an extremity of the minor axis. The argument now follows closely that in (b), namely, the circle with diameter LM meets UV in points X_1 and X_2 which are the required foci.

The polar line of a focus of a conic is called a *directrix* of the conic associated with the focus. The directrix of a parabola is, then, perpendicular to the axis of the parabola; the directrices of a central conic, not a circle, are perpendicular to the transverse axis of the conic.

We conclude this brief study of the conics by stating a number of theorems, some of which may be new to the reader.

Theorem 14.22. Perpendicular tangents to a parabola meet on the directrix.

For a proof, see Problem 14.3.

Theorem 14.23. If two tangents to a parabola are perpendicular, the join of their points of contact is on the focus.

Theorem 14.24. The segment AP of a tangent to a parabola between the directrix and the point of tangency subtends a right angle at the focus.

Theorem 14.25. The orthocenter of a triangle circumscribed about a parabola is on the directrix.

For a proof, see Problem 14.4.

Theorem 14.26. The line joining a focus of a conic to the points of intersection of any two tangents makes equal angles with the lines joining the focus and the points of contact of the tangents.

Solved Problems

14.1. Prove: The altitudes of a triangle are concurrent.

The theorem is trivial if the triangle is a right triangle; suppose, then, the triangle PQR (see Fig. 14-11) is oblique.

Let $PQ \cdot l_\infty = \bar{A}_\infty$, $QR \cdot l_\infty = \bar{B}_\infty$, $RP \cdot l_\infty = \bar{C}_\infty$. Then the perpendicular on P to QR and the perpendicular on R to PQ meet l_∞ in $\bar{B}'_\infty$ and $\bar{A}'_\infty$ respectively. Let $P\bar{B}'_\infty \cdot R\bar{A}'_\infty = S$. By Theorem 6.3, page 66, the sides of the complete quadrangle $PQRS$ meet l_∞ in three reciprocal pairs of an involution; call this involution $\mathcal{J}$. By construction, two of these pairs are reciprocal pairs of the involution $\mathcal{A}$ whence, by Theorem 6.2, page 65, $\mathcal{J} = \mathcal{A}$. Then SQ, being on $\bar{C}'_\infty$, is perpendicular to PR and we have the theorem.

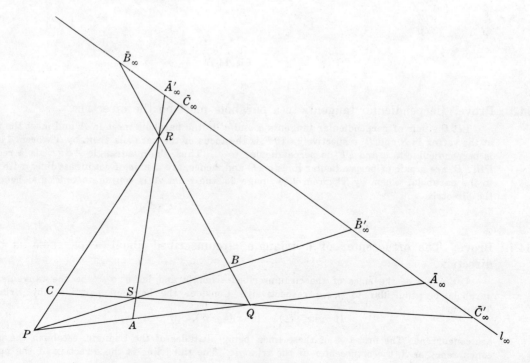

Fig. 14-11

14.2. Let P, R be distinct points and let O be any point on the perpendicular bisector p of PR. Show that there is a unique circle having O as center and PR as a chord.

Let $p \cdot PR = C$. The problem is trivial if $O = C$; suppose O is not on PR as in Fig. 14-12. Denote by C the unique circle on P and having O as center. (To construct this circle, let $OP \cdot l_\infty = \bar{P}_\infty$, construct Q such that $H(O, \bar{P}_\infty; P, Q)$, and proceed as in the section: Circles.) If C is on R, the problem is solved. Suppose, instead, that C meets PR again in $R' \neq R$. By Theorem 14.7, O (on p) is also on the perpendicular bisector of PR'. But then $H(R, P; C, A_\infty)$ and $H(R', P; C, A_\infty)$. Hence, $R' = R$ and C has PR as a chord.

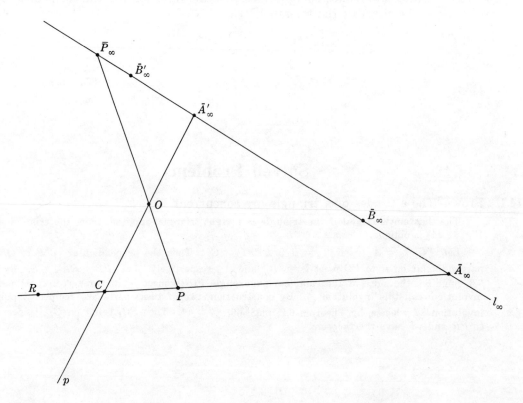

Fig. 14-12

14.3. Prove: Perpendicular tangents to a parabola meet on the directrix.

Let the pair of perpendicular tangents a and a' to the parabola meet in A and meet the tangent at the vertex in B and B' respectively. If F is the focus of the parabola then, by Theorem 14.21, FB is perpendicular to a and FB' is perpendicular to a'. Thus, the quadrangle $ABFB'$ is a rectangle; FB, FB' are a pair of perpendicular lines on F and, hence, are a pair of conjugate lines with respect to the parabola. Then, by Theorem 9.14', page 95, the point A is conjugate to F and hence is on the directrix.

14.4. Prove: The orthocenter of a triangle circumscribed about a parabola is on the directrix.

Let p, q, r be the sides of the circumscribed triangle and let p', q', r' be the tangents to the parabola perpendicular to p, q, r respectively. Consider the hexagon $pqrr'l_\infty p'$; by Brianchon's Theorem

$$(p \cdot q)(r' \cdot l_\infty), \quad (q \cdot r)(l_\infty \cdot p'), \quad (r \cdot r')(p' \cdot p)$$

are concurrent. The first two of these lines, being altitudes of the triangle, establish the point of concurrency as S, the orthocenter of the triangle. The third line is the directrix of the parabola; hence S is on the directrix as required.

Supplementary Problems

14.5.   Prove: The lines joining the orthocenter of a given oblique triangle to one of its vertices and to the ideal point on the side opposite that vertex are a reciprocal pair in the circular involution on the orthocenter.

14.6.   Given an oblique triangle PQR with orthocenter S, prove:
  (a) The points P, Q, R are respectively the orthocenters of the triangles QRS, PRS, PQS.
  (b) The pedal triangle of each of the triangles PQR, QRS, PRS, PQS is the diagonal triangle of the quadrangle $PQRS$.

14.7.   Prove: The perpendicular bisectors of the sides of a triangle are concurrent. The point of concurrency O is called the *circumcenter* of the triangle.
  Hint. Let the midpoints of the sides PQ, QR, RP of the triangle be D, E, F respectively. The perpendicular bisectors of the triangle PQR are the altitudes of its *medial triangle DEF*.

14.8.   Prove: Theorems $14.6, \ldots, 14.11$.

14.9.   Given in Fig. 14-13 a circle with center O, an inscribed triangle ABC, and the tangent t at A. Show that (bt) and (ac) are equal.

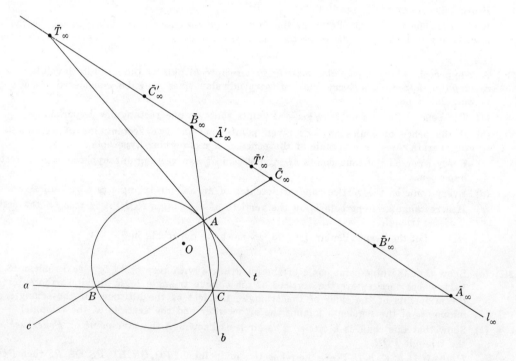

Fig. 14-13

14.10.   Prove: The polar line with respect to a circle of any point P is perpendicular to the diameter on P.

14.11.   Prove: The orthocenter of any self-polar triangle with respect to a circle coincides with the center of the circle.

14.12.   Prove Theorems $14.14, 14.15, 14.16$.

14.13.   In Fig. 14-11 locate the midpoints D and E of PQ and QR respectively. Verify that the triangles $RD\bar{A}'_\infty$ and $PE\bar{B}'_\infty$ are perspective from a point. Conclude that the centroid G, the circumcenter O, and the orthocenter S of the triangle PQR are on a line. This line is known as the *Euler line* of the triangle.

14.14. Let C be a circle, C be its center, and A be any other interior point; let p be any line on A and P be its pole with respect to C. Show that, generally, the pair of conjugate lines p and AP are not perpendicular; hence A is never a focus of C.

14.15. Given two circles C, with center C, and C', with center C', intersecting in P and Q. Prove:
(a) The tangents to C (C') at P and Q meet on the line of centers $c = CC'$.
(b) If the two tangents at P are perpendicular, they are singly on the centers C and C' as also are the tangents at Q.

14.16. Two intersecting circles whose tangents at each of their common points are perpendicular are called *orthogonal*. If the circles of Problem 14.15 are orthogonal, prove: Any diameter DD' of C (C') meets C' (C) in R, S such that $H(D, D'; R, S)$.

14.17. Prove: If C and C' are two circles meeting in P and Q, if a line d on the center of C meets C in D, D' and C' in R, S and if $H(D, D'; R, S)$, then C and C' are orthogonal.

14.18. Prove: If S is the orthocenter of the triangle PQR, then the circle C on the segment joining two vertices as diameter and the circle C' on the segment joining S and the third vertex as diameter are orthogonal.

14.19. Prove: Theorems 14.23, 14.24, 14.26.
Hint. For Theorem 14.26. Let F be the focus and d the associated directrix; t, t' be the tangents intersecting in U; T, T' be the respective points of contact; $TT' \cdot d = K$. Then K is the pole of FU.

14.20. Any hyperbola whose asymptotes meet l_∞ in a reciprocal pair of the absolute involution is called a *rectangular hyperbola*. Every pair of perpendicular lines may be considered as a degenerate rectangular hyperbola. Prove:
(a) The pencil of conics on four general points contains one rectangular hyperbola.
(b) If the pencil of conics on four given points contains two rectangular hyperbolas (including degenerate cases), every conic of the pencil is a rectangular hyperbola.
(c) Every conic on the four points of intersection of two rectangular hyperbolas is a rectangular hyperbola.
(d) Every conic on the vertices and orthocenter of an oblique triangle is a rectangular hyperbola.
(e) If a rectangular hyperbola is on the vertices of a triangle PQR, it is also on the orthocenter S of the triangle.
Hint. Let the perpendicular on P to QR meet the hyperbola in T.

14.21. (a) Show that the nine-point conic of the quadrangle with respect to l_∞ (see Problem 13.37, page 141), whose vertices are the vertices of the oblique triangle PQR and its orthocenter S, is on the midpoints of the sides of the triangle, the feet of the altitudes of the triangle, and the midpoints of the segments joining the orthocenter and the vertices of the triangle.
(b) Show that this conic is a circle. The circle is known as the *nine-point* or *Feuerbach circle* of the triangle PQR.
Hint. If D, E, F, T, U, V are the respective midpoints of PQ, QR, RP, PS, QS, RS then $DUVF$ and $TDEV$ are rectangles having DV as common diagonal.

Chapter 15

Analytic Projective Geometry

INTRODUCTION

In Chapter 1 attention was called to the fact that the Greeks interpreted arithmetic and algebra in terms of geometry. It is interesting to note that we have now come a full half turn, the tendency today being to solve any problem in geometry by algebraic means. In this and the next two chapters, we investigate some of the uses of algebraic methods in plane projective geometry.

In Chapter 7 we saw that a model of certain finite projective geometries could be built by defining an ordered triple of numbers to be a point, and a linear equation in three variables to be a line. In each case, the components of the triples and the coefficients in the equations were elements of a certain finite set. We shall proceed in a similar fashion except that the set of numbers with which we work is the set of all real numbers.

DEFINITIONS

We begin by defining in algebraic terms the basic ingredients of any plane geometry, namely, point, line, and the "on" relation. In so doing, it is essential we keep in mind that in our geometry point and line are to be dual elements.

A *point* is an ordered triple of real numbers, $(x_1, x_2, x_3) \neq (0, 0, 0)$, with the convention that (x_1, x_2, x_3) and $(\lambda x_1, \lambda x_2, \lambda x_3)$, where $\lambda \neq 0$, are the same point.

A *line* is an ordered triple of real numbers, $[X_1, X_2, X_3] \neq [0, 0, 0]$, with the convention that $[X_1, X_2, X_3]$ and $[\lambda X_1, \lambda X_2, \lambda X_3]$, where $\lambda \neq 0$, is the same line.

Thus $(1, 2, 3)$, i.e., "parentheses $1, 2, 3$" is a point while $[1, 2, 3]$, i.e., "brackets $1, 2, 3$" is a line.

The point $X: (x_1, x_2, x_3)$ is said to be *on* the line $x: [X_1, X_2, X_3]$ and, dually, the line $[X_1, X_2, X_3]$ is said to be *on* the point (x_1, x_2, x_3) provided

$$\{Xx\} = \{xX\} = x_1 X_1 + x_2 X_2 + x_3 X_3 = 0 \qquad (1)$$

Example 15.1.

(a) $(6, 8, -10), (3, 4, -5), (-1, -4/3, 5/3)$ are one and the same point.

(b) $[2, 1, 2], [6, 3, 6], [-1, -\frac{1}{2}, -1]$ are one and the same line.

(c) The point $P: (3, 4, -5)$ is on the line $p: [2, 1, 2]$ since

$$x_1 X_1 + x_2 X_2 + x_3 X_3 = 3 \cdot 2 + 4 \cdot 1 + (-5) \cdot 2 = 0$$

The numbers x_1, x_2, x_3 are called the *coordinates* of the point (x_1, x_2, x_3); the numbers X_1, X_2, X_3 are called the *coordinates* of the line $[X_1, X_2, X_3]$. When (x_1, x_2, x_3) is a variable point on the fixed line $[X_1, X_2, X_3]$, (1) may be written as

$$X_1 x_1 + X_2 x_2 + X_3 x_3 = 0$$

and is called the *equation* of the line.   When $[X_1, X_2, X_3]$ is a variable line on the fixed point (x_1, x_2, x_3),

$$x_1 X_1 + x_2 X_2 + x_3 X_3 = 0$$

is called the *equation* of the point.

Example 15.2.

(a) The equation of the line $[1, 2, 3]$ is $x_1 + 2x_2 + 3x_3 = 0$; the coordinates of the line $2x_1 - 4x_2 + 5x_3 = 0$ are $[2, -4, 5]$.

(b) The equation of the point $(2, -1, 0)$ is $2X_1 - X_2 = 0$; the coordinates of the point $X_1 - X_3 = 0$ are $(1, 0, -1)$.

Clearly, the dual of any discussion is obtained essentially by the simultaneous interchange of

      (i)   small and capital letters,

     (ii)   parentheses and brackets.

In the familiar Cartesian coordinate system each number pair (x, y) defines a unique point of the plane and, conversely, each point is defined by a unique number pair.   On the contrary, each linear equation $ax + by + c = 0$, where not both a and b are zero, defines a unique line while this same line is represented by a class of linear equations $\lambda ax + \lambda by + \lambda c = 0$, $\lambda \neq 0$, of which $ax + by + c = 0$ is one representation.   In the analytic geometry under discussion in this chapter, both points and lines have multiple representations.   However, in the interest of brevity we shall continue, as in Example 15.2, to write "the coordinates" and "the equation", meaning thereby some one of the class of representations of the point and line.

It should be expected, of course, that the analytic geometry of point and line triples will provide a model of the real synthetic geometry studied in the earlier chapters.   Verifications of a number of the axioms of Chapter 7 will be made in the Examples and Solved Problems of this chapter; verification of the remaining axioms are left for the reader to supply.

COLLINEAR POINTS AND CONCURRENT LINES

Let two points Y and Z be given by $Y: (y) = (y_1, y_2, y_3)$ and $Z: (z) = (z_1, z_2, z_3)$ respectively.   By definition, Y has an infinitude of representations $\alpha(y) = (\alpha y_1, \alpha y_2, \alpha y_3)$, where α is any non-zero real number, and similarly for Z.   Suppose, among these representations, we find

$$\alpha(y) = (\alpha y_1, \alpha y_2, \alpha y_3) = (\beta z_1, \beta z_2, \beta z_3) = \beta(z) \qquad (2)$$

Then Y and Z are coincident points; moreover, writing $\lambda = \alpha$ and $\mu = -\beta$, we have as a condition for coincidence the existence of non-zero real numbers λ, μ such that

$$\lambda(y) + \mu(z) = (0, 0, 0) = 0 \qquad (3)$$

If, on the other hand, no relation *(3)* exists, the points Y and Z are distinct.

Consider, next, three distinct points $Y: (y)$, $Z: (z)$, $W: (w)$.   Suppose, among their representation, we find $\alpha(y)$, $\beta(z)$, $\gamma(w)$ such that

$$\alpha(y) + \beta(z) + \gamma(w) = (0, 0, 0) = 0 \qquad (4)$$

that is,

$$\alpha y_1 + \beta z_1 + \gamma w_1 = 0$$

$$\alpha y_2 + \beta z_2 + \gamma w_2 = 0 \qquad (5)$$

$$\alpha y_3 + \beta z_3 + \gamma w_3 = 0$$

Although (5) always has the trivial solution $\alpha = \beta = \gamma = 0$, it is a theorem of algebra that the system will have a non-trivial solution if and only if the determinant

$$\begin{vmatrix} y_1 & z_1 & w_1 \\ y_2 & z_2 & w_2 \\ y_3 & z_3 & w_3 \end{vmatrix} = 0 \tag{6}$$

Now (6) implies

$$\begin{vmatrix} y_1 & y_2 & y_3 \\ z_1 & z_2 & z_3 \\ w_1 & w_2 & w_3 \end{vmatrix} = 0 \tag{7}$$

which, in turn, assures the existence of numbers X_1, X_2, X_3, not all zero, such that

$$X_1 y_1 + X_2 y_2 + X_3 y_3 = 0$$

$$X_1 z_1 + X_2 z_2 + X_3 z_3 = 0$$

$$X_1 w_1 + X_2 w_2 + X_3 w_3 = 0$$

But these are simply the conditions that the points $(y), (z), (w)$ be on the line $[X_1, X_2, X_3]$. Thus, we have

Theorem 15.1. A necessary and sufficient condition that the distinct points $(y), (z), (w)$ be collinear is

$$\begin{vmatrix} y_1 & y_2 & y_3 \\ z_1 & z_2 & z_3 \\ w_1 & w_2 & w_3 \end{vmatrix} = 0$$

There follows

Theorem 15.2. The line determined by the distinct points (y) and (z) has equation

$$\begin{vmatrix} x_1 & x_2 & x_3 \\ y_1 & y_2 & y_3 \\ z_1 & z_2 & z_3 \end{vmatrix} = \begin{vmatrix} y_2 & y_3 \\ z_2 & z_3 \end{vmatrix} x_1 + \begin{vmatrix} y_3 & y_1 \\ z_3 & z_1 \end{vmatrix} x_2 + \begin{vmatrix} y_1 & y_2 \\ z_1 & z_2 \end{vmatrix} x_3 = 0$$

and coordinates $\left[\begin{vmatrix} y_2 & y_3 \\ z_2 & z_3 \end{vmatrix}, \begin{vmatrix} y_3 & y_1 \\ z_3 & z_1 \end{vmatrix}, \begin{vmatrix} y_1 & y_2 \\ z_1 & z_2 \end{vmatrix} \right]$.

Example 15.3.

(a) The points $P: (2, 1, -3)$, $Q: (4, -2, 4)$, $R: (10, -1, 0)$ are distinct (check this) and are collinear since

$$\begin{vmatrix} 2 & 1 & -3 \\ 4 & -2 & 4 \\ 10 & -1 & 0 \end{vmatrix} = \begin{vmatrix} 12 & 1 & -3 \\ -16 & -2 & 4 \\ 0 & -1 & 0 \end{vmatrix} = 0$$

(b) The line joining P and Q has equation

$$\begin{vmatrix} x_1 & x_2 & x_3 \\ 2 & 1 & -3 \\ 4 & -2 & 4 \end{vmatrix} = \begin{vmatrix} 1 & -3 \\ -2 & 4 \end{vmatrix} x_1 + \begin{vmatrix} -3 & 2 \\ 4 & 4 \end{vmatrix} x_2 + \begin{vmatrix} 2 & 1 \\ 4 & -2 \end{vmatrix} x_3$$

$$= -2x_1 - 20x_2 - 8x_3 = 0$$

Another representation is $x_1 + 10x_2 + 4x_3 = 0$.

If, on the contrary, no relation (4) exists for the points Y, Z, W, they are non-collinear.

Finally, consider any four points $Y:(y)$, $Z:(z)$, $U:(u)$, $W:(w)$ of the plane. We show there always exist real numbers $\alpha, \beta, \gamma, \delta$, not all zero, such that

$$\alpha(y) + \beta(z) + \gamma(u) + \delta(w) = (0,0,0) = \mathbf{0} \tag{8}$$

Suppose some two of the points, say Y and Z, are coincident; then (8) holds for $\alpha \neq 0$, $\beta \neq 0$, $\gamma = \delta = 0$. Suppose, next, that no two of the points are coincident but Y, Z, U are collinear; then (8) holds for $\alpha \neq 0$, $\beta \neq 0$, $\gamma \neq 0$, $\delta = 0$. Lastly, suppose no three of the points are collinear; then, since

$$\begin{vmatrix} y_1 & z_1 & u_1 \\ y_2 & z_2 & u_2 \\ y_3 & z_3 & u_3 \end{vmatrix} \neq 0$$

the system of equations

$$\alpha y_1 + \beta z_1 + \gamma u_1 = -\delta w_1$$

$$\alpha y_2 + \beta z_2 + \gamma u_2 = -\delta w_2$$

$$\alpha y_3 + \beta z_3 + \gamma u_3 = -\delta w_3$$

has a unique solution α, β, γ for any $\delta \neq 0$, and (8) holds.

Dually, two lines $y:[Y]$ and $z:[Z]$ are coincident provided there exist non-zero real numbers λ, μ such that

$$\lambda[Y] + \mu[Z] = [0,0,0] = \mathbf{0} \tag{3'}$$

while three distinct lines $y:[Y]$, $z:[Z]$, $w:[W]$ are concurrent provided there exist non-zero real numbers α, β, γ such that

$$\alpha[Y] + \beta[Z] + \gamma[W] = \mathbf{0} \tag{4'}$$

It is left for the reader to obtain

Theorem 15.1'. A necessary and sufficient condition that the distinct lines $[Y], [Z], [W]$ be concurrent is

$$\begin{vmatrix} Y_1 & Y_2 & Y_3 \\ Z_1 & Z_2 & Z_3 \\ W_1 & W_2 & W_3 \end{vmatrix} = 0$$

and

Theorem 15.2'. The point determined by the distinct lines $[Y]$ and $[Z]$ has equation

$$\begin{vmatrix} X_1 & X_2 & X_3 \\ Y_1 & Y_2 & Y_3 \\ Z_1 & Z_2 & Z_3 \end{vmatrix} = \begin{vmatrix} Y_2 & Y_3 \\ Z_2 & Z_3 \end{vmatrix} X_1 + \begin{vmatrix} Y_3 & Y_1 \\ Z_3 & Z_1 \end{vmatrix} X_2 + \begin{vmatrix} Y_1 & Y_2 \\ Z_1 & Z_2 \end{vmatrix} X_3 = 0$$

and coordinates $\left(\begin{vmatrix} Y_2 & Y_3 \\ Z_2 & Z_3 \end{vmatrix}, \begin{vmatrix} Y_3 & Y_1 \\ Z_3 & Z_1 \end{vmatrix}, \begin{vmatrix} Y_1 & Y_2 \\ Z_1 & Z_2 \end{vmatrix} \right)$.

See Problem 15.1.

PENCILS OF POINTS AND LINES

In Problem 15.2, we prove

Theorem 15.3. If (y) and (z) are distinct points, then $\lambda(y) + \mu(z) = (\lambda y + \mu z)$, where λ, μ are any non-zero real numbers, is another point on the line determined by (y) and (z).

Let (y), (z), $(\lambda y + \mu z)$ be three distinct collinear points. Since $\lambda \neq 0$,

$$(\lambda y + \mu z) \quad \text{and} \quad \left(y + \frac{\mu}{\lambda} z\right) = (y + vz)$$

are the same point. Consider $(y + vz) = (y) + v(z)$ with (y) and (z) fixed and v a real parameter. When $v = 0$, $(y + vz)$ is the point (y); when $v \neq 0$, $(y + vz)$ is a point distinct from (y) and from (z); when $v_1 \neq v_2$ and $v_1 \cdot v_2 \neq 0$, the points (y), (z), $(y + v_1 z)$, $(y + v_2 z)$ are four distinct collinear points. Moreover, if $W \neq Z$ is on the line YZ it follows from (6) that $(w) = (y) + v(z)$ for some real value of v. Let us accept the convention: $(y + vz) = (z)$ when $v = \infty$. We have

Theorem 15.4.   If (y) and (z) are distinct points then, as v varies over the extended real number system (the set of all real numbers together with ∞), $(y + vz)$ defines the pencil of points on the line determined by (y) and (z).

Dually, we have

Theorem 15.3'.   If $[Y]$ and $[Z]$ are distinct lines, then $\lambda Y + \mu Z$, where λ, μ are any non-zero real numbers, is another line on the point determined by $[Y]$ and $[Z]$.

and

Theorem 15.4'.   If $[Y]$ and $[Z]$ are distinct lines then, as v varies over the extended real number system, $[Y + vZ]$ defines the pencil of lines on the point determined by $[Y]$ and $[Z]$.

Example 15.4.

(a) Locate the point of intersection T of the line r, joining the points $(3, 1, -2)$ and $(1, -5, 3)$, and the line $s: 2x_1 - 3x_2 - 4x_3 = 0$. (b) Find the equation of the line p on the point $R: (1, 2, -2)$ and the point of intersection of the lines $2x_1 - 3x_2 + 7z_3 = 0$ and $5x_1 + 2x_2 = 0$.

(a) Since T is on r, we have $T: (3 + \lambda, 1 - 5\lambda, -2 + 3\lambda)$ for some value of λ. Since T is also on s, then $2(3 + \lambda) - 3(1 - 5\lambda) - 4(-2 + 3\lambda) = 11 + 5\lambda = 0$ and $\lambda = -11/5$. Thus,

$$T: (3 - 11/5, 1 + 11, -2 - 33/5) = (4/5, 12, -43/5)$$

or $T: (4, 60, -43)$.

The problem can also be solved by writing the equation of r and solving simultaneously with the equation of s.

(b) The equation of any line concurrent with the two given lines is of the form $(2 + 5\lambda)x_1 + (-3 + 2\lambda)x_2 + 7x_3 = 0$. Since R is on the required line,

$$(2 + 5\lambda) \cdot 1 + (-3 + 2\lambda) \cdot 2 + 7(-2) = -18 + 9\lambda = 0 \quad \text{and} \quad \lambda = 2$$

Then $(2 + 5 \cdot 2)x_1 + (-3 + 2 \cdot 2)x_2 + 7x_3 = 12x_1 + x_2 + 7x_3 = 0$ is the equation of p.

See also Problem 15.3.

ANALYTIC PROOFS

Suppose we are concerned with distinct points on a line. By Theorem 15.3, we may take (y), (z), $\lambda(y) + \mu(z)$ to be three arbitrary points of the line. When more than three points are needed, it is simpler to take them as (y), (z), $(y) + \alpha(z)$, $(y) + \beta(z)$, $\ldots$, $(y) + \theta(z)$. When only three points are needed, a further simplification is possible. Consider the set $P: (y)$, $Q: (z)$, $R: (w) = \lambda(y) + \mu(z)$. Now $\lambda(y) = (\lambda y)$ is just another representation of the point P and $\mu(z)$ is just another representation of the point Q. Thus it is always possible to choose new representations $(y^*) = (\lambda y)$ for P and $(z^*) = (\mu z)$ for Q so that

$$(w) = (y^*) + (z^*)$$

or, with (y^*), (z^*) as before and $(w^*) = (-w)$ so that

$$(y^*) + (z^*) + (w^*) = (0, 0, 0) = 0$$

as best suits our purpose.

Example 15.5.

Given the collinear points $P\colon (y) = (2, 1, -3)$, $Q\colon (z) = (4, -2, 4)$, $R\colon (w) = (10, -1, 0)$ of Example 15.3, choose representations (y^*) and (z^*) for P and Q such that $R\colon (y^*) + (z^*)$.

Set $\lambda(y) + \mu(z) = (w)$ and solve $\begin{cases} 2\lambda + 4\mu = 10 \\ \lambda - 2\mu = -1 \\ -3\lambda + 4\mu = 0 \end{cases}$ for $\lambda = 2$, $\mu = 3/2$. Then $(y^*) = (2y) = (4, 2, -6)$

and $(z^*) = (3z/2) = (6, -3, 6)$ are the required representations.

We may, then, begin a demonstration by assuming $P\colon (y)$, $Q\colon (z)$, $R\colon (w)$ three distinct collinear points whose coordinates satisfy

$$\text{(i)} \qquad (w) = (y) + (z) = (y + z)$$

or satisfy

$$\text{(ii)} \qquad (y) + (z) + (w) = (0, 0, 0) = \mathbf{0}$$

Similarly, if P, Q, R, S are four points, no three of which are collinear, it follows from (8) that representations $(y), (z), (u), (w)$ respectively may be chosen so that

$$\text{(iii)} \qquad (y) + (z) + (u) + (w) = (0, 0, 0) = \mathbf{0}$$

or such that

$$\text{(iv)} \qquad (w) = (y) + (z) + (u)$$

With respect to lines, the dual of each of (i)-(iv) obtains.

Example 15.6.

Given a triangle ABC and a point P not on any side. Let $AP \cdot BC = A'$, $BP \cdot CA = B'$, $CP \cdot AB = C'$; $BC \cdot B'C' = A''$, $CA \cdot C'A' = B''$, $AB \cdot A'B' = C''$. Show that A'', B'', C'' are collinear.

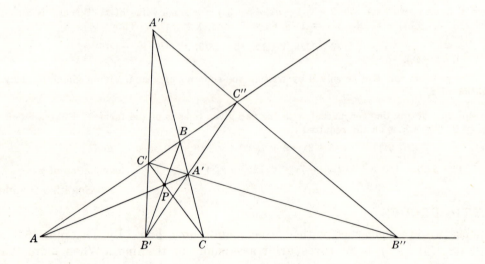

Fig. 15-1

Take $A\colon (a)$, $B\colon (b)$, $C\colon (c)$ so that $P\colon (a + b + c)$; then $A'\colon (b + c)$, $B'\colon (c + a)$, $C'\colon (a + b)$. Since A'' is on BC, its coordinates are of the form $\lambda(b) + \mu(c)$; since A'' is also on $B'C'$, we can only have $A''\colon (b - c)$. Similarly, $B''\colon (c - a)$ and $C''\colon (a - b)$. Collinearity follows from

$$\begin{vmatrix} A'' \\ B'' \\ C'' \end{vmatrix} = \begin{vmatrix} b_1 - c_1 & b_2 - c_2 & b_3 - c_3 \\ c_1 - a_1 & c_2 - a_2 & c_3 - a_3 \\ a_1 - b_1 & a_2 - b_2 & a_3 - b_3 \end{vmatrix} = 0$$

or $A'' + B'' + C'' = (b - c) + (c - a) + (a - b) = \mathbf{0}$.

In Problem 15.4 we verify Axiom 6, page 76, (Descartes' Two-Triangle Theorem). In Problem 15.5 we give another proof and in Problem 15.6 investigate multiply perspective triangles (see Problem 3.34, page 44).

Axiom 7, page 76, is verified in Problem 15.7. In Problem 15.8 we prove

Theorem 15.5. The harmonic conjugate of $D: (a) + \lambda(b)$ with respect to $A: (a)$ and $B: (b)$ is $E: (a) - \lambda(b)$.

Dually, we have

Theorem 15.5'. The harmonic conjugate of $d: [A] + \lambda[B]$ with respect to $a: [A]$ and $b: [B]$ is $e: [A] - \lambda[B]$.

Example 15.7.

Given the collinear points $A: (a) = (1, 2, 3)$, $B: (b) = (2, -1, -2)$, $C: (c) = (8, 1, 0)$, locate the harmonic conjugate D of C with respect to A and B.

If C is given by the representation $(a) + \lambda(b)$ then, by Theorem 15.5, a representation of D is $(a) - \lambda(b)$. Set $(c) = \alpha(a) + \beta(b)$ and solve $\begin{cases} \alpha + 2\beta = 8 \\ 2\alpha - \beta = 1 \\ 3\alpha - 2\beta = 0 \end{cases}$ for $\alpha = 2$, $\beta = 3$; then $C: (a) + \dfrac{\beta}{\alpha}(b) =$ $(a) + \frac{3}{2}(b)$ and $D: (a) - \frac{3}{2}(b) = (-2, 7/2, 6)$. Another representation is $D: (-4, 7, 12)$.

CROSS RATIO

On the line determined by $A: (a)$ and $B: (b)$, take four points $P: (a) + \alpha(b)$, $Q: (a) + \beta(b)$, $R: (a) + \gamma(b)$, $S: (a) + \delta(b)$. We define the *cross ratio* of these points in terms of the parameters $\alpha, \beta, \gamma, \delta$, as

$$(P, Q; R, S) \;=\; (\alpha, \beta; \gamma, \delta) \;=\; \frac{(\alpha - \gamma)(\beta - \delta)}{(\alpha - \delta)(\beta - \gamma)} \tag{9}$$

Now there is no essential difference *in form* between (9) above and (5), page 20, of Chapter 2. For, if A, B, C, D are at distances a, b, c, d respectively from a fixed point O on the line, then (5) becomes

$$(A, B; C, D) \;=\; \frac{AC \cdot BD}{AD \cdot BC} \;=\; \frac{(c - a)(d - b)}{(d - a)(c - b)}$$

As a consequence, Theorems 2.2-2.5 hold here without change. Moreover, when the points P, Q, R are held fixed while $S \to B$ (i.e. $\delta \to \infty$), we obtain

$$(P, Q; R, B) \;=\; \frac{\alpha - \gamma}{\beta - \gamma}$$

Finally, if $R = A$, $S = B$, $P: (a) + \alpha(b)$, $Q: (a) - \alpha(b)$, (i.e. $H(P, Q; A, B)$ by Problem 15.8), then $(P, Q; A, B) = -1$.

PROJECTIVITIES

The Fundamental Theorem, page 77, is equivalent to

A projectivity between two pencils of points is completely determined by three distinct pairs of corresponding points.

and it is this form which we now verify.

We begin with the most primitive projectivity — the elementary perspectivity between a pencil of lines on a point P and the pencil of points obtained when the pencil of lines is cut by any line p, not on P. By the dual of (i), page 160, we may select any three lines

q, r, s on P and choose representations so that $q: [Y]$, $r: [Z]$, $s: [Y + Z]$. Let $p \cdot q = Q$, $p \cdot r = R$, $p \cdot s = S$. Since Q, R, S are collinear, we may choose representations [see (i)] so that $Q: (y)$, $R: (z)$, $S: (y + z)$. We have then

$$Q \text{ on } q \text{ so that } \quad \{Yy\} = 0,$$

$$R \text{ on } r \text{ so that } \quad \{Zz\} = 0,$$

$$S \text{ on } s \text{ so that } \quad \{(Y + Z)(y + z)\} = \{Yy\} + \{Zz\} + \{Yz\} + \{Zy\}$$

$$= \{Yz\} + \{Zy\} = 0$$

Let $t: [Y + \lambda Z]$ be any other line on P. We are required to show that $p \cdot t = T: (y + \lambda z)$. This follows since

$$\{(Y + \lambda Z)(y + \lambda z)\} = \{Yy\} + \lambda(\{Yz\} + \{Zy\}) + \lambda^2\{Zz\} = 0$$

Now join the points Q, R, S to any point $W \neq P$ and not on p, and cut by any line w not on W in the points (y'), (z'), $(y' + z')$ respectively. We leave for the reader to show that the join of any other point $(y + \lambda z)$ on p and W is met by w in the point $(y' + \lambda z')$. Thus,

$$(y), (z), (y + z), (y + \lambda z) \;\overline{\overline{\wedge}}\; (y'), (z'), (y' + z'), (y' + \lambda z')$$

Moreover, it is clear that another projection and section would result in

$$(y), (z), (y + z), (y + \lambda z) \;\overline{\overline{\wedge}}\; (y''), (z''), (y'' + z''), (y'' + \lambda z'')$$

and so $$(y'), (z'), (y' + z'), (y' + \lambda z') \;\overline{\wedge}\; (y''), (z''), (y'' + z''), (y'' + \lambda z'')$$

It is clear from the above discussion that in a projectivity between two pencils of points

$$(A, B, C, D, \ldots) \;\overline{\wedge}\; (A', B', C', D', \ldots)$$

on distinct lines or on the same line, matters can always be arranged so that the parameter defining any point, say P, in terms of A and B is the same as that defining P', the correspondent of P, in terms of A' and B'; in brief, a projectivity preserves parameters. It follows, then, that projectivities preserve cross ratios; that is,

$$(P, Q, R, S) \;\overline{\wedge}\; (P', Q', R', S')$$

implies $$(P, Q; R, S) = (P', Q'; R', S')$$

Conversely, suppose $(P, Q; R, S) = (P', Q'; R', S')$. We may always choose representations $R: (r)$, $S: (s)$ so that $Q: (r + s)$; and $R': (r')$, $S': (s')$ so that $Q': (r' + s')$. Suppose $P: (r + \lambda s)$. Since $(P, Q; R, S) = (\lambda, 1; 0, \infty) = \lambda$, then $(P', Q'; R', S') = \lambda$ and $P': (r' + \lambda s')$. Thus the correspondence between P, Q, R, S and P', Q', R', S' is a projectivity and we have

$$(P, Q; R, S) = (P', Q'; R', S')$$

implies $$(P, Q, R, S) \;\overline{\wedge}\; (P', Q', R', S')$$

SEPARATION

Let $A: (a)$, $B: (b)$, $D: (a + \lambda b)$, $E: (a + \mu b)$ be four distinct points on a line. We define:

The pair of points D, E is said to *separate* the pair A, B if and only if $\lambda/\mu < 0$.

The verification of Axioms 8 and 9, page 76, are now immediate.

For Axiom 10, page 76, consider the distinct collinear points $A: (a)$, $B: (b)$, $D: (a + \lambda b)$, $E_1: (a + \mu b)$, $E_2: (a + \nu b)$. We are given:

(i) D, E_1 separate A, B so that $\lambda/\mu < 0$,

(ii) B, E_2 separate A, E_1.

For (ii) it is necessary to express the coordinates of B, E_2 as linear combinations of those of A and E_1. We find

$$B: \big(a - 1(a + \mu b)\big)$$

$$E_2: \left(a + \frac{v}{\mu - v}(a + \mu b)\right)$$

and so $(v - \mu)/v < 0$.

We are required to show that D, E_2 separate A, B; that is, that $\lambda/v < 0$. Since $\lambda/\mu < 0$, either $\lambda > 0$, $\mu < 0$ or $\lambda < 0$, $\mu > 0$. Suppose $\lambda > 0$, $\mu < 0$; then $(v - \mu)/v < 0$ implies $v < 0$, $|v| < |\mu|$ and so $\lambda/v < 0$. Suppose $\lambda < 0$, $\mu > 0$; then $(v - \mu)/v < 0$ implies $v > 0$, $v < \mu$ and so $\lambda/v < 0$. Thus, Axiom 10 is verified.

Solved Problems

15.1. Given the lines $p: 3x_1 - 3x_2 - x_3 = 0$, $q: 7x_1 - 11x_2 - 5x_3 = 0$, $r: 10x_1 - 11x_2 - 4x_3 = 0$, (a) show they are concurrent, (b) locate the point of concurrency.

(a) The lines are concurrent since
$$\begin{vmatrix} 3 & -3 & -1 \\ 7 & -11 & -5 \\ 10 & -11 & -4 \end{vmatrix} = \begin{vmatrix} 0 & 0 & -1 \\ -8 & 4 & -5 \\ -2 & 1 & -4 \end{vmatrix} = 0.$$

(b) **First Solution.** $\begin{vmatrix} X_1 & X_2 & X_3 \\ 3 & -3 & -1 \\ 7 & -11 & -5 \end{vmatrix} = 4X_1 + 8X_2 - 12X_3 = 0.$ The point of concurrency is $(4, 8, -12)$ or $(1, 2, -3)$.

Second Solution. Eliminating the term in x_2 between the equations of q and r, we have $3x_1 + x_3 = 0$. Take $x_1 = 1$, $x_3 = -3$; then $x_2 = 2$ and the required point is $(1, 2, -3)$ as before.

15.2. Prove: If (y) and (z) are distinct points, then $\lambda(y) + \mu(z)$, where $\lambda \cdot \mu \neq 0$, is another point on the line determined by (y) and (z).

Suppose $\lambda(y) + \mu(z) = v(y)$. Then
$$\mu(z) = v(y) - \lambda(y) = (v - \lambda)(y)$$

and so $\mu(z)$ and $(v - \lambda)(y)$ are representations of the same point, contrary to the assumption that (y) and (z) are distinct points. Similarly, $\lambda(y) + \mu(z) \neq v(z)$. Thus (y), (z) and $\lambda(y) + \mu(z)$ are distinct points. Since

$$\begin{vmatrix} y_1 & y_2 & y_3 \\ z_1 & z_2 & z_3 \\ \lambda y_1 + \mu z_1 & \lambda y_2 + \mu z_2 & \lambda y_3 + \mu z_3 \end{vmatrix} = 0$$

these points are collinear.

15.3. Show that the point $P: (y)$ and the point of intersection Q of the lines $r: [Z]$ and $s: [W]$ are on the line $t: [\{Wy\}Z - \{Zy\}W]$.

The problem is trivial if either r or s is on P. Suppose neither r nor s is on P; then t is a third line on $r \cdot s$. Take $t: [Z] + \lambda[W] = [Z + \lambda W]$, $\lambda \neq 0$. Now t will be on P provided $\{[Z + \lambda W]y\} = \{Zy\} + \lambda\{Wy\} = 0$ or $\lambda = -\dfrac{\{Zy\}}{\{Wy\}}$. Then $t: \left[Z - \dfrac{\{Zy\}}{\{Wy\}}W\right]$ or $[\{Wy\}Z - \{Zy\}W]$ as required.

Note. The equation of t has the form

$$\{Wy\}\{Zx\} - \{Zy\}\{Wx\} = 0 \quad \text{or} \quad \{Wy\}\{Zx\} = \{Zy\}\{Wx\}$$

15.4. Verify Axiom 6, page 76: If two triangles are perspective from a point, they are perspective from a line.

Consider in Fig. 15-2 the triangles ABC and $A'B'C'$ perspective from the point P. Take $P:(p)$, $A:(a)$, $B:(b)$, $C:(c)$; then $A':(p)+\lambda(a)$, $B':(p)+\mu(b)$, $C':(p)+\nu(c)$. Let $R=AB\cdot A'B'$, $S=BC\cdot B'C'$, $T=CA\cdot C'A'$. Since R is on the line AB, the coordinates of R are, by Theorem 15.3, page 158, some linear combination of (a) and (b); since R is also on $A'B'$, this linear combination must be $\lambda(a)-\mu(b)$. Then $R:\lambda(a)-\mu(b)$ and similarly $S:\mu(b)-\nu(c)$ and $T:\nu(c)-\lambda(a)$. Now

$$\nu(c)-\lambda(a) \;=\; -1\big(\mu(b)-\nu(c)\big)\;+\;(-1)\big(\lambda(a)-\mu(b)\big)$$

that is, the coordinates of T can be written as a linear combination of those of R and S. Thus R, S, T are collinear and the triangles are perspective from the line RST.

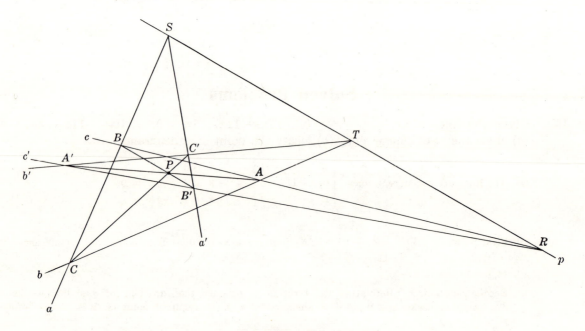

Fig. 15-2

15.5. Prove: Desargues' Two-Triangle Theorem.

Using Fig. 15-2, take $A:(a)$, $B:(b)$, $C:(c)$; $c'=A'B':\{Wx\}=W_1x_1+W_2x_2+W_3x_3=0$, $a'=B'C':\{Yx\}=0$, $b'=C'A':\{Zx\}=0$. The equation of an arbitrary line on the point A' is given by $\{Wx\}+\lambda\{Zx\}=0$ and, by Problem 15.3, it will be on A provided

$$\{Za\}\,\{Wx\} \;=\; \{Wa\}\,\{Zx\}$$

Similarly, the equations of the lines BB' and CC' are

$$\{Wb\}\,\{Yx\} \;=\; \{Yb\}\,\{Wx\}$$

and

$$\{Yc\}\,\{Zx\} \;=\; \{Zc\}\,\{Yx\}$$

respectively. The condition that these equations have a common solution is found by eliminating x_1, x_2, x_3. Forming the product

$$\{Za\}\,\{Wx\}\cdot\{Wb\}\,\{Yx\}\cdot\{Yc\}\,\{Zx\} \;=\; \{Wa\}\,\{Zx\}\cdot\{Yb\}\,\{Wx\}\cdot\{Zc\}\,\{Yx\}$$

and removing the common factors, we have as the condition that $A, B, C \,\overline{\overline{\wedge}}\, A', B', C'$

(i) $$\{Za\}\,\{Wb\}\,\{Yc\} \;=\; \{Wa\}\,\{Yb\}\,\{Zc\}$$

Now take $A:\{aU\}=a_1U_1+a_2U_2+a_3U_3=0$, $B:\{bU\}=0$, $C:\{cU\}=0$; $c'=A'B':[W]$, $a'=B'C':[Y]$, $b'=C'A':[Z]$. The equation of an arbitrary point on the line $a=BC$ is given by $\{bU\}+\lambda\{cU\}=0$ and, by the dual of Problem 15.2, will be on a', that is, will be the point S, provided

$$\{cY\}\{bU\} \;=\; \{bY\}\{cU\}$$

Similarly, the equations of the points R and T are

$$\{bW\}\{aU\} \;=\; \{aW\}\{bU\}$$

and

$$\{aZ\}\{cU\} \;=\; \{cZ\}\{aU\}$$

respectively. The condition that these points be collinear, that is, that $a,b,c \barwedge a',b',c'$ is found by eliminating U_1, U_2, U_3 to be

(i′) $$\{cY\}\{bW\}\{aZ\} \;=\; \{bY\}\{aW\}\{cZ\}$$

Since (i) and (i′) are identical, we have: If two triangles are perspective from a point (line), they are perspective from a line (point).

15.6. Show that the triangles ABC and $A'B'C'$ of Problem 15.5 are perspective: (1) in the order $A,B,C \barwedge B',C',A'$ provided

$$\{Wa\}\{Yb\}\{Zc\} \;=\; \{Ya\}\{Zb\}\{Wc\}$$

and (2) in the order $A,B,C \barwedge C',A',B'$ provided

$$\{Ya\}\{Zb\}\{Wc\} \;=\; \{Za\}\{Wb\}\{Yc\}$$

(1) The equation of an arbitrary line on B' is given by $\{Yx\} + \lambda\{Wx\} = 0$ and it will be on A provided

$$\{Wa\}\{Yx\} \;=\; \{Ya\}\{Wx\}$$

Similarly, the equations of the lines BC' and CA' are

$$\{Yb\}\{Zx\} \;=\; \{Zb\}\{Yx\}$$

and

$$\{Zc\}\{Wx\} \;=\; \{Wc\}\{Zx\}$$

respectively. Thus $A,B,C \barwedge B',C',A'$ provided

$$\{Wa\}\{Yb\}\{Zc\} \;=\; \{Ya\}\{Zb\}\{Wc\}$$

It will be left for the reader to show this is also the condition that $a,b,c \barwedge b',c',a'$.

(2) The equation of an arbitrary line on C' is given by $\{Yx\} + \lambda\{Zx\} = 0$ and it will be on A provided

$$\{Za\}\{Yx\} \;=\; \{Ya\}\{Zx\}$$

Similarly, the equations of the lines BA' and CB' are

$$\{Wb\}\{Zx\} \;=\; \{Zb\}\{Wx\}$$

and

$$\{Yc\}\{Wx\} \;=\; \{Wc\}\{Yx\}$$

respectively. Thus $A,B,C \barwedge C',A',B'$ provided

$$\{Za\}\{Wb\}\{Yc\} \;=\; \{Ya\}\{Zb\}\{Wc\}$$

It will be left for the reader to show this is also the condition that $a,b,c \barwedge c',a',b'$.

15.7. Verify Axiom 7: The diagonal points of a complete quadrangle are never collinear.

Consider in Fig. 15-3 the complete quadrangle $PQRS$ with diagonal points A, B, C. Let representations $P\!:\!(p)$, $Q\!:\!(q)$, $R\!:\!(r)$, $S\!:\!(s)$ be chosen so that [see (iii), page 160]

$$(p) + (q) + (r) + (s) \;=\; \mathbf{0}$$

From

$$(p) + (q) \;=\; -\big((r) + (s)\big)$$

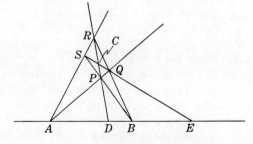

Fig. 15-3

it follows that the points $(p) + (q) = (p + q)$ and $(r) + (s) = (r + s)$ are identical. Since this point is on both the line PQ and the line RS, it must be the diagonal point A. Similarly, $(p + s)$ and $(q + r)$ define the diagonal point B while $(p + r)$ and $(s + q)$ define the diagonal point C.

Suppose A, B, C are collinear. Then there exist numbers λ, μ, ν, not all zero, such that

$$\lambda(r + s) + \mu(q + r) + \nu(s + q) = 0$$

and, hence, such that

$$(\mu + \nu)(q) + (\lambda + \mu)(r) + (\nu + \lambda)(s) = 0$$

Since not all of λ, μ, ν are zero, it follows that not all of $(\mu + \nu)$, $(\lambda + \mu)$, $(\nu + \lambda)$ are zero; hence the points Q, R, S must be collinear. But this is impossible and, so, the diagonal points are never collinear.

15.8. Prove: The harmonic conjugate of $D: (a) + \lambda(b)$ with respect to $A:(a)$ and $B:(b)$ is $E: (a) - \lambda(b)$.

Refer to Fig. 15-3 in which $H(A, B; D, E)$ is established by the complete quadrangle $PQRS$. Take $Q:(q)$ and $P:(a) + \mu(q)$. Since R is on BQ it has coordinates of the form $\alpha(b) + \beta(q)$; since R is also on DP, we have $R: \lambda(b) - \mu(q)$. Since S is on both BP and AR, we have $S: (a) - \lambda(b) + \mu(q)$. Finally, since E is on both QS and AB, then $E: (a) - \lambda(b)$ as required.

Supplementary Problems

15.9. Determine which of the following

      (1) $(1, 0, -1), (1, -2, 1), (3, -2, -1)$ (3) $(1, 2, 1), (3, 5, 2), (3, 4, 1)$

      (2) $(1, 1, 1), (1, 2, 0), (3, 4, 1)$ (4) $(1, 0, 1), (1, 1, 0), (0, 1, 1)$

are collinear triples of points. If collinear, obtain an equation of their common line; if non-collinear, obtain equations of the three lines determined by them.

Ans. (1) $x_1 + x_2 + x_3 = 0$

      (2) $2x_1 - x_2 - x_3 = 0$, $3x_1 - 2x_2 - x_3 = 0$, $2x_1 - x_2 - 2x_3 = 0$

15.10. Determine which of the following

      (1) $[1, 3, -2], [2, -2, 1], [11, 1, -2]$ (3) $[1, 3, 1], [2, 1, 3], [0, 1, 0]$

      (2) $[3, 2, -1], [1, 0, -2], [4, 1, 3]$ (4) $[1, 1, -1], [1, -1, 1], [1, -1, -1]$

are concurrent triples of lines. If concurrent, obtain coordinates of their common point; if non-concurrent, obtain coordinates of the three points determined by them.

Ans. (1) $(1, 5, 8)$; (3) $(8, -1, -5), (1, 0, -1), (3, 0, -2)$

15.11. Given the points $A: (1, 2, 3)$, $B: (2, 4, 3)$, $C: (1, 2, -2)$.

(i)   Show that A, B, C are collinear.

(ii) Take other representations $(a), (b), (c)$ of A, B, C so that $(a) + (b) + (c) = 0$.

(iii) Take other representations so that $(a) + (b) = (c)$.

Ans. (ii) $(a) = (7, 14, 21)$, $(b) = (-10, -20, -15)$, $(c) = (3, 6, -6)$

      (iii) $(a) = (-7/3, -14/3, -7)$, $(b) = (10/3, 20/3, 5)$

15.12. Obtain other representations $[P]$ and $[Q]$ for the lines $p: [1, 3, -2]$ and $q: [2, -2, 1]$ so that $r: [R] = [11, 1, -2]$ is given by $[R] = [P] + [Q]$.   *Ans.* $[P] = [3, 9, -6]$, $[Q] = [8, -8, 4]$

15.13. Given the points $A: (1, 2, 3)$, $B: (2, 5, -6)$, $C: (6, -7, 2)$, $D: (4, 4, -1)$.

(i)   Show that no three are collinear;

(ii) Take other representations $(a), (b), (c), (d)$ respectively so that $(a) + (b) + (c) + (d) = 0$.

Ans. $(a) = (4, 8, 12)$, $(b) = (6, 15, -18)$, (c) $(6, -7, 2)$, $(d) = (-16, -16, 4)$

15.14. Express the coordinates $[1, 1, 1]$ of the line u as a linear combination of those of the non-concurrent lines $[1, 1, 0], [1, 0, 1], [0, 1, 1]$.

15.15. Show that the lines $x_1 + 2x_2 + 3x_3 = 0$, $4x_1 + 3x_2 - x_3 = 0$, $3x_1 - 5x_2 + 4x_3 = 0$ form a triangle. Write an equation of the line joining each vertex of this triangle to the point $(1, 1, 1)$.

    Ans. $3x_1 + x_2 - 4x_3 = 0$, $5x_1 - 18x_2 + 13x_3 = 0$, $8x_1 - 17x_2 + 9x_3 = 0$

15.16. In Fig. 15-1, take $A: (1, 0, 0)$, $B: (0, 1, 0)$, $C: (0, 0, 1)$, $P: (1, 1, 1)$; obtain coordinates of all other points and equations of all lines of the figure.

    Partial Ans. $C': (1, 1, 0)$, $C'': (1, -1, 0)$, $p = A''B''C'': x_1 + x_2 + x_3 = 0$

    Note. The line p is the polar line (see Problem 3.11, page 43) of P with respect to the triangle ABC.

15.17. Given the distinct points $A_1: (1, 0, 0)$, $A_2: (0, 1, 0)$, $A_3: (1, 1, 1)$, $A_4: (1, a^2, a)$, $A_5: (1, b^2, b)$, $A_6: (1, c^2, c)$.

    (i)   Obtain $A_1A_2 \cdot A_4A_5 = R: (1, -ab, 0)$, $A_2A_3 \cdot A_5A_6 = S: (1, b+c-bc, 1)$ and $RS: abx_1 + x_2 - (b+c-bc+ab)x_3 = 0$.

    (ii) Show that R, S and $T = A_3A_4 \cdot A_6A_1$ are collinear.

15.18. Prove the special case of the Theorem of Pappus (see Problem 3.4, page 40).

    Hint. Take $A_1: (a)$, $B_2: (b)$, $P: (c)$, $O: (a+b+c)$. Then $C_3: (a-b)$, $C_2: (a+c-b)$, $C_1: (a+2c-b)$.

15.19. In Fig. 4-3, page 48, take $P: (1, 0, 0)$, $Q: (0, 1, 0)$, $R: (0, 0, 1)$, $S: (1, 1, 1)$. (*a*) Obtain coordinates of all other points and equations of all lines of the figure. (*b*) Show that the diagonal triangle is perspective with each of the four triangles determined by P, Q, R, S. For each perspective pair locate the center and axis of perspectivity.

    Partial Ans. (*a*) $C: (1, 0, 1)$, $E: (1, 2, 1)$, $F: (1, -1, 0)$, $AC: x_1 - x_2 - x_3 = 0$.

                 (*b*) The four points P, Q, R, S and the four lines defined in Theorem 4.14, page 48.

15.20. Prove the Theorem of Pappus.

    Hint. In Fig. 2-9', take $A_1: (a)$, $B_2: (b)$, $A_3: (c)$, $C_2: (a)+(b)+(c)$; $A_2: (a)+\rho(c)$, $B_1: (a)+(b)+\sigma(c)$. Then $C_1: \sigma(b)+(\sigma-\rho)(c)$ and $C_3: (\rho-\sigma)(a)+\rho(b)$.

15.21. Show that the lines $a: x_1 - x_2 - x_3 = 0$, $b: 3x_1 - x_2 + x_3 = 0$ and $c: x_2 + 2x_3 = 0$ are concurrent. Obtain the harmonic conjugate d of the line c with respect to a and b.   *Ans.* $3x_1 - 2x_2 - x_3 = 0$

15.22. In Problem 15.21, let $Q = a \cdot b$. If $C \neq Q$ is any point on c, call the line d the *polar line* of the point C with respect to a and b. Find the polar line of $P: (1, 1, 2)$ with respect to a and b.

    Ans. $x_1 + x_2 + 3x_3 = 0$

15.23. In Problem 15.18 show that p is the polar line of O with respect to r and s.

15.24. Using the results of Problems 15.5 and 15.6, show that if the triangles are perspective in any two of the orders

$$ABC \barwedge A'B'C', \quad ABC \barwedge B'C'A', \quad ABC \barwedge C'A'B'$$

they are also perspective in the third order.

15.25. Find the conditions that the triangles of Problem 15.5 be perspective in the orders (i) $ABC \barwedge A'C'B'$, (ii) $ABC \barwedge B'A'C'$, (iii) $ABC \barwedge C'B'A'$.

    Partial Ans. (i) $\{Za\} \cdot \{Yb\} \cdot \{Wc\} = \{Wa\} \cdot \{Zb\} \cdot \{Yc\}$

15.26. Verify that if the triangles of Problem 15.5 are perspective in any five orders, they are fully perspective, that is, are perspective in the six possible orders.

15.27. Write out in full the dual of the last three sections of this chapter.

15.28. Given $A: (a)$, $B: (b)$, $D: (a+\lambda b)$ and $E: (a+\mu b)$, obtain $(A, B; D, E) = \lambda/\mu$. Thus, separation has been defined in terms of cross ratio.

15.29. In each of the following

        (i) $A: (1, 1, 0)$, $B: (1, 0, 1)$, $D: (2, 1, 1)$, $E: (0, 1, -1)$

        (ii) $A: (1, 1, 0)$, $B: (1, 0, 1)$, $D: (2, 1, 1)$, $E: (3, 2, 1)$

        (iii) $A: (2, 1, 4)$, $B: (3, 5, -1)$, $D: (1, 1, 1)$, $E: (20, 31, -2)$

        (iv) $A: (1, -2, 3)$, $B: (-2, 3, 1)$, $D: (5, -7, -6)$, $E: (2, -5, 13)$

determine whether D, E separates A, B.   *Ans.* Separates: (i).

Chapter 16

Coordinate Systems
and Projective Transformations

INTRODUCTION

In the preceding chapter a geometry of number triples, identified as points, was introduced as a model of plane projective geometry. Although it was found convenient to speak of the number triples as coordinates of the points, it must be recognized that there is, at least at the moment, a difference between these number triples and the familiar coordinates of a point. In our geometry, the number triple *is* the point; in Plane Analytic Geometry, the coordinates of a point are assigned relative to a previously established coordinate system. We propose now to set up suitable coordinate systems on the projective line and in the projective plane. For the purpose of clarity, let us think of the resulting coordinate(s) of a point as its *relative coordinate(s)*. We shall later discover that a coordinate system can be established in the plane so that the number triples of Chapter 15 — call them *absolute coordinates* — are, themselves, relative coordinates.

The first step in constructing a geometry is to choose a fundamental element. The reader's knowledge of geometry is probably limited to point geometry, i.e., geometries in which the point is the fundamental element. In this geometry, the line is said to be one-dimensional, the plane is two-dimensional and ordinary space is three-dimensional. In our geometry of number triples, we recognize, in view of the duality principle, both the point and line as fundamental elements. We saw in Chapter 15 that the points on a line can be given by means of a single parameter; thus, point geometry on a line is one-dimensional. Dually, the lines on a point can be given by means of a single parameter; thus, line geometry on a point is also one-dimensional. Later we shall see that both point and line geometry on the plane are two dimensional.

The nature of the geometry under consideration is determined by the selection of a set of transformations on the coordinates of the fundamental element. In metric plane geometry, we deal with the set of transformations known as rigid motions — translations and rotations — since under these transformations both lengths of segments and measures of angles are invariant. In our geometry we deal with the set of transformations, called projective transformations, under which cross ratio is invariant.

PROJECTIVE COORDINATES ON A LINE (POINT)

Consider in Fig. 16-1 the line o of the metric plane. The familiar coordinate system on the line is established by selecting on o a point O from which all measurements along the line are to be made, a unit of measure, and a sense of (positive) direction along o. Essentially, this consists in selecting two points O and U on o to which coordinates 0 and 1 respectively are assigned. The non-homogeneous coordinate x of any third point X on o is

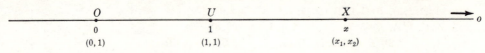

Fig. 16-1

then the directed distance of X from O. In turn, a homogeneous coordinate system on o may be established by assigning coordinates $(0, 1)$ to O, $(1, 1)$ to U, and, to any other point X with non-homogeneous coordinate x, any pair of coordinates (x_1, x_2) satisfying $x_1/x_2 = x$.

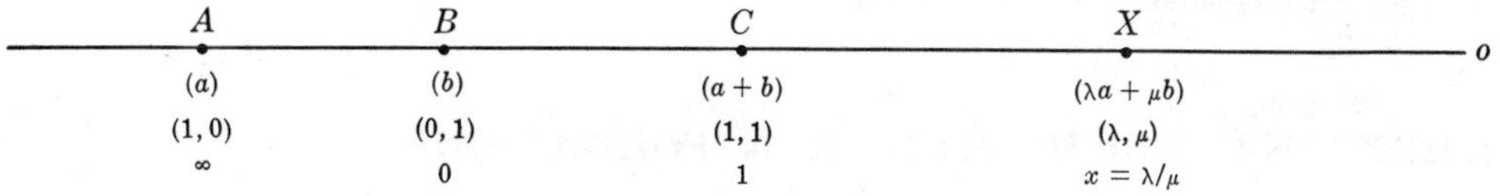

A	B	C	X
(a)	(b)	$(a + b)$	$(\lambda a + \mu b)$
$(1, 0)$	$(0, 1)$	$(1, 1)$	(λ, μ)
∞	0	1	$x = \lambda/\mu$

Fig. 16-2

Consider next (see Fig. 16-2) the line o in the projective plane. On o take two distinct points $A: (a) = (a_1, a_2, a_3)$ and $B: (b) = (b_1, b_2, b_3)$. With the understanding that these representations shall remain fixed, any choice of non-zero numbers λ, μ defines a unique point $X: (x) = (\lambda a + \mu b)$ on o. Conversely, for the representation (x) of X, we recover the pair λ, μ. Now suppose another representation, say (αx), of X be used. Then, from $(\alpha x) = (\alpha \lambda a + \alpha \mu b)$, we obtain a new pair $\alpha \lambda, \alpha \mu$. However, since one pair is a constant multiple of the other, this causes no concern. The above calculations have been made subject to the restriction that the representations of A and B remain fixed. To show the necessity of this restriction, we need only take new representations $A: (\alpha a)$ and $B: (\beta b)$ and see, for the particular choice λ, μ above, that while $(\alpha \lambda a + \beta \mu b)$ is a point on o, it is not the point X. In practice, the fixing of the representations of the base points A and B is effected by selecting a third point $C: (c)$ on o and choosing representations (a^*) and (b^*) for A and B so that $(a^*) + (b^*) = (c)$. We now define:

> With respect to a coordinate system on a line o consisting of the distinct points $A: (a)$, $B: (b)$, $C: (c)$ subject to $(a) + (b) = (c)$, the *relative homogeneous projective coordinates* of any point $X: (x)$ on o are (λ, μ) where $(x) = \lambda(a) + \mu(b)$ and the *abscissa* or *relative non-homogeneous projective coordinate* of the same point X is $x = \lambda/\mu$.

Essentially, a system of homogeneous projective coordinates on a line is established by (i) selecting three distinct points (*reference points*) on the line, (ii) assigning coordinates $(1, 0)$ and $(0, 1)$ in that order to two of the points (*base points*) A and B, (iii) assigning coordinates $(1, 1)$ to the remaining point (*unit point*) C. These same points establish a non-homogeneous coordinate system on the line in which A, B, C have the respective coordinates $\infty, 0, 1$.

Example 16.1.

If $A: (2, 1, -3)$, $B: (4, -2, 4)$, $C: (10, -1, 0)$ are taken as reference points on the line o, find the homogeneous projective coordinates and the non-homogeneous projective coordinate of $D: (8, 2, -7)$ on o.

First, choose new representations (see Example 15.5, page 160) $A: (a) = (4, 2, -6)$ and $B: (b) = (6, -3, 6)$

so that $C: (a + b)$. Now set $\lambda(a) + \mu(b) = (8, 2, -7)$ and solve $\begin{cases} 4\lambda + 6\mu = 8 \\ 2\lambda - 3\mu = 2 \\ -6\lambda + 6\mu = -7 \end{cases}$ for $\lambda = 3/2$, $\mu = 1/3$.

Then, in relative homogeneous coordinates, we have $D: (3/2, 1/3)$; another representation is $D: (9, 2)$. In either case, the non-homogeneous coordinate is $\dfrac{\lambda}{\mu} = \dfrac{3/2}{1/3} = \dfrac{9}{2}$.

In Problem 16.1 we prove

Theorem 16.1. The cross ratio of any four points on a line is independent of the coordinate system established on the line.

Dually, a system of homogeneous projective coordinates on a point O is established by assigning coordinates $[1, 0]$ and $[0, 1]$ to two base lines a and b on O and the coordinates $[1, 1]$ to a third line c on O. These same lines also establish a non-homogeneous system in which a, b, c have the respective coordinates $\infty, 0, 1$. For an example, the reader has only to write out the dual of Example 16.1 in which $A: (2, 1, -3)$ becomes $a: [2, 1, -3]$, etc.   In view of this, we shall restrict attention in the next three sections to coordinate systems on lines. The task of writing the dual of these sections together with all examples and problems will be left as an exercise for the reader.

ONE-DIMENSIONAL PROJECTIVE TRANSFORMATIONS

In the projective plane take any two distinct lines.   On one of them select three distinct points P, Q, R and suppose, relative to a coordinate system previously established on the line, the non-homogeneous coordinates are found to be $P: p, Q: q, R: r$.   Likewise, on the other line suppose three distinct points P', Q', R' selected whose non-homogeneous coordinates relative to a previously established coordinate system are found to be $P': p'$, $Q': q', R'': r'$.   Let $X: x$ and $X': x'$ be any other pair of corresponding points in the projectivity $(P, Q, R) \barwedge (P', Q', R')$.   Then,

$$(P, Q; R, X) = (P', Q'; R', X') \tag{1}$$

and so

$$\frac{(p-r)(q-x)}{(p-x)(q-r)} = \frac{(p'-r')(q'-x')}{(p'-x')(q'-r')} \tag{2}$$

We now solve (2) for x' in terms of x.   Putting $\dfrac{p-r}{q-r} = s$ and $\dfrac{p'-r'}{q'-r'} = s'$, we have $s\dfrac{q-x}{p-x} = s'\dfrac{q'-x'}{p'-x'}$ whence

$$s(x'x - qx' - p'x + p'q) = s'(x'x - px' - q'x + pq')$$

$$[(s - s')x + ps' - qs]x' = (sp' - s'q')x + pq's' - p'qs$$

and

$$x' = \frac{ax + b}{cx + d} \tag{3}$$

where $a = sp' - s'q', \ b = pq's' - p'qs, \ c = s - s', \ d = ps' - qs$.

In terms of homogeneous coordinates, we have, by setting $x = x_1/x_2$ and $x' = x'_1/x'_2$,

$$\frac{x'_1}{x'_2} = \frac{ax_1 + bx_2}{cx_1 + dx_2} \tag{4}$$

or

$$\begin{aligned} \rho x'_1 &= ax_1 + bx_2 \\ \rho x'_2 &= cx_1 + dx_2 \end{aligned}, \quad \rho \neq 0 \tag{5}$$

Equations (3), (4), (5) are variously said to define a *mapping* of one pencil of points upon another, or a *projective transformation* which carries the points on one line into the points on another, or a *projectivity* between two distinct pencils of points.   Dually, we have

$$X' = \frac{AX + B}{CX + D} \tag{3'}$$

$$\frac{X'_1}{X'_2} = \frac{AX_1 + BX_2}{CX_1 + DX_2} \tag{4'}$$

$$\begin{aligned} \rho X'_1 &= AX_1 + BX_2 \\ \rho X'_2 &= CX_1 + DX_2 \end{aligned}, \quad \rho \neq 0 \tag{5'}$$

as equations of the projective transformation in non-homogeneous and homogeneous coordinates which carries the lines on one point O into the lines on another point O'.

Equation (*3*) is also of the form generally called a *linear transformation*. Is then every linear transformation

$$\tau\colon x' = \frac{ax + b}{cx + d} \tag{3''}$$

with not all a, b, c, d equal to zero, a projective transformation, that is, a transformation having an inverse and preserving cross ratios? We show that (*3''*) is projective provided it carries distinct points into distinct points.

Let $P\colon p$ and $Q\colon q$ be any two distinct points on line o and $P'\colon p' = \dfrac{ap + b}{cp + d}$ and $Q'\colon q' = \dfrac{aq + b}{cp + d}$ be their respective correspondents on another line o' or on the same line o. From

$$p' - q' = \frac{(ad - bc)(p - q)}{(cp + d)(cq + d)}$$

it follows, since $p - q \neq 0$, that $p' - q' \neq 0$ if and only if $ad - bc \neq 0$.

Now consider (*3''*) with the added restriction $ad - bc \neq 0$. Then it may be solved for $x = \dfrac{dx' - b}{-cx' + a}$ which, by replacement in (*3''*), is easily shown to be the inverse of (*3''*). Finally, let $P\colon p, P'\colon p'; Q\colon q, Q'\colon q'; R\colon r, R'\colon r'; S\colon s, S'\colon s'$ be any four distinct pairs of corresponding points. From

$$p' - r' = \frac{(ad - bc)(p - r)}{(cp + d)(cr + d)}, \qquad p' - s' = \frac{(ad - bc)(p - s)}{(cp + d)(cs + d)}$$

$$q' - r' = \frac{(ad - bc)(q - r)}{(cq + d)(cr + d)}, \qquad q' - s' = \frac{(ad - bc)(q - s)}{(cq + d)(cs + d)}$$

there follows by a simple calculation

$$(P', Q'; R', S') = \frac{(p' - r')(q' - s')}{(p' - s')(q' - r')} = \frac{(p - r)(q - s)}{(p - s)(q - r)} = (P, Q; R, S)$$

We have proved

Theorem 16.2. Every linear transformation

$$\tau\colon x' = \frac{ax + b}{cx + d}, \qquad \text{where} \quad ad - bc \neq 0$$

is a projective transformation.

Example 16.2.

Given the points $P\colon 2, Q\colon 3, R\colon 4$ on the line o and the points $P'\colon 3, Q'\colon 4, R'\colon 2$ on another line o'. (*a*) Determine the linear transformation which carries P, Q, R into P', Q', R' respectively. (*b*) Locate the point S' on o' into which $S\colon -1$ is carried. (*c*) Locate the point T on o which is carried into $T'\colon 0$ on o'. (*d*) Is the transformation projective?

(*a*) Since $x = 2$ is carried into $x' = 3$, we have using (*3''*)

$$3 = \frac{2a + b}{2c + d} \qquad \text{or} \qquad 2a + b - 6c - 3d = 0 \tag{i}$$

Similarly, we obtain

$$3a + b - 12c - 4d = 0 \tag{ii}$$

and

$$4a + b - 8c - 2d = 0 \tag{iii}$$

By subtracting (i) from (ii) and (iii), we obtain

$$a - 6c - d = 0 \qquad \text{and} \qquad 2a - 2c + d = 0$$

which, when added, yields $3a - 8c = 0$. Take $a = 8$, $c = 3$; then $d = -10$ and $b = -28$. The required transformation is $x' = \dfrac{8x - 28}{3x - 10}$.

(*b*) When $x = -1$, $x' = \dfrac{-8 - 28}{-3 - 10} = \dfrac{36}{13}$.

(c) When $x' = 0$, $8x - 28 = 0$ and $x = \frac{7}{2}$.

(d) Since $ad - bc = 4 \neq 0$, the transformation is projective.

In Problem 16.2 we prove

Theorem 16.3. If the joins of three pairs of corresponding points of two projective pencils are concurrent, the pencils are perspective.

PROJECTIVITIES ON A LINE. NON-HOMOGENEOUS COORDINATES

Suppose now that the two lines o and o' of the above section coincide. There are then two interpretations of the projective transformation

$$\tau: \quad x' = \frac{ax + b}{cx + d}, \quad ad - bc \neq 0 \tag{6}$$

In the first, we have a pencil of points on the line o referred to two distinct coordinate systems. Each point X on o has two coordinates — x when referred to one system and x' when referred to the other — and (6) merely changes each x into its proper x'. Under this interpretation, (6) is called a *passive* or *alias* transformation or a *transformation of coordinates*.

In the second, we have two superposed pencils of points on o referred to the same coordinate system. Then (6) carries a point $X: x$ on o into, generally, another point $X': x'$ on o. Under this interpretation, (6) is called an *active* or *alibi* transformation.

As an alibi transformation, (6) is the analytic definition of a projectivity on a line; frequently, it is called a *collineation* on a line.

We now investigate the possibility that the collineation (6) between two superposed pencils carries some point M into itself. Such a point M, if one exists, is called a *double (fixed, self-corresponding*, or *invariant) point* of the projectivity. The double points, if any, are obtained by setting $x' = x$ in (6) and solving the resulting quadratic equation in x.

First, we consider the special cases which arise when $c = 0$. The quadratic equation is then in reality a linear equation so that $x = \infty$ is always a root and $M: \infty$ is always a double point.

(i) When $c = 0$, $b \neq 0$, and $a \neq d$, we have $(d - a)x = b$; there are two distinct double points: $M: \infty$ and $N: \dfrac{b}{d - a}$.

(ii) When $c = 0$, $b \neq 0$, and $a = d$, the double points M and N of (i) coincide. There is a single double point $M: \infty$.

(iii) When $c = b = 0$ and $a \neq d$, then $(d - a)x = 0$. There are two double points $M: \infty$ and $N: 0$.

(iv) When $c = b = 0$ and $a = d$, *every* point of the line is a double point. This projectivity is known as the *identity*.

Otherwise, when $x' = x$, (6) may be written as

$$cx^2 + (d - a)x - b = 0, \quad c \neq 0$$

whose roots are

$$x_1 = \frac{a - d + \sqrt{(d - a)^2 + 4bc}}{2c} \quad \text{and} \quad x_2 = \frac{a - d - \sqrt{(d - a)^2 + 4bc}}{2c}$$

There are three cases:

Case 1. $(d-a)^2 + 4bc > 0$. The roots are real and distinct. There are two distinct double points; τ is called a *hyperbolic projectivity*.

Case 2. $(d-a)^2 + 4bc = 0$. The roots are real and coincident. There is one double point; τ is called a *parabolic projectivity*.

Case 3. $(d-a)^2 + 4bc < 0$. The roots are complex numbers. There is no real double point; τ is called an *elliptic projectivity*.

Thus, aside from the identity transformation, τ is either hyperbolic, parabolic, or elliptic.

See Problem 16.3.

In Problem 16.4, we prove

Theorem 16.4.   The cross ratio K of the two double elements of a hyperbolic projectivity and a pair of corresponding elements is independent of the pair of corresponding elements used.

The real constant K is called the *characteristic invariant* of the hyperbolic projectivity. It follows that any hyperbolic projectivity can be written in the *canonical form*

$$\frac{x' - x_2}{x' - x_1} = K \frac{x - x_2}{x - x_1}$$

where x_1, x_2 are the double points and K is the characteristic invariant.

In Problem 16.5, canonical forms for parabolic projectivities

$$x' = x + p, \quad p = b/a, \quad \text{when } c = 0$$

and

$$\frac{1}{x' - x_1} = \frac{1}{x - x_1} + p, \quad p = \frac{-c}{cx_1 - a}, \quad \text{when } c \neq 0$$

are obtained.

Example 16.3.

Reduce to canonical form: (a) $x' = \dfrac{-5x + 6}{2x - 4}$,   (b) $x' = \dfrac{3x - 8}{2x - 5}$.

(a) The projectivity is hyperbolic with double points $M: 3/2$, $N = -2$. Using the pair of corresponding points $P: 1$; $P': -1/2$, we find

$$K = (M, N; P, P') = (3/2, -2; 1, -1/2) = 1/8$$

The canonical form is $\dfrac{x' + 2}{x' - 3/2} = \dfrac{1}{8} \dfrac{x + 2}{x - 3/2}$.

(b) The projectivity is parabolic with double point $M: 2$.   The canonical form is $\dfrac{1}{x' - 2} = \dfrac{1}{x - 2} - 2$.

INVOLUTIONS ON A LINE

Let τ be a projective transformation of a line onto itself whose inverse is $\tau^{-1} = \tau$. (See Problem 16.13(f) for an example.)   Then $\tau \cdot \tau^{-1} = \tau^{-1} \cdot \tau = \tau^2 = \mathcal{I}$, the identity, and the projectivity τ is called a *quadratic involution*. More often, it is called, simply, an *involution*. If p, p' is a pair of corresponding points in an involution, so also is p', p; we will therefore speak of p, p' as a *reciprocal pair* of the involution. (See Chapter 6.)   In Problem 16.6 we prove

Theorem 16.5.   The projective transformation

$$x' = \frac{ax + b}{cx + d} \tag{3}$$

is an involution if and only if $a + d = 0$.

The equation of an involution may be written in the form

$$fxx' + g(x + x') + h = 0, \quad g^2 - fh \neq 0 \tag{7}$$

and, hence, is completely determined by giving any two reciprocal pairs of its points.

Example 16.4.

Obtain the equation of the involution having x_1, x_1'; x_2, x_2' as two reciprocal pairs.

Substituting each pair in (7) yields

$$fx_1x_1' + g(x_1 + x_1') + h = 0$$
$$fx_2x_2' + g(x_2 + x_2') + h = 0$$

Eliminating the coefficients between (7) and the above equations, the required equation in determinant form is

$$\begin{vmatrix} xx' & x + x' & 1 \\ x_1x_1' & x_1 + x_1' & 1 \\ x_2x_2' & x_2 + x_2' & 1 \end{vmatrix} = 0 \tag{8}$$

The double points of an involution are found by writing $x' = x$ in (7) and solving the resulting equation. There will be either two distinct real roots (the involution is *hyperbolic*), two coincident real roots (the involution is *parabolic*) or two imaginary roots (the involution is *elliptic*). Parabolic involutions are generally ignored (see Problem 16.26). A hyperbolic involution is completely determined when its double points or one double point and another reciprocal pair are known.

Finally, an involution is completely determined when the quadratic equation giving its double points is known. For, suppose this equation is

$$ax^2 + 2bx + c = 0, \quad b^2 - ac \neq 0 \tag{9}$$

Now consider the involution $axx' + b(x + x') + c = 0$ \hfill (10)

which is hyperbolic or elliptic according as the roots of (9) are real or complex. Equation (10) can be derived from (9) by the simple rule:

   Replace x^2 by xx' and $2x$ by $x + x'$.

Equation (10) is called the *polarized form* of (9).

Example 16.5.

Derive the equation of the involution from the equation $3x^2 + 5x - 6 = 0$ giving its double points.

By polarizing the given equation, we have

$$3xx' + \frac{5}{2}(x + x') - 6 = 0 \quad \text{or} \quad 6xx' + 5(x + x') - 12 = 0$$

It is not difficult to prove

Theorem 16.6. The pairs of points x, x' harmonic to two given points x_1, x_2 are reciprocal pairs of an involution having x_1, x_2 as double points.

PROJECTIVITIES ON A LINE. HOMOGENEOUS COORDINATES

The study of projectivities on a line and in the projective plane (also, in projective spaces of higher dimensions) is greatly simplified by the use of homogeneous coordinates. In this section we restudy projectivities on a line using

$$\begin{cases} \rho x_1' = ax_1 + bx_2 \\ \rho x_2' = cx_1 + dx_2 \end{cases}, \quad (ad - bc \neq 0) \tag{5}$$

Again, looking ahead to future work, we propose a change of notation and rewrite (5) as

$$\begin{cases} \rho x_1' = a_{11}x_1 + a_{12}x_2 \\ \rho x_2' = a_{21}x_1 + a_{22}x_2 \end{cases}, \quad (a_{11}a_{22} - a_{12}a_{21} \neq 0) \tag{11}$$

To investigate the possible double points, we replace x_1' by x_1, x_2' by x_2 in (11) and consider the resulting equations

$$\begin{cases} (a_{11} - \rho)x_1 + \quad a_{12}x_2 = 0 \\ \quad a_{21}x_1 + (a_{22} - \rho)x_2 = 0 \end{cases} \tag{12}$$

Now (12) will have non-trivial solutions if and only if

$$\begin{vmatrix} a_{11} - \rho & a_{12} \\ a_{21} & a_{22} - \rho \end{vmatrix} = 0 \tag{13}$$

or

$$\rho^2 - (a_{11} + a_{22})\rho + (a_{11}a_{22} - a_{12}a_{21}) = 0$$

Equation (13) is called the *characteristic equation* of the projectivity, and its roots ρ_1, ρ_2 are called the *characteristic roots*. A non-identity projectivity (5) is hyperbolic, parabolic, or elliptic according as the roots ρ_1, ρ_2 are real and distinct, real and coincident or imaginary.

Example 16.6.

Examine the projectivity $\begin{cases} x_1' = 2x_1 + 5x_2 \\ x_2' = 2x_1 - x_2 \end{cases}$ for double points.

The characteristic equation is $\begin{vmatrix} 2 - \rho & 5 \\ 2 & -1 - \rho \end{vmatrix} \equiv \rho^2 - \rho - 12 = 0$; the characteristic roots are $\rho_1 = 4$, $\rho_2 = -3$. The double points of the projectivity are given when ρ is replaced by $\rho_1 = 4$ and $\rho_2 = -3$ in either of the equations

$$(2 - \rho)x_1 + \quad 5x_2 = 0$$
$$2x_1 + (-1 - \rho)x_2 = 0$$

Using the first equation, we find:

$$(2 - 4)x_1 + 5x_2 = -2x_1 + 5x_2 = 0; \quad \text{the double point is } (5, 2),$$
$$(2 + 3)x_1 + 5x_2 = 5(x_1 + x_2) = 0; \quad \text{the double point is } (1, -1).$$

The reader will verify that the same results obtain when the second equation is used. Thus the projectivity is hyperbolic.

By Theorem 16.5 any involution on a line has the form

$$\begin{aligned} \rho x_1' = a_{11}x_1 + a_{12}x_2 \\ \rho x_2' = a_{21}x_1 - a_{11}x_2 \end{aligned}, \quad a_{11}^2 + a_{12}a_{21} \neq 0$$

with invariant points satisfying

$$a_{21}x_1^2 - 2a_{11}x_1x_2 - a_{12}x_2^2 = 0$$

The involution is hyperbolic or elliptic according as $a_{11}^2 + a_{12}a_{21} > 0$ or $a_{11}^2 + a_{12}a_{21} < 0$.

HOMOGENEOUS POINT COORDINATES IN THE PLANE

In establishing a coordinate system on a line, use was made of any three of its points $A: (a)$, $B: (b)$, $C: (a + b)$. Relative to this coordinate system, the homogeneous coordinates of any fourth point $P: (p)$ on the line were obtained as the ordered pair of numbers (p_1, p_2) satisfying $(p) = p_1(a) + p_2(b)$. The basic points of this coordinate system were the points A and B; the role of the third point C was solely that of determining for all time fixed representations (absolute coordinates) for A and B. A simple extension of the above will be used to establish a coordinate system in the projective plane.

In the plane, take any four points A_1, A_2, A_3, A_4, no three of which are collinear. Choose representations of the first three points so that we have A_1: (a), A_2: (b), A_3: (c), A_4: $(a + b + c)$. Note that the role of A_4 is simply to fix representations for A_1, A_2, A_3. These three points are vertices of a triangle, called the *triangle of reference*; the point A_4 is called the *unit point*. The coordinate system, thus defined, will be indicated by $A_1, A_2, A_3; A_4$. In terms of this coordinate system, we define the *relative coordinates* as A_1: $(1, 0, 0)$, A_2: $(0, 1, 0)$, A_3: $(0, 0, 1)$, A_4: $(1, 1, 1)$ and Y: (α, β, γ) for any point Y: (y), where $(y) = \alpha(a) + \beta(b) + \gamma(c)$.

Example 16.7.

Take four points with absolute coordinates A_1: $(1, 2, 3)$, A_2: $(2, -3, 4)$, A_3: $(4, 5, -6)$, A_4: $(11, 9, -5)$. (The reader will verify that no three of these points are collinear.) Set up the coordinate system $A_1, A_2, A_3; A_4$ and find:

(1) The relative coordinates of P with absolute coordinates $(2, 2, 1)$.

(2) The absolute coordinates of Q with relative coordinates $(4, -2, 1)$.

(3) The relations existing between the absolute coordinates (x_1'', x_2'', x_3'') and the relative coordinates (x_1, x_2, x_3) of an arbitrary point X.

First, we fix the representations A_1: (a), A_2: (b), A_3: (c) so that A_4: $(a + b + c)$. Set

$$\lambda(1, 2, 3) + \mu(2, -3, 4) + \nu(4, 5, -6) = (11, 9, -5)$$

and solve the system
$$\lambda + 2\mu + 4\nu = 11$$
$$2\lambda - 3\mu + 5\nu = 9$$
$$3\lambda + 4\mu - 6\nu = -5$$

to obtain $\lambda = 1$, $\mu = 1$, $\nu = 2$. Then A_1: $(1, 2, 3)$, A_2: $(2, -3, 4)$, A_3: $(8, 10, -12)$; A_4: $(11, 9, -5)$ satisfy all requirements for triangle of reference and unit point.

(1) Set $\alpha(1, 2, 3) + \beta(2, -3, 4) + \gamma(8, 10, -12) = (2, 2, 1)$ and solve the system

$$\alpha + 2\beta + 8\gamma = 2$$
$$2\alpha - 3\beta + 10\gamma = 2$$
$$3\alpha + 4\beta - 12\gamma = 1$$

to obtain $\alpha = 37/60$, $\beta = 7/40$, $\gamma = 31/240$. Then, in relative coordinates, we have P: $(37/60, 7/40, 31/240)$; another representation is P: $(148, 42, 31)$.

(2) The relative coordinates of Q are: $\alpha = 4$, $\beta = -2$, $\gamma = 1$; the absolute coordinates were

$$4(1, 2, 3) - 2(2, -3, 4) + 1(8, 10, -12) = (8, 24, -8)$$

Another representation is Q: $(1, 3, -1)$.

(3) Identifying (x_1, x_2, x_3) with $(4, -2, 1)$ of (2), we have a representation Q: $(\rho x_1'', \rho x_2'', \rho x_3'')$ given by

$$1 \cdot x_1 + 2 \cdot x_2 + 8 \cdot x_3 = \rho x_1''$$
$$2 \cdot x_1 - 3 \cdot x_2 + 10 \cdot x_3 = \rho x_2''$$
$$3 \cdot x_1 + 4 \cdot x_2 - 12 \cdot x_3 = \rho x_3''$$

in which the multipliers of x_1, x_2, x_3 in the first, second, third equality are respectively the first, second, third (absolute) coordinates of the vertices of the reference triangle.

As in metric geometry, an analytic proof may be simplified by a proper choice of the coordinate system.

Example 16.8.

(a) For a proof of Desargues' Two-Triangle Theorem (Problem 15.4), take one of the triangles (see Fig. 16-3 below) as reference triangle and the center of perspectivity as unit point.

(b) For a proof of the Theorem of Pappus, take A_1: $(1, 0, 0)$, B_2: $(0, 1, 0)$ and A_3: $(0, 0, 1)$ and C_2: $(1, 1, 1)$ in Fig. 2-9', page 28.

(c) In the study of a complete quadrangle, the vertices may be taken as the vertices of the reference triangle and the unit point or (see Problem 16.41) the diagonal triangle may be taken as the reference triangle and the unit point as any one of the vertices of the quadrangle.

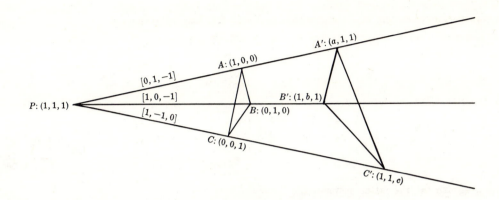

Fig. 16-3

RELATION BETWEEN ABSOLUTE AND RELATIVE COORDINATES

In what is to follow it is assumed that the reader is acquainted with the algebra of matrices; the minimum required in this book is treated briefly in the Appendix. In order to use matrix algebra most efficiently, a change of notation for the vertices of the reference triangle is necessary. For the reference triangle defined above, we write $A_1: (a_{11}, a_{21}, a_{31})$, $A_2: (a_{12}, a_{22}, a_{32})$, $A_3: (a_{13}, a_{23}, a_{33})$ where the second subscript indicates the point and the first subscript indicates the particular coordinate of that point. Then the absolute coordinates (x_1'', x_2'', x_3'') and relative coordinates (x_1, x_2, x_3) of an arbitrary point X satisfy [see Example 16.7(c)]

$$\rho x_1'' = a_{11}x_1 + a_{12}x_2 + a_{13}x_3$$

$$\rho x_2'' = a_{21}x_1 + a_{22}x_2 + a_{23}x_3$$

$$\rho x_3'' = a_{31}x_1 + a_{32}x_2 + a_{33}x_3$$

which, in matrix notation, is

$$\rho \begin{bmatrix} x_1'' \\ x_2'' \\ x_3'' \end{bmatrix} = \begin{bmatrix} a_{11} & a_{12} & a_{13} \\ a_{21} & a_{22} & a_{23} \\ a_{31} & a_{32} & a_{33} \end{bmatrix} \cdot \begin{bmatrix} x_1 \\ x_2 \\ x_3 \end{bmatrix} = \mathbf{A} \cdot \begin{bmatrix} x_1 \\ x_2 \\ x_3 \end{bmatrix} \tag{14}$$

where $\mathbf{A} = [a_{ij}]$, $(i, j = 1, 2, 3)$, and $\rho \neq 0$ is a proportionality factor. By means of (14) we pass from relative coordinates to absolute coordinates.

Since A_1, A_2, A_3 are non-collinear, $|\mathbf{A}| \neq 0$; hence $\mathbf{A}^{-1}$ exists. Multiplying (14) on the left by $\mathbf{A}^{-1}$, we have

$$\rho \mathbf{A}^{-1} \cdot \begin{bmatrix} x_1'' \\ x_2'' \\ x_3'' \end{bmatrix} = \begin{bmatrix} x_1 \\ x_2 \\ x_3 \end{bmatrix}$$

or

$$\sigma \begin{bmatrix} x_1 \\ x_2 \\ x_3 \end{bmatrix} = \mathbf{A}^{-1} \cdot \begin{bmatrix} x_1'' \\ x_2'' \\ x_3'' \end{bmatrix} \tag{15}$$

by which we pass from absolute coordinates to relative coordinates.

Equations (14) may be written down as soon as a coordinate system has been established. Note that *the elements of the columns of* $\mathbf{A}$ *are the absolute coordinates of the vertices* A_1, A_2, A_3 *of the reference triangle* in that order.

Example 16.9.

For the coordinate system of Example 16.7, (*14*) and (*15*) become respectively

$$\rho \begin{bmatrix} x_1'' \\ x_2'' \\ x_3'' \end{bmatrix} = \begin{bmatrix} 1 & 2 & 8 \\ 2 & -3 & 10 \\ 3 & 4 & -12 \end{bmatrix} \cdot \begin{bmatrix} x_1 \\ x_2 \\ x_3 \end{bmatrix} = \mathbf{A} \begin{bmatrix} x_1 \\ x_2 \\ x_3 \end{bmatrix}$$

and

(i)
$$\sigma \begin{bmatrix} x_1 \\ x_2 \\ x_3 \end{bmatrix} = \frac{1}{240} \begin{bmatrix} -4 & 56 & 44 \\ 54 & -36 & 6 \\ 17 & 2 & -7 \end{bmatrix} \cdot \begin{bmatrix} x_1'' \\ x_2'' \\ x_3'' \end{bmatrix} = \mathbf{A}^{-1} \begin{bmatrix} x_1'' \\ x_2'' \\ x_3'' \end{bmatrix} = \frac{1}{|\mathbf{A}|} (\mathrm{adj}\, \mathbf{A}) \begin{bmatrix} x_1'' \\ x_2'' \\ x_3'' \end{bmatrix}$$

By letting σ absorb $|\mathbf{A}|$, the latter assumes the form

(ii)
$$\sigma \begin{bmatrix} x_1 \\ x_2 \\ x_3 \end{bmatrix} = (\mathrm{adj}\, \mathbf{A}) \begin{bmatrix} x_1'' \\ x_2'' \\ x_3'' \end{bmatrix}$$

COORDINATES RELATIVE TO TWO COORDINATE SYSTEMS

Consider in the projective plane, a second coordinate system with reference triangle $B_1 B_2 B_3$ and unit point B_4. From the section above, it follows that the absolute coordinates (x_1'', x_2'', x_3'') and relative coordinates (x_1', x_2', x_3') with respect to this system of the point X satisfy

$$\rho \begin{bmatrix} x_1'' \\ x_2'' \\ x_3'' \end{bmatrix} = \begin{bmatrix} b_{11} & b_{12} & b_{13} \\ b_{21} & b_{22} & b_{23} \\ b_{31} & b_{32} & b_{33} \end{bmatrix} \cdot \begin{bmatrix} x_1' \\ x_2' \\ x_3' \end{bmatrix} = \mathbf{B} \begin{bmatrix} x_1' \\ x_2' \\ x_3' \end{bmatrix} \tag{16}$$

and

$$\sigma \begin{bmatrix} x_1' \\ x_2' \\ x_3' \end{bmatrix} = \mathbf{B}^{-1} \begin{bmatrix} x_1'' \\ x_2'' \\ x_3'' \end{bmatrix} \tag{17}$$

Combining (*14*) and (*16*), we obtain the relation

$$\tau' \mathbf{B} \begin{bmatrix} x_1' \\ x_2' \\ x_3' \end{bmatrix} = \mathbf{A} \begin{bmatrix} x_1 \\ x_2 \\ x_3 \end{bmatrix}$$

from which follow

$$\tau' \begin{bmatrix} x_1' \\ x_2' \\ x_3' \end{bmatrix} = \mathbf{B}^{-1} \cdot \mathbf{A} \begin{bmatrix} x_1 \\ x_2 \\ x_3 \end{bmatrix} = \mathbf{C} \begin{bmatrix} x_1 \\ x_2 \\ x_3 \end{bmatrix} \tag{18}$$

giving the coordinates (x_1', x_2', x_3') of X relative to the coordinate system $B_1, B_2, B_3; B_4$ when the coordinates (x_1, x_2, x_3) relative to the coordinate system $A_1, A_2, A_3; A_4$ are known and

$$\tau \begin{bmatrix} x_1 \\ x_2 \\ x_3 \end{bmatrix} = \mathbf{A}^{-1} \cdot \mathbf{B} \begin{bmatrix} x_1' \\ x_2' \\ x_3' \end{bmatrix} = \mathbf{D} \begin{bmatrix} x_1' \\ x_2' \\ x_3' \end{bmatrix} \tag{19}$$

giving the coordinates (x_1, x_2, x_3) of the point X relative to the coordinate system $A_1, A_2, A_3; A_4$ when the coordinates (x_1', x_2', x_3') relative to the coordinate system $B_1, B_2, B_3; B_4$ are known.

Consider in (*18*) the matrix $\mathbf{C} = \mathbf{B}^{-1}\cdot\mathbf{A}$. The columns of $\mathbf{A}$ are the absolute coordinates of the vertices of the reference triangle $A_1A_2A_3$ and the columns of $\mathbf{C}$ are, then, the coordinates of the vertices of this same triangle relative to the coordinate system $B_1, B_2, B_3; B_4$. A similar interpretation of $\mathbf{D} = \mathbf{A}^{-1}\cdot\mathbf{B}$ in (*19*) may be given. We conclude:

When the coordinates of a point relative to a coordinate system $\mathcal{S}_1$ are known, the coordinates of the same point relative to another coordinate system $\mathcal{S}_2$ are found by multiplying the known coordinates on the left by the matrix whose columns are the coordinates of the vertices of the reference triangle of $\mathcal{S}_1$ relative to the coordinate system $\mathcal{S}_2$.

PROJECTIVE TRANSFORMATIONS IN THE PLANE

The equations

$$\rho\begin{bmatrix} x_1' \\ x_2' \\ x_3' \end{bmatrix} = \begin{bmatrix} e_{11} & e_{12} & e_{13} \\ e_{21} & e_{22} & e_{23} \\ e_{31} & e_{32} & e_{33} \end{bmatrix}\cdot\begin{bmatrix} x_1 \\ x_2 \\ x_3 \end{bmatrix}, \qquad |\mathbf{E}| = [e_{ij}] \neq 0 \tag{20}$$

and

$$\sigma\begin{bmatrix} x_1 \\ x_2 \\ x_3 \end{bmatrix} = \mathbf{E}^{-1}\begin{bmatrix} x_1' \\ x_2' \\ x_3' \end{bmatrix} \tag{21}$$

may be interpreted either (i) as a means of passing from the relative coordinates (x_1, x_2, x_3) of a point X, with respect to a coordinate system $\mathcal{S}_1$, to its absolute coordinates (x_1', x_2', x_3') and vice versa or (ii) as a means of passing from the relative coordinates (x_1, x_2, x_3) of a point X, referred to a coordinate system $\mathcal{S}_1$, to the relative coordinates (x_1', x_2', x_3') of the same point, referred to a coordinate system $\mathcal{S}_2$, and vice versa. These two interpretations become the same by noting that there exists a coordinate system — call it the natural coordinate system — with respect to which the absolute coordinates of a point are themselves relative coordinates. This coordinate system is defined by $A_1: (1, 0, 0)$, $A_2: (0, 1, 0)$, $A_3: (0, 0, 1)$; $A_4: (1, 1, 1)$ in absolute coordinates. Any point $X: (x_1, x_2, x_3)$ in absolute coordinates may be written as $x_1(1, 0, 0) + x_2(0, 1, 0) + x_3(0, 0, 1)$ and, hence, referred to this (the natural) coordinate system has relative coordinates (x_1, x_2, x_3).

The effect then of (*20*) and (*21*) is simply to change the name (coordinates) of each point in the plane. In view of this, they are called *passive* or *alias* transformations.

As with its counterpart (*5*), a second interpretation of (*20*) is possible. In this interpretation, we have distinct points $X: (x_1, x_2, x_3)$ and $X': (x_1', x_2', x_3')$ whose coordinates are relative to the *same* coordinate system. The effect of (*20*) is to carry the point X into the point X'; the effect of (*21*) is to carry the point X' back into X. With this interpretation, (*20*) and (*21*) are called *active* transformations, or *alibi* transformations. We may now write (*20*) and (*21*) more briefly as

$$\rho X'^T = \mathbf{E}X^T \quad \text{and} \quad \sigma X^T = \mathbf{E}^{-1}X'^T$$

This cumbersome notation could have been avoided by adopting from the beginning vector notation in which the coordinates of X are given as $\begin{pmatrix} x_1 \\ x_2 \\ x_3 \end{pmatrix}$ rather than by the familiar (x_1, x_2, x_3). Instead, we shall agree to drop the transpose signs, recalling always that in the equations of a transformation written in matrix notation, the coordinates of a point enter naturally as a 3×1 matrix or as a column of a 3-square matrix. With this understanding we write

$$\rho X' = \mathbf{E}X \tag{20}$$

and

$$\sigma X = \mathbf{E}^{-1}X' \tag{21}$$

There can be no misunderstanding by so doing; literally, $\mathbf{E}X = [e_{ij}] \cdot [x_1, x_2, x_3]$ is meaningless.

In Problem 16.7 we prove

Theorem 16.7.   The transformation *(20)* preserves collinearity.

In Problem 16.46 the reader is asked to prove

Theorem 16.8.   The transformation *(20)* preserves cross ratio.

Example 16.10.

Relative to a coordinate system $\mathcal{S}_1$, take $A_1:(a_{11}, a_{21}, a_{31})$, $A_2:(a_{12}, a_{22}, a_{32})$, $A_3:(a_{13}, a_{23}, a_{33})$ as reference triangle and $A_4:\left(\sum_j a_{1j},\ \sum_j a_{2j},\ \sum_j a_{3j}\right)$ as unit point of a new system $\mathcal{S}_2$. When referred to $\mathcal{S}_2$ let A_1 have coordinates $(1, 0, 0)$, A_2 have coordinates $(0, 1, 0)$ and A_3 have coordinates $(0, 0, 1)$. Also, let an arbitrary point $X:(x_1, x_2, x_3)$, referred to $\mathcal{S}_1$, have coordinates (x'_1, x'_2, x'_3) when referred to $\mathcal{S}_2$. (*a*) Obtain equations *(20)* and *(21)* effecting the change from one system to the other. (*b*) Show that the effect of a second transformation of the form $\tau X'' = \begin{bmatrix} a & 0 & 0 \\ 0 & b & 0 \\ 0 & 0 & c \end{bmatrix} X' = \mathbf{D}X'$, $|\mathbf{D}| \neq 0$, on *(20)* is to retain the reference triangle while changing the unit point.

(*a*) Since the effect of *(21)* is to carry $(1, 0, 0)$ into (a_{11}, a_{21}, a_{31}), $(0, 1, 0)$ into (a_{12}, a_{22}, a_{32}) and $(0, 1, 0)$ into (a_{13}, a_{23}, a_{33}), we find for *(21)*

$$(\text{i}) \qquad \sigma X = \begin{bmatrix} a_{11} & a_{12} & a_{13} \\ a_{21} & a_{22} & a_{23} \\ a_{31} & a_{32} & a_{33} \end{bmatrix} X' = \mathbf{A}X' = \mathbf{E}^{-1}X'$$

and obtain for *(20)*

$$(\text{ii}) \qquad \rho X' = \mathbf{A}^{-1}X = \mathbf{E}X$$

(*b*) Applying the inverse $\tau'X' = \mathbf{D}^{-1}X''$ to (i), we find

$$(\text{iii}) \qquad \sigma X = \mathbf{E}^{-1}\mathbf{D}^{-1}X'' = \begin{bmatrix} a_{11}/a & a_{12}/b & a_{13}/c \\ a_{21}/a & a_{22}/b & a_{23}/c \\ a_{31}/a & a_{32}/b & a_{33}/c \end{bmatrix} X''$$

Call this new coordinate system $\mathcal{S}_3$. Now (iii) carries the vertices of the reference triangle of $\mathcal{S}_3$ back into $(a_{11}/a, a_{21}/a, a_{31}/a)$, $(a_{12}/b, a_{22}/b, a_{32}/b)$, $(a_{13}/c, a_{23}/c, a_{33}/c)$ respectively which are merely different representations of the vertices of the reference triangle relative to $\mathcal{S}_1$. Thus, $\tau X'' = \mathbf{D}X'$ is a transformation which leaves unchanged the vertices of the reference triangle but, since it changes their representations, selects a new unit point.

In summary, *(20)* is a point transformation having an inverse and preserving both collinearity and cross ratio; as such, it belongs to the class called projective transformations. Dually, there is a line transformation having an inverse and preserving both concurrency and cross ratio. We now show that this latter transformation is induced by *(20)* and is, in effect, identical with it.

Consider the line $x: X_1x_1 + X_2x_2 + X_3x_3 = 0$ which in matrix notation is

$$[X_1, X_2, X_3] \cdot \begin{bmatrix} x_1 \\ x_2 \\ x_3 \end{bmatrix} = 0 \tag{22}$$

The effect of *(20)* is to carry x into the line x' whose equation is

$$[X_1, X_2, X_3] \cdot \rho \mathbf{E}^{-1} \cdot \begin{bmatrix} x_1' \\ x_2' \\ x_3' \end{bmatrix} = [X_1', X_2', X_3'] \cdot \begin{bmatrix} x_1' \\ x_2' \\ x_3' \end{bmatrix} = 0$$

The induced line transformation is given by

$$\tau[X_1', X_2', X_3'] = [X_1, X_2, X_3] \cdot \mathbf{E}^{-1}$$

which, by taking the transpose of both sides, assumes the more familiar form

$$\tau \begin{bmatrix} X_1' \\ X_2' \\ X_3' \end{bmatrix} = (\mathbf{E}^{-1})^T \begin{bmatrix} X_1 \\ X_2 \\ X_3 \end{bmatrix} \tag{23}$$

or, more briefly, $\tau x' = (\mathbf{E}^{-1})^T x$.   Its inverse is given, in turn, by

$$\tau' x = \mathbf{E}^T x' \tag{24}$$

Example 16.11.

The projective transformation

$$\rho X' = \mathbf{E} X = \begin{bmatrix} 1 & 3 & 5 \\ 1 & -2 & 3 \\ 2 & 4 & -3 \end{bmatrix} X$$

carries the line $l\colon 4x_1 + 3x_2 - 5x_3 = 0$ onto

$$l'\colon [4,3,-5] \cdot \mathbf{E}^{-1} X' = [4,3,-5] \cdot \begin{bmatrix} -6 & 29 & 19 \\ 9 & -13 & 2 \\ 8 & 2 & -5 \end{bmatrix} \begin{bmatrix} x_1' \\ x_2' \\ x_3' \end{bmatrix} = [-37, 67, 107] \begin{bmatrix} x_1' \\ x_2' \\ x_3' \end{bmatrix}$$

$$= -37x_1' + 67x_2' + 107x_3' = 0$$

or, dropping the primes since we are thinking of this as an alibi transformation, $37x_1 - 67x_2 - 107x_3 = 0$.

By (*23*) the induced line transformation becomes

$$\sigma \begin{bmatrix} X_1' \\ X_2' \\ X_3' \end{bmatrix} = \begin{bmatrix} -6 & 9 & 8 \\ 29 & -13 & 2 \\ 19 & 2 & -5 \end{bmatrix} \begin{bmatrix} X_1 \\ X_2 \\ X_3 \end{bmatrix}$$

which, of course, yields the same line l'.

Projective transformations of the type (*20*) and (*23*) are called *collineations*. Later we shall consider other projective transformations, called *correlations*, which carry points into lines and lines into points.

In Problem 16.49 the reader is asked to verify

Theorem 16.9.   A projective transformation [collineation of the form (*20*)] is uniquely determined when four pairs of corresponding points, no three points of either set being collinear, are given.

CANONICAL FORM OF A COLLINEATION

Consider the collineation

$$\rho X' = \mathbf{E} X, \quad |\mathbf{E}| \neq 0 \tag{20}$$

If $\mathbf{E} = I$, the effect of the transformation is to carry each point of the plane and also each line of the plane into itself. The transformation is known as the *identity*. Suppose $\mathbf{E} \neq I$. We raise the question: Are there individual points and lines of the plane which remain fixed under the transformation? Such points and lines, if any, are called *double* (*self-corresponding, invariant, fixed*) points and lines of the transformation. The double points of *(20)* are those points for which $X' = X$; hence they are those points satisfying $\rho X = \mathbf{E} X$ or

$$(\mathbf{E} - \rho I)X \;=\; \begin{bmatrix} e_{11} - \rho & e_{12} & e_{13} \\ e_{21} & e_{22} - \rho & e_{23} \\ e_{31} & e_{32} & e_{33} - \rho \end{bmatrix} \begin{bmatrix} x_1 \\ x_2 \\ x_3 \end{bmatrix} \;=\; 0$$

The system of equations

$$\begin{cases} (e_{11} - \rho)x_1 + e_{12}x_2 + e_{13}x_3 = 0 \\ e_{21}x_1 + (e_{22} - \rho)x_2 + e_{23}x_3 = 0 \\ e_{31}x_1 + e_{32}x_2 + (e_{33} - \rho)x_3 = 0 \end{cases} \tag{25}$$

will have a non-trivial solution if and only if

$$\phi(\rho) \;=\; \begin{vmatrix} e_{11} - \rho & e_{12} & e_{13} \\ e_{21} & e_{22} - \rho & e_{23} \\ e_{31} & e_{32} & e_{33} - \rho \end{vmatrix} \;=\; 0 \tag{26}$$

This is the so-called characteristic equation of the matrix $\mathbf{E}$ which will also be called the *characteristic equation* of the collineation *(20)*. The equation has either three real roots or one real and a pair of conjugate imaginary roots. We shall be concerned with collineations having real characteristic roots. Let ρ_i be one of the roots. The matrix $\mathbf{E} - \rho_i I$ is then singular and the system of equations *(25)* with ρ replaced by ρ_i will have a non-trivial solution, say, $X^i : (x_{i_1}, x_{i_2}, x_{i_3})$ which is a double point of the transformation associated with the characteristic root ρ_i.

In Problem 16.8, we prove

Theorem 16.10. If $\rho_1, X^1; \rho_2, X^2; \rho_3, X^3$ are the characteristic roots and associated double points of *(20)*, then $\rho_1 \neq \rho_2 \neq \rho_3$ implies X^1, X^2, X^3 non-collinear.

Corresponding to each double point X^i of a collineation *(20)* there is a double line whose existence follows when the argument above is applied to the induced collineation *(23)*. In this connection, recall that if ρ_i is a characteristic root of *(20)*, then $1/\rho_i$ is a characteristic root of *(23)*.

Suppose X^1 is a double point of *(20)* and take it as the point $A_1 : (1, 0, 0)$ of a new triangle of reference. The condition that A_1 be fixed requires that the first equation of *(20)* reduce to $\rho x_1' = e_{11}x_1$, i.e. $e_{12} = e_{13} = 0$. Thus when the characteristic roots of *(20)* are real and distinct and the associated double points are taken as vertices of a new reference triangle, *(20)* reduces to

$$(i) \qquad \rho x_1' = e_{11}x_1, \qquad \rho x_2' = e_{22}x_2, \qquad \rho x_3' = e_{33}x_3$$

The transformation which effects this reduction is $\tau X = \mathbf{R}Y$, in which the columns of $\mathbf{R}$ are the coordinates of X^1, X^2, X^3. Applying this alias transformation to *(20)*, we obtain

$$\rho \mathbf{R}Y' = \mathbf{E}\mathbf{R}Y$$

or, reverting to the notation in (i),

$$\rho X' = \mathbf{R}^{-1}\mathbf{E}\mathbf{R}X$$

The matrices $\mathbf{E}$ and $\mathbf{R}^{-1}\mathbf{ER}$, being similar, have the same characteristic roots; hence, $e_{11} = \rho_1$, $e_{22} = \rho_2$, $e_{33} = \rho_3$.

Then (i) becomes $\qquad \rho x_1' = \rho_1 x_1, \quad \rho x_2' = \rho_2 x_2, \quad \rho x_3' = \rho_3 x_3 \qquad\qquad (27)$

called a *canonical form* of (20).

Example 16.12.

Show that the collineation $\rho X' = \begin{bmatrix} 1 & 0 & -1 \\ 1 & 2 & 1 \\ 2 & 2 & 3 \end{bmatrix} X$ has distinct double points and obtain the canonical form.

The characteristic equation is

$$\phi(\rho) = \begin{vmatrix} 1-\rho & 0 & -1 \\ 1 & 2-\rho & 1 \\ 2 & 2 & 3-\rho \end{vmatrix} = 6 - 11\rho + 6\rho^2 - \rho^3 = 0$$

and the characteristic roots are $\rho_1 = 1$, $\rho_2 = 2$, $\rho_3 = 3$. When $\rho = \rho_1 = 1$, (25) becomes

$$\begin{cases} \qquad\quad -\ x_3 = 0 \\ x_1 + \ x_2 + \ x_3 = 0 \\ 2x_1 + 2x_2 + 2x_3 = 0 \end{cases}$$

and $X^1 : (1, -1, 0)$ is the associate double point. When $\rho = \rho_2 = 2$, (25) becomes

$$\begin{cases} -x_1 \qquad\quad -\ x_3 = 0 \\ x_1 \qquad\quad +\ x_3 = 0 \\ 2x_1 + 2x_2 + \ x_3 = 0 \end{cases}$$

and $X^2 : (2, -1, -2)$ is the associate double point. When $\rho = \rho_3 = 3$, (25) becomes

$$\begin{cases} -2x_1 \qquad\quad -\ x_3 = 0 \\ x_1 - \ x_2 + \ x_3 = 0 \\ 2x_1 + 2x_2 \qquad\quad = 0 \end{cases}$$

and $X^3 : (1, -1, -2)$ is the associate double point.

The double lines are $x^1 = X^2 X^3 : 2x_2 - x_3 = 0;$ $x^2 = X^3 X^1 : x_1 + x_2 = 0$ and $x^3 = X^1 X^2 : 2x_1 + 2x_2 + x_3 = 0$.

Take $\mathbf{R} = \begin{bmatrix} 1 & 2 & 1 \\ -1 & -1 & -1 \\ 0 & -2 & -2 \end{bmatrix}$; then $\mathbf{R}^{-1} = \dfrac{1}{2}\begin{bmatrix} 0 & -2 & 1 \\ 2 & 2 & 0 \\ -2 & -2 & -1 \end{bmatrix}$ and

$$\rho X' = \mathbf{R}^{-1}C\mathbf{R}X = \frac{1}{2}\begin{bmatrix} 0 & -2 & 1 \\ 2 & 2 & 0 \\ -2 & -2 & -1 \end{bmatrix} \cdot \begin{bmatrix} 1 & 0 & -1 \\ 1 & 2 & 1 \\ 2 & 2 & 3 \end{bmatrix} \cdot \begin{bmatrix} 1 & 2 & 1 \\ -1 & -1 & -1 \\ 0 & -2 & -2 \end{bmatrix} X = \begin{bmatrix} 1 & 0 & 0 \\ 0 & 2 & 0 \\ 0 & 0 & 3 \end{bmatrix} X$$

and the canonical form is $\rho x_1' = x_1, \quad \rho x_2' = 2x_2, \quad \rho x_3' = 3x_3$.

See also Problem 16.9.

PLANAR HOMOLOGIES AND ELATIONS

Suppose the characteristic equation (26) of (20) has a simple root ρ_1 and a root ρ_2 of multiplicity two. Two cases arise, each leading to a distinct canonical form, according as the rank of the coefficient matrix of (25) with ρ replaced by ρ_2 is two or one. The case when the rank is two is illustrated in Problem 16.10. When the rank is one, the collineation is called a *planar homology*. An illustration is given in

Example 16.13.

Obtain the canonical form of the collineation $X' = \begin{bmatrix} 2 & 2 & 1 \\ 1 & 3 & 1 \\ 1 & 2 & 2 \end{bmatrix} X$.

The characteristic equation is

$$\phi(\rho) \;=\; \begin{bmatrix} 2-\rho & 2 & 1 \\ 1 & 3-\rho & 1 \\ 1 & 2 & 2-\rho \end{bmatrix} \;=\; (5-\rho)(1-\rho)^2 \;=\; 0$$

and the characteristic roots are $\rho_1 = 5$, $\rho_2 = 1$, $\rho_3 = 1$.

When $\rho = \rho_1 = 5$, (25) becomes

$$\begin{cases} -3x_1 + 2x_2 + x_3 = 0 \\ x_1 - 2x_2 + x_3 = 0 \\ x_1 + 2x_2 - 3x_3 = 0 \end{cases}$$

and X': $(1,1,1)$ is the associate double point.

When $\rho = \rho_2 = 1$, (25) becomes

(i)
$$\begin{cases} x_1 + 2x_2 + x_3 = 0 \\ x_1 + 2x_2 + x_3 = 0 \\ x_1 + 2x_2 + x_3 = 0 \end{cases}$$

with coefficient matrix $\begin{bmatrix} 1 & 2 & 1 \\ 1 & 2 & 1 \\ 1 & 2 & 1 \end{bmatrix}$ of rank $r = 1$. Then every point on the line x^1: $x_1 + 2x_2 + x_3 = 0$ is a

double point; also x^1 and the pencil of lines joining X^1 (not on x^1) to the points of x^1 are double lines. There is a difference in these double lines, however; x^1 *is also a line of double points.*

As vertices of a new reference triangle (see Fig. 16-4) take X^1 and any two distinct points as X^2: $(1, 0, -1)$ and X^3: $(1, -1, 1)$ on x^1. Then

$$\mathbf{R} = \begin{bmatrix} 1 & 1 & 1 \\ 1 & 0 & -1 \\ 1 & -1 & 1 \end{bmatrix}, \qquad \mathbf{R}^{-1} = \frac{1}{4}\begin{bmatrix} 1 & 2 & 1 \\ 2 & 0 & -2 \\ 1 & -2 & 1 \end{bmatrix}, \qquad \rho X' = \mathbf{R}^{-1}\mathbf{E}\mathbf{R}X = \begin{bmatrix} 5 & 0 & 0 \\ 0 & 1 & 0 \\ 0 & 0 & 1 \end{bmatrix}X$$

and the required canonical form is

$$\rho x_1' = \rho_1 x_1, \qquad \rho x_2' = \rho_2 x_2, \qquad \rho x_3' = \rho_2 x_3$$

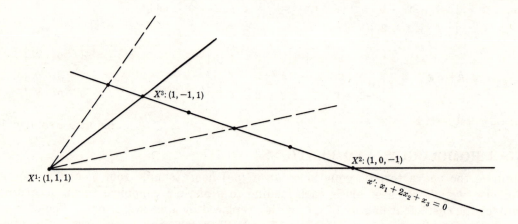

Fig. 16-4

A planar homology, then, is a collineation (not the identity) which has a double line k consisting entirely of double points and a pencil of double lines on a double point K, not on k. The line k is called the *axis* and the point K is called the *center* of the homology.

Next, suppose the characteristic equation (*26*) of (*20*) has a root ρ_1 of multiplicity three. Again, two cases arise according as the rank of the coefficient matrix of (*25*), with ρ replaced by ρ_1, is two or one. The case when the rank is two is illustrated in Problem 16.11. When the rank is one, the collineation is called an *elation*. An illustration is given in

Example 16.14.

Obtain a canonical form of the collineation $\rho X' = \begin{bmatrix} 2 & 2 & 3 \\ -2 & -3 & -6 \\ 1 & 2 & 4 \end{bmatrix} X.$

The characteristic equation is

$$\phi(\rho) = \begin{bmatrix} 2-\rho & 2 & 3 \\ -2 & -3-\rho & -6 \\ 1 & 2 & 4-\rho \end{bmatrix} = (1-\rho)^3 = 0$$

and the characteristic roots are $\rho_1 = \rho_2 = \rho_3 = 1$.

When $\rho = \rho_1 = 1$, (*25*) becomes

$$\begin{cases} x_1 + 2x_2 + 3x_3 = 0 \\ -2x_1 - 4x_2 - 6x_3 = 0 \\ x_1 + 2x_2 + 3x_3 = 0 \end{cases}$$

of rank $r = 1$. There is then a line of double points x^3: $x_1 + 2x_2 + 3x_3 = 0$. From the induced collineation

$\sigma x' = \begin{bmatrix} 0 & 2 & -1 \\ -2 & 5 & -2 \\ -3 & 6 & -2 \end{bmatrix} x,$ or following the note to Problem 16.9, it is found that there is a pencil of double

lines on X^1: $(1, -2, 1)$.

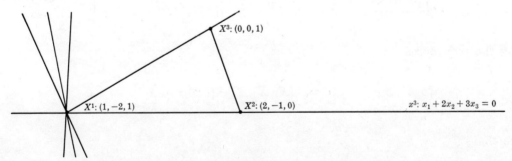

Fig. 16-5

As vertices of a new triangle of reference take X^1, any other point as X^2: $(2, -1, 0)$ on x^3, and any other point of the plane as X^3: $(0, 0, 1)$ such that X^1, X^2, X^3 are non-collinear. Then

$$\mathbf{R} = \begin{bmatrix} 1 & 2 & 0 \\ -2 & -1 & 0 \\ 1 & 0 & 1 \end{bmatrix}, \qquad \mathbf{R}^{-1} = \frac{1}{3}\begin{bmatrix} -1 & -2 & 0 \\ 2 & 1 & 0 \\ 1 & 2 & 3 \end{bmatrix}$$

and we obtain the canonical form

$$\rho x_1' = x_1 + 3x_3, \qquad \rho x_2' = x_2, \qquad \rho x_3' = x_3$$

An elation, then, is a collineation (not the identity) which has a line k of double points called its *axis* and a pencil of double lines on K (a point on k) called its *center*.

CORRELATIONS

The equations

$$\rho \begin{bmatrix} X_1 \\ X_2 \\ X_3 \end{bmatrix} = \mathbf{C} \begin{bmatrix} x_1 \\ x_2 \\ x_3 \end{bmatrix}, \qquad \mathbf{C} = [c_{ij}], \quad (i,j = 1,2,3), \quad |\mathbf{C}| \neq 0 \qquad (28)$$

define a one-to-one correspondence, between the points $X = (x_1, x_2, x_3)$ and the lines $x: [X_1, X_2, X_3]$ of the plane, called a *correlation*. We may regard (28) then as a transformation which carries the point X of the projective plane into the line x of this plane. Its inverse, which carries x back into X, is given by

$$\sigma \begin{bmatrix} x_1 \\ x_2 \\ x_3 \end{bmatrix} = \mathbf{C}^{-1} \begin{bmatrix} X_1 \\ X_2 \\ X_3 \end{bmatrix} \qquad (29)$$

Let $P: (y)$ and $Q: (z)$ be distinct points and denote their respective correspondents under (28) as $p: [Y]$ and $q: [Z]$. It follows easily that the correspondent of $R: (y + \lambda z)$, another point on the line PQ is $r: [Y + \lambda Z]$, another line on the point $p \cdot q$. The correlation (28) then carries a line s of points into a point S of lines. (We leave for the reader to verify that under (28) cross ratio is also preserved and, thus, correlations are projective transformations.) In general, see Problem 16.54, the point S and the line s are not correspondents. Hence, paralleling our study of collineations, the correlation (28) induces another transformation (correlation), generally distinct from (28), which carries the lines of the plane into points of the plane.

To find the equation of this induced correlation, suppose it carries the line $\{X'x\} = 0$ into the point $\{Xx'\} = 0$. (Here, the primes denote fixed coordinates.) Under (28), $\{X'x\} = 0$ is carried into

$$[X_1', X_2', X_3'] \cdot \begin{bmatrix} x_1 \\ x_2 \\ x_3 \end{bmatrix} = [X_1', X_2', X_3'] \cdot \mathbf{C}^{-1} \cdot \begin{bmatrix} X_1 \\ X_2 \\ X_3 \end{bmatrix} = 0$$

Taking the transposes

$$[x_1, x_2, x_3] \cdot \begin{bmatrix} X_1' \\ X_2' \\ X_3' \end{bmatrix} = [X_1, X_2, X_3] \cdot (\mathbf{C}^{-1})^T \cdot \begin{bmatrix} X_1' \\ X_2' \\ X_3' \end{bmatrix} = 0$$

we see that

$$\sigma \begin{bmatrix} x_1' \\ x_2' \\ x_3' \end{bmatrix} = (\mathbf{C}^{-1})^T \begin{bmatrix} X_1' \\ X_2' \\ X_3' \end{bmatrix}$$

or, upon dropping the primes,

$$\sigma \begin{bmatrix} x_1 \\ x_2 \\ x_3 \end{bmatrix} = (\mathbf{C}^{-1})^T \begin{bmatrix} X_1 \\ X_2 \\ X_3 \end{bmatrix} \qquad (30)$$

defines the induced transformation. Its inverse is

$$\tau \begin{bmatrix} X_1 \\ X_2 \\ X_3 \end{bmatrix} = \mathbf{C}^T \begin{bmatrix} x_1 \\ x_2 \\ x_3 \end{bmatrix} \qquad (31)$$

A correlation *(28)* whose matrix **C** is symmetric is called a *polar correlation* or *polarity*. Since for a polarity *(29)* and *(30)* are identical, the polarity not only carries a point P into a line p but also carries the lines x of the pencil on P into the points X of the pencil on p.

The importance of a polarity as

$$\rho \begin{bmatrix} X_1 \\ X_2 \\ X_3 \end{bmatrix} = \mathbf{C} \begin{bmatrix} x_1 \\ x_2 \\ x_3 \end{bmatrix}, \qquad \sigma \begin{bmatrix} x_1 \\ x_2 \\ x_3 \end{bmatrix} = \mathbf{C}^{-1} \begin{bmatrix} X_1 \\ X_2 \\ X_3 \end{bmatrix} \tag{32}$$

where $\mathbf{C} = [c_{ij}]$, $c_{ij} = c_{ji}$, $|\mathbf{C}| \neq 0$, is due to the fact that, whereas a point P does not generally lie on its correspondent, the locus of all points X which *do* lie on their correspondents is a curve of order two, called a *point conic*. In other words, every point to line polarity defines a point conic and, dually, every line to point polarity defines a *line conic*. For an example, see Problem 16.12.

Conics will be considered in some detail in the next chapter. There we shall use definitions in keeping with their introduction in Chapter 8.

Solved Problems

16.1. Prove: The cross ratio of any four points on a line o is independent of the coordinate system established on the line.

Let P, Q, R, S be the four points and suppose, relative to the coordinate system consisting of $A: (a)$, $B: (b)$; $C: (a+b)$ on o, we find $P: p$, $Q: q$, $R: r$, $S: s$ while, relative to another coordinate system on o consisting of $A', B'; C'$ we have $P: p'$, $Q: q'$, $R: r'$, $S: s'$. Suppose, relative to the first coordinate system, $A': u$ and $B': v$. Then for the point P we have

$$P: (pa+b) \qquad \text{and} \qquad P: \big(p'(ua+b) + (va+b)\big) = \left(\frac{p'u+v}{1+p'} a + b \right)$$

Thus, $p = \dfrac{p'u+v}{1+p'}$ and, similarly, $q = \dfrac{q'u+v}{1+q'}$, $r = \dfrac{r'u+v}{1+r'}$, $s = \dfrac{s'u+v}{1+s'}$. Then

$$p - r = \frac{p'u+v}{1+p'} - \frac{r'u+v}{1+r'} = \frac{(u-v)(p'-r')}{(1+p')(1+r')}$$

and, similarly,

$$p - s = \frac{(u-v)(p'-s')}{(1+p')(1+s')}, \qquad q - s = \frac{(u-v)(q'-s')}{(1+q')(1+s')}, \qquad q - r = \frac{(u-v)(q'-r')}{(1+q')(1+r')}$$

Finally,

$$(p, q; r, s) = \frac{(p-r)(q-s)}{(p-s)(q-r)} = \frac{(p'-r')(q'-s')}{(p'-s')(q'-r')} = (p', q'; r', s')$$

as required.

16.2. Prove: If the joins of three pairs of corresponding points of two projective pencils are concurrent, the pencils are perspective.

Let $P: p$, $P': p'$; $Q: q$, $Q': q'$; $R: r$, $R': r'$ be three pairs of corresponding points whose joins $a = PP'$, $b = QQ'$, $c = RR'$ are on the point O. Let $S: s$, $S': s'$ be any other pair of corresponding points of the projectivity and let $OS = d$, $OS' = d'$. Since

$$(a, b; c, d) = (p, q; r, s) = (p', q'; r', s') = (a, b; c, d')$$

it follows that $d = d'$ and, hence, S and S' are perspective from O.

16.3. Obtain the equation of the projectivity which carries the points $1, 3, \infty$ into the points $5, 4, \infty$ respectively.

The required equation has the form $x' = \dfrac{ax + b}{cx + d}$. When $x = \infty$, $x' = a/c = \infty$ and $c = 0$. The equation has the form $dx' - ax - b = 0$. Since $x' = 5$ when $x = 1$ and $x' = 4$ when $x = 3$, we have

$$5d - a - b = 0 \quad \text{and} \quad 4d - 3a - b = 0$$

By subtraction, $d + 2a = 0$; take $d = -2$, $a = 1$. Then $b = 5d - a = -11$ and the required equation is $x' = \dfrac{x - 11}{-2}$.

16.4. Prove: The cross ratio K of the two double points of a hyperbolic projectivity and a pair of corresponding points is independent of the pair of corresponding points selected.

Consider the hyperbolic projectivity

$$x' = \frac{ax + b}{cx + d} \tag{3}$$

whose double points are given by

$$M: x_1 = \frac{a - d + \theta}{2c} \quad \text{and} \quad N: x_2 = \frac{a - d - \theta}{2c}$$

where $\theta = \sqrt{(d - a)^2 + 4bc}$. Let $P: p$ and $P': p'$ be a pair of corresponding points. Then

$$K = (M, N; P, P') = (x_1, x_2; p, p') = \frac{x_1 - p}{x_1 - p'} \cdot \frac{x_2 - p'}{x_2 - p} = \frac{pp' - x_1 p' - x_2 p + x_1 x_2}{pp' - x_1 p - x_2 p' + x_1 x_2}$$

From (3), we obtain $pp' = \dfrac{ap - dp' + b}{c}$ while $x_1 x_2 = -b/c$. Then

$$K = \frac{\dfrac{ap - dp' + b}{c} - p' \dfrac{a - d + \theta}{2c} - p \dfrac{a - d - \theta}{2c} - \dfrac{b}{c}}{\dfrac{ap - dp' + b}{c} - p \dfrac{a - d + \theta}{2c} - p' \dfrac{a - d - \theta}{2c} - \dfrac{b}{c}} = \frac{(p - p')(a + d + \theta)}{(p - p')(a + d - \theta)} = \frac{a + d + \theta}{a + d - \theta}$$

Clearly, the value of K depends solely upon the parameters a, b, c, d defining (3); in particular, it is independent of the pair P, P' of corresponding points used in obtaining it.

16.5. Derive the canonical forms

$$x' = x + p, \qquad p = b/a \qquad \text{when } c = 0,$$

$$\frac{1}{x' - x_1} = \frac{1}{x - x_1} + p, \quad p = \frac{-c}{cx_1 - a} \qquad \text{when } c \neq 0,$$

for parabolic projectivities.

When $c = 0$, then $b \neq 0$ and $a = d$. The equation of the projectivity is $x' = x + b/a$ or

$$x' = x + p, \quad p = b/a$$

When $c \neq 0$, we have Case 2, page 173. The double point is $x_1 = \dfrac{a - d}{2c}$; hence, $a - d = 2cx_1$. Since $(d - a)^2 + 4bc = 0$, we have $b = -cx_1^2$. Then

$$x' = \frac{ax + b}{cx + d} = \frac{ax - cx_1^2}{cx + a - 2cx_1}$$

$$xx' + \frac{a}{c}x' - 2x'x_1 - \frac{a}{c}x + x_1^2 = 0$$

$$(x' - x)(x_1 - a/c) - (x' - x_1)(x - x_1) = 0$$

$$\frac{x' - x}{(x' - x_1)(x - x_1)} = \frac{1}{x_1 - a/c}$$

and
$$\frac{1}{x'-x_1} - \frac{1}{x-x_1} = \frac{-1}{x_1 - a/c} = p$$

Thus, as required,
$$\frac{1}{x'-x_1} = \frac{1}{x-x_1} + p, \qquad p = \frac{-c}{cx_1 - a}$$

16.6. Prove: The projective transformation $\tau: x' = \dfrac{ax+b}{cx+d}$ of a line onto itself is an involution if and only if $a+d=0$.

Suppose τ is an involution which carries $P: p$ into $P': p'$, where $P' \neq P$. Then $\tau^{-1}: x = \dfrac{ax'+b}{cx'+d}$ carries P' back into P and we have
$$cpp' + dp' - ap - b = 0$$
and
$$cpp' - ap' + dp - b = 0$$
By subtraction, $(a+d)(p'-p) = 0$. Now $p'-p \neq 0$ since $P' \neq P$; hence, $a+d=0$ as required.

Conversely, suppose $a+d=0$ so that $\tau: x' = \dfrac{ax+b}{cx-a}$. We leave for the reader to show that τ^2 is the identity and, hence, τ is an involution.

PROJECTIVE COORDINATES IN THE PLANE

16.7. Prove: The transformation $\rho X' = \mathbf{E}X$, $|\mathbf{E}| \neq 0$, preserves collinearity.

Let (y), (z), $(y+\lambda z)$ be distinct collinear points. Under the transformation they are carried respectively into
$$\mathbf{E}Y, \quad \mathbf{E}Z, \quad \mathbf{E}\begin{bmatrix} y_1+\lambda z_1 \\ y_2+\lambda z_2 \\ y_3+\lambda z_3 \end{bmatrix} = \mathbf{E}\begin{bmatrix} y_1 \\ y_2 \\ y_3 \end{bmatrix} + \lambda \mathbf{E}\begin{bmatrix} z_1 \\ z_2 \\ z_3 \end{bmatrix} = \mathbf{E}Y + \lambda \mathbf{E}Z$$
which, in turn, are clearly collinear points.

16.8. Prove: If ρ_1, X^1; ρ_2, X^2; ρ_3, X^3 are the characteristic roots and associated double points of $\rho X' = \mathbf{E}X$, then $\rho_1 \neq \rho_2 \neq \rho_3$ implies X^1, X^2, X^3 non-collinear.

Suppose $\rho_1 \neq \rho_2 \neq \rho_3$ but X_1, X_2, X_3 collinear. Then there exist constants $\alpha_1, \alpha_2, \alpha_3$ not all zero such that
$$\text{(i)} \qquad \alpha_1 X_1 + \alpha_2 X_2 + \alpha_3 X_3 = 0$$

Multiplying (i) on the left by $\mathbf{E}$ and using $\mathbf{E}X_i = \rho_i X_i$, $(i=1,2,3)$, we have
$$\text{(ii)} \qquad \alpha_1 \mathbf{E}X_1 + \alpha_2 \mathbf{E}X_2 + \alpha_3 \mathbf{E}X_3 = \alpha_1 \rho_1 X_1 + \alpha_2 \rho_2 X_2 + \alpha_3 \rho_3 X_3 = 0$$

Multiplying (ii) on the left by $\mathbf{E}$, we obtain
$$\text{(iii)} \qquad \alpha_1 \rho_1^2 X_1 + \alpha_2 \rho_2^2 X_2 + \alpha_3 \rho_3^2 X_3 = 0$$

Write (i), (ii), (iii) in matrix notation as
$$\text{(iv)} \qquad \begin{bmatrix} 1 & 1 & 1 \\ \rho_1 & \rho_2 & \rho_3 \\ \rho_1^2 & \rho_2^2 & \rho_3^2 \end{bmatrix} \cdot \begin{bmatrix} \alpha_1 X_1 \\ \alpha_2 X_2 \\ \alpha_3 X_3 \end{bmatrix} = \mathbf{B}\begin{bmatrix} \alpha_1 X_1 \\ \alpha_2 X_2 \\ \alpha_3 X_3 \end{bmatrix} = 0$$

Now $|\mathbf{B}| = (\rho_1-\rho_2)(\rho_2-\rho_3)(\rho_3-\rho_1) \neq 0$ and, hence, $\mathbf{B}^{-1}$ exists. Multiplying (iv) on the left by $\mathbf{B}^{-1}$, we have
$$\mathbf{B}^{-1}\mathbf{B} \cdot \begin{bmatrix} \alpha_1 X_1 \\ \alpha_2 X_2 \\ \alpha_3 X_3 \end{bmatrix} = \mathbf{I} \cdot \begin{bmatrix} \alpha_1 X_1 \\ \alpha_2 X_2 \\ \alpha_3 X_3 \end{bmatrix} = \begin{bmatrix} \alpha_1 X_1 \\ \alpha_2 X_2 \\ \alpha_3 X_3 \end{bmatrix} = 0$$

But this requires $\alpha_1 = \alpha_2 = \alpha_3 = 0$, contrary to our assumption. Hence, X_1, X_2, X_3 are non-collinear.

16.9. Examine the line transformation induced by the point transformation $\rho X' = \mathbf{E}X = \begin{bmatrix} 1 & 0 & -1 \\ 1 & 2 & 1 \\ 2 & 2 & 3 \end{bmatrix} X$ of Example 16.12 for double elements.

Here $\mathbf{E}^{-1} = \begin{bmatrix} \frac{2}{3} & -\frac{1}{3} & \frac{1}{3} \\ -\frac{1}{6} & \frac{5}{6} & -\frac{1}{3} \\ -\frac{1}{3} & -\frac{1}{3} & \frac{1}{3} \end{bmatrix}$; the induced line transformation is

$$\sigma x' = (\mathbf{E}^{-1})^T x = \begin{bmatrix} \frac{2}{3} & -\frac{1}{6} & -\frac{1}{3} \\ -\frac{1}{3} & \frac{5}{6} & -\frac{1}{3} \\ \frac{1}{3} & -\frac{1}{3} & \frac{1}{3} \end{bmatrix} x$$

The system of equations corresponding to (25), page 182, is

(i)
$$(\tfrac{2}{3} - \sigma)X_1 - \tfrac{1}{6}X_2 - \tfrac{1}{3}X_3 = 0$$
$$-\tfrac{1}{3}X_1 + (\tfrac{5}{6} - \sigma)X_2 - \tfrac{1}{3}X_3 = 0$$
$$\tfrac{1}{3}X_1 - \tfrac{1}{3}X_2 + (\tfrac{1}{3} - \sigma)X_3 = 0$$

which will have a non-trivial solution if and only if

$$\phi'(\sigma) = \begin{vmatrix} \tfrac{2}{3} - \sigma & -\tfrac{1}{6} & -\tfrac{1}{3} \\ -\tfrac{1}{3} & \tfrac{5}{6} - \sigma & -\tfrac{1}{3} \\ \tfrac{1}{3} & -\tfrac{1}{3} & \tfrac{1}{3} - \sigma \end{vmatrix} = 0$$

The roots of $\phi'(\sigma) = 0$ are the reciprocals $\sigma_1 = 1$, $\sigma_2 = \tfrac{1}{2}$, $\sigma_3 = \tfrac{1}{3}$ of the roots of the characteristic equation $\phi(\rho) = 0$ of $\mathbf{E}$.

For $\sigma = \sigma_1 = 1$, (i) becomes
$$\begin{cases} -2X_1 - X_2 - 2X_3 = 0 \\ -2X_1 - X_2 - 2X_3 = 0 \\ X_1 - X_2 - 2X_3 = 0 \end{cases}$$

and x^1: $[0, 2, -1]$ or x^1: $2x_2 - x_3 = 0$ is the associate double line.

Similarly, for $\sigma = \sigma_2 = \tfrac{1}{2}$ and $\sigma = \sigma_3 = \tfrac{1}{3}$, we find x^2: $x_1 + x_2 = 0$ and x^3: $2x_1 + 2x_2 + x_3 = 0$ as respective double lines.

Note. From $\rho X = \mathbf{E}X$, where ρ and X are characteristic root and associated double point of $\mathbf{E}$, we obtain $\tfrac{1}{\rho}X = \mathbf{E}^{-1}X$; the converse is also true. Hence X is an associated double point of $\mathbf{E}^{-1}$ if and only if X is an associated double point of $\mathbf{E}$. Dually, x is an associated double line of $(\mathbf{E}^T)^{-1}$ if and only if x is an associated double line of $\mathbf{E}^T$. Since the characteristic roots of $\mathbf{E}^T$ are those of $\mathbf{E}$, the double lines obtained above may be found more easily from $(\mathbf{E}^T - \rho\mathbf{I})x = 0$ for the characteristic roots of $\mathbf{E}$. For example, from $(\mathbf{E}^T - \rho\mathbf{I})x = \begin{bmatrix} 1-\rho & 1 & 2 \\ 0 & 2-\rho & 2 \\ -1 & 1 & 3-\rho \end{bmatrix}\begin{bmatrix} X_1 \\ X_2 \\ X_3 \end{bmatrix} = 0$

for the characteristic root $\rho = \rho_3 = 3$, we find x^3: $[2, 2, 1]$ as before.

16.10. Obtain the canonical form of $\rho X' = \begin{bmatrix} 2 & 1 & -1 \\ 0 & 2 & -1 \\ -3 & -2 & 3 \end{bmatrix} X$.

The characteristic equation is

$$\phi(\rho) = \begin{bmatrix} 2-\rho & 1 & -1 \\ 0 & 2-\rho & -1 \\ -3 & -2 & 3-\rho \end{bmatrix} = (5-\rho)(1-\rho)^2 = 0$$

and the characteristic roots are $\rho_1 = 5$, $\rho_2 = \rho_3 = 1$.

When $\rho = \rho_1 = 5$, (25), page 182, becomes

$$\begin{cases} -3x_1 + x_2 - x_3 = 0 \\ \qquad\;\; -3x_2 - x_3 = 0 \\ -3x_1 - 2x_2 - 2x_3 = 0 \end{cases}$$

and $X^1 : (-4, -3, 9)$ is the associate double point.

When $\rho = \rho_2 = 1$, (25) becomes

$$\begin{cases} x_1 + x_2 - x_3 = 0 \\ \qquad\;\; x_2 - x_3 = 0 \\ -3x_1 - 2x_2 + 2x_3 = 0 \end{cases}$$

of rank $r = 2$ and $X^2 : (0, 1, 1)$ is the associate double point.

For this transformation there are two double points X^1 and X^2 with X^2 counted twice. Dually, there are two double lines — $x^3 = X^1X^2 : 3x_1 - x_2 + x_3 = 0$ and $x^1 : x_1 + x_2 - x_3 = 0$ — the latter being identified by the use of the induced line transformation

$$\sigma x' = \begin{bmatrix} 4 & 3 & 6 \\ -1 & 3 & 1 \\ 1 & 2 & 4 \end{bmatrix} x \quad \text{or from} \quad \sigma x' = \mathbf{E}^T x = \begin{bmatrix} 2 & 0 & -3 \\ 1 & 2 & -2 \\ -1 & -1 & 3 \end{bmatrix} x$$

As vertices of a new reference triangle take X^1, X^2 and any other point, say, $X^3 : (1, -2, -1)$ on x^1. Then

$$\mathbf{R} = \begin{bmatrix} -4 & 0 & 1 \\ -3 & 1 & -2 \\ 9 & 1 & -1 \end{bmatrix}, \quad \mathbf{R}^{-1} = \frac{1}{16}\begin{bmatrix} -1 & -1 & 1 \\ 21 & 5 & 11 \\ 12 & -4 & 4 \end{bmatrix}$$

and we obtain the canonical form

$$\rho x_1' = 5x_1, \quad \rho x_2' = x_2 - x_3, \quad \rho x_3' = x_3$$

16.11. Obtain the canonical form of $\quad \rho X' = \begin{bmatrix} 0 & 1 & 0 \\ 0 & 0 & 1 \\ 1 & -3 & 3 \end{bmatrix} X.$

The characteristic equation is

$$\phi(\rho) = \begin{bmatrix} -\rho & 1 & 0 \\ 0 & -\rho & 1 \\ 1 & -3 & 3 - \rho \end{bmatrix} = (1 - \rho)^3$$

and the characteristic roots are $\rho_1 = \rho_2 = \rho_3 = 1$.

When $\rho = \rho_1 = 1$, (25), page 182, becomes

$$\begin{cases} -x_1 + x_2 \qquad\;\; = 0 \\ \qquad\; -x_2 + x_3 = 0 \\ x_1 - 3x_2 + 2x_3 = 0 \end{cases}$$

of rank $r = 2$ and $X^1 : (1, 1, 1)$ is the associate double point.

Thus there is a single double point counted three times and a single double line $x^3 : x_1 - 2x_2 + x_3 = 0$ counted three times. Since X^1 is on x^3, take as vertices of a new triangle of reference X^1, another point as $X^2 : (2, 1, 0)$ on x^3, and any other point as $X^3 : (0, 0, 1)$ such that X^1, X^2, X^3 are non-collinear. Then

$$\mathbf{R} = \begin{bmatrix} 1 & 2 & 0 \\ 1 & 1 & 0 \\ 1 & 0 & 1 \end{bmatrix}, \quad \mathbf{R}^{-1} = \begin{bmatrix} -1 & 2 & 0 \\ 1 & -1 & 0 \\ 1 & -2 & 1 \end{bmatrix}$$

and the canonical form is

$$\rho x_1' = x_1 - x_2 + 2x_3, \quad \rho x_2' = x_2 - x_3, \quad \rho x_3' = x_3$$

16.12. Derive the equations of (a) the point conic and (b) the line conic defined by the polarity

$$\rho \begin{bmatrix} X_1 \\ X_2 \\ X_3 \end{bmatrix} = \mathbf{C} \begin{bmatrix} x_1 \\ x_2 \\ x_3 \end{bmatrix} \quad \text{and its inverse} \quad \sigma \begin{bmatrix} x_1 \\ x_2 \\ x_3 \end{bmatrix} = \mathbf{C}^{-1} \begin{bmatrix} X_1 \\ X_2 \\ X_3 \end{bmatrix}$$

(a) *The Point Conic.*

The polarity carries the point (x_1, x_2, x_3) into the line $[X_1, X_2, X_3]$. Since we require the point to be on the line, we must have $X_1x_1 + X_2x_2 + X_3x_3 = 0$. Now, under the polarity,

$$[x_1, x_2, x_3] \begin{bmatrix} X_1 \\ X_2 \\ X_3 \end{bmatrix} = 0 \quad \text{is carried into} \quad [x_1, x_2, x_3] \cdot \mathbf{C} \cdot \begin{bmatrix} x_1 \\ x_2 \\ x_3 \end{bmatrix} = 0$$

which, when expanded, becomes

$$[x_1, x_2, x_3] \cdot \begin{bmatrix} c_{11} & c_{12} & c_{13} \\ c_{12} & c_{22} & c_{23} \\ c_{13} & c_{23} & c_{33} \end{bmatrix} \cdot \begin{bmatrix} x_1 \\ x_2 \\ x_3 \end{bmatrix} = [x_1, x_2, x_3] \begin{bmatrix} c_{11}x_1 + c_{12}x_2 + c_{13}x_3 \\ c_{12}x_1 + c_{22}x_2 + c_{23}x_3 \\ c_{13}x_1 + c_{23}x_2 + c_{33}x_3 \end{bmatrix}$$

$$= c_{11}x_1^2 + c_{22}x_2^2 + c_{33}x_3^2 + 2c_{12}x_1x_2 + 2c_{13}x_1x_3 + 2c_{23}x_2x_3 = 0$$

This is the equation of the point conic defined by the polarity.

(b) *The Line Conic.*

The inverse of the polarity carries the line $[X_1, X_2, X_3]$ into the point (x_1, x_2, x_3). Since the line is to be on the point, we must again have $X_1x_1 + X_2x_2 + X_3x_3 = 0$. Under the inverse of the polarity

$$[X_1, X_2, X_3] \begin{bmatrix} x_1 \\ x_2 \\ x_3 \end{bmatrix} = 0 \quad \text{is carried into} \quad [X_1, X_2, X_3] \mathbf{C}^{-1} \begin{bmatrix} X_1 \\ X_2 \\ X_3 \end{bmatrix} = 0$$

the equation of the line conic.

Note that the line equation of a given point conic

$$[x_1, x_2, x_3] \mathbf{C} \begin{bmatrix} x_1 \\ x_2 \\ x_3 \end{bmatrix} = 0 \quad \text{is} \quad [X_1, X_2, X_3] \mathbf{C}^{-1} \begin{bmatrix} X_1 \\ X_2 \\ X_3 \end{bmatrix} = 0$$

and the point equation of a given line conic

$$[X_1, X_2, X_3] \mathbf{C} \begin{bmatrix} X_1 \\ X_2 \\ X_3 \end{bmatrix} = 0 \quad \text{is} \quad [x_1, x_2, x_3] \mathbf{C}^{-1} \begin{bmatrix} x_1 \\ x_2 \\ x_3 \end{bmatrix} = 0$$

Supplementary Problems

16.13. Determine the equation of the transformation which carries

(a) $0, 1, 2$ into $0, 4, 3$ respectively.
(b) $0, 1, 2$ into $2, 1, 3$ respectively.
(c) $0, 2, 5$ into $9, 1, 4$ respectively.
(d) $0, 1, \infty$ into $1, \infty, 0$ respectively.
(e) $2, 3, 4$ into $2, 3, 5$ respectively.
(f) $1, 2, \infty$ into $-1, -5, 3$ respectively.

Ans. (a) $x' = \dfrac{12x}{5x - 2}$ (d) $x' = \dfrac{1}{1 - x}$

(b) $x' = \dfrac{7x - 8}{3x - 4}$ (e) $x' = \dfrac{x + 6}{-x + 6}$

(c) $x' = \dfrac{5x - 9}{x - 1}$ (f) $x' = \dfrac{3x - 1}{x - 3}$

16.14. Obtain the real double points for the transformations of Problem 16.13 and thus show: $(a), (b), (e), (f)$ are hyperbolic; (c) is parabolic; (d) is elliptic.

16.15. The transformation in Problem 16.13(e) has $M:2$ and $N:3$ as double points and carries $P:4$ into $P':5$. Find two other pairs of corresponding Q, Q'; R, R' and show that

$$(M, N; P, P') = (M, N; Q, Q') = (M, N; R, R')$$

16.16. Obtain canonical forms for the hyperbolic and parabolic projectivities of Problem 16.13.

> *Ans.* (a) $\dfrac{x'}{x' - 14/5} = -6 \dfrac{x}{x - 14/5}$ (e) $\dfrac{x' - 2}{x' - 3} = \dfrac{3}{4} \dfrac{x - 2}{x - 3}$
>
> (b) $\dfrac{x' - 1}{x' - 8/3} = -4 \dfrac{x - 1}{x - 8/3}$ (f) $\dfrac{x' - 3 + 2\sqrt{2}}{x' - 3 - 2\sqrt{2}} = -\dfrac{x - 3 + 2\sqrt{2}}{x - 3 - 2\sqrt{2}}$
>
> (c) $\dfrac{1}{x' - 3} = \dfrac{1}{x - 3} + \dfrac{1}{2}$

16.17. Given the projective transformation $\tau: x' = \dfrac{ax + b}{cx + d}$, obtain the inverse transformation τ^{-1} and verify that $\tau \cdot \tau^{-1} = \tau^{-1} \cdot \tau$ is the identity transformation.

16.18. Write the inverse of each transformation of Problem 16.13.

> *Partial Ans.* (a) $x = \dfrac{2x'}{5x' - 12}$, (b) $x = \dfrac{4x' - 8}{3x' - 7}$, (f) $x = \dfrac{3x' - 1}{x' - 3}$

16.19. Given the projective transformations $\tau_1: x' = \dfrac{ax + b}{cx + d}$ and $\tau_2: x' = \dfrac{ex + f}{gx + h}$, show: (a) $\tau_2 \cdot \tau_1$ is a projective transformation. (b) $\tau_1 \cdot \tau_2$ is a projective transformation, (c) generally, $\tau_1 \cdot \tau_2 \neq \tau_2 \cdot \tau_1$.

> *Hint.* For (a) write $\tau_1: x' = \dfrac{ax + b}{cx + d}$ and $\tau_2: x'' = \dfrac{ex' + f}{gx' + h}$ and obtain $\tau_2 \cdot \tau_1: x'' = \dfrac{a'x + b'}{c'x + d'}$.
> Recall that this is a projective transformation if and only if $a'd' - b'c' \neq 0$.

16.20. Show that for $\tau_2 \cdot \tau_1$ in Problem 16.19, $a'd' - b'c' = (ad - bc)(eh - fg)$.

16.21. Suppose τ_1 and τ_2 are hyperbolic projectivities having the same double points M, N and with characteristic invariants K_1 and K_2 respectively. Show, using the canonical forms, that $\tau_2 \cdot \tau_1$ is hyperbolic with double points M, N and characteristic invariant $K_2 \cdot K_1$. Hence, $\tau_1 \cdot \tau_2 = \tau_2 \cdot \tau_1$.

16.22. Consider the products of two parabolic projectivities τ_1 and τ_2 having the same double point M and parameters p_1 and p_2 respectively.

16.23. Determine the involution having $1, -1$; $-2, 3$ as two of its reciprocal pairs.
> *Ans.* $xx' + 5(x + x') + 1 = 0$.

16.24. Suppose two reciprocal pairs of an involution are given by the roots of the quadratic equations (i) $a_1x^2 + b_1x + c_1 = 0$ and (ii) $a_2x^2 + b_2x + c_2 = 0$. Denote by x_1, x_1' the roots of (i); then

$$x_1x_1' = c_1/a_1 \text{ and } x_1 + x_1' = -b_1/a_1. \text{ Obtain the equation of the involution as } \begin{vmatrix} xx' & x + x' & 1 \\ c_1 & -b_1 & a_1 \\ c_2 & -b_2 & a_2 \end{vmatrix} = 0.$$

16.25. Obtain the equation of the involution determined by the reciprocal pairs given by $2x^2 - x - 1 = 0$ and $x^2 + 3x + 2 = 0$.   *Ans.* $7xx' + 5(x + x') + 1 = 0$.

16.26. What happens if the quadratic equations in Problem 16.24 have a root in common?

16.27. Determine the involution whose double points are $1, -2$.   *Ans.* $2xx' + (x + x') - 4 = 0$.

16.28. Prove Theorem 16.6, page 174.
> *Hint.* Show that $H(x_1, x_2; x, x')$ implies $2xx' - (x_1 + x_2)(x + x') + 2x_1x_2 = 0$.

16.29. Given two pairs of points p, p'; q, q' on a line, prove that there exists a unique pair x_1, x_2 on the line which separate harmonically each of the given pairs.

16.30. Given the pairs $3, 18$; $-4, -2/3$ on a line, locate the pair whose existence is established in Problem 16.29.   *Ans.* $-2, 6$.

16.31. By means of Problem 16.24, show that in the involution determined by $q_1: a_1x^2 + b_1x + c_1 = 0$ and $q_2: a_2x^2 + b_2x + c_2 = 0$ any other reciprocal pair is given by $q_1 + \lambda q_2$ for some unique value of λ.

PROJECTIVE COORDINATES IN THE PLANE

16.32. Prove Desargues' Two-Triangle Theorem using Fig. 16-3.

16.33. Prove the Theorem of Pappus when the coordinate system is chosen as in Example 16.8(b), page 176.

16.34. In Fig. 3-1(a), page 30, take $A: (1, 0, 0)$, $B: (0, 1, 0)$, $C: (0, 0, 1)$ and $P: (1, 1, 1)$. Find relative coordinates of Q, R, S and the equations of all sides of the complete quadrangle.

16.35. Prove: If three distinct collinear points, one on each side of the diagonal triangle of a complete quadrilateral, are marked, their harmonic conjugates with respect to the vertices of the quadrilateral are collinear.

Hint. Using Fig. 3-1(b), page 31, take $a: [1, 0, 0]$, $b: [0, 1, 0]$, $c: [0, 0, 1]$ and $p: [1, 1, 1]$.

16.36. Given a triangle ABC and two points A' and B', locate C' such that $AA', BC', B'C$ are concurrent and also AC', BB', CA' are concurrent. (a) Show that AB', BA', CC' are then concurrent. (b) What additional condition insures AB', BC', CA' concurrent?

Hint. In Fig. 16-3, page 177, interchange the labels C' and B' so that AA', BC', CB' are concurrent in P and AC', BB', CA' are concurrent provided $abc = 1$.   *Ans.* (b) $a = c$.

16.37. P and Q are distinct points on a fixed line r while A and B are distinct points not on r. Let $R = AP \cdot BQ$ and $S = AQ \cdot BP$. Show that $T = AB \cdot RS$ is a fixed point, i.e. is independent of the choice of P and Q.

Hint. Take $r: x_1 - x_2 = 0$, $P: (1, 1, a)$, $Q: (1, 1, b)$; $A: (1, 0, 0)$, $B: (0, 1, 0)$.

16.38. State and prove the dual of Problem 16.37.

16.39. Prove: If X is any point on the side PR and Y is any point on the side QS of a quadrangle $PQRS$, then $T = QX \cdot PY$, $U = SX \cdot RY$ and $B = PS \cdot QR$ are also collinear.

16.40. (a) Show that $p: x_1/a_1 + x_2/a_2 + x_3/a_3 = 0$ is the equation of the polar line of the point $P: (a)$ with respect to the reference triangle.

(b) Show that the polar lines of $P: (a)$ with respect to the pairs of sides of the reference triangle form a second triangle.

(c) Show that the two triangles in (b) are perspective from P and p.

16.41. Prove the theorem of Problem 4.2, page 51; also the theorem in the note.

Hint. Take the reference triangle as ABC and take A', B', C' on the line $ax_1 + bx_2 + cx_3 = 0$.

16.42. If $H(A, B; C, D)$ and $H(A', B'; C', D)$ on distinct lines, then AA', BB', CC' are concurrent.

16.43. Prove: A projective transformation of the form (*20*), page 179, is uniquely determined when four pairs of corresponding points, no three points of either set being collinear, are given.

Hint. Equations (*20*) have eight essential constants since the equations may be divided by any one of the e_{ij}. One of the coordinates of any point may be taken as 1; hence, transforming a point A into its correspondent A' gives two independent relations among the eight essential constants in (*20*). Of course, it may happen that the given pairs of points are such that the e_{ij} selected above must be zero. In this case, we eventually reach contradictory relations and must begin anew with some other e_{ij}. (Why can we be sure of a solution?)

16.44. Take $B_1: (1, 1, 2)$, $B_2: (3, -2, 4)$, $B_3: (5, 3, -3)$; $B_4: (9, 2, 3)$ in absolute coordinates and write the equations of the transformations (16) and (17). Combine these equations with those of Example 16.7, page 176, to obtain

$$\tau' \begin{bmatrix} x_1' \\ x_2' \\ x_3' \end{bmatrix} = \begin{bmatrix} 109 & -23 & 14 \\ -11 & 65 & -82 \\ -3 & -10 & 144 \end{bmatrix} \begin{bmatrix} x_1 \\ x_2 \\ x_3 \end{bmatrix} = \mathbf{C} \begin{bmatrix} x_1 \\ x_2 \\ x_3 \end{bmatrix} \qquad (19)$$

and

$$\tau \begin{bmatrix} x_1 \\ x_2 \\ x_3 \end{bmatrix} = \begin{bmatrix} 140 & 52 & 16 \\ 30 & 258 & 144 \\ 5 & 19 & 112 \end{bmatrix} \begin{bmatrix} x_1' \\ x_2' \\ x_3' \end{bmatrix} = \mathbf{D} \begin{bmatrix} x_1' \\ x_2' \\ x_3' \end{bmatrix} \qquad (20)$$

Verify that the columns of $\mathbf{C}$ are the coordinates of the reference triangle $A_1 A_2 A_3$ relative to the coordinate system $B_1, B_2, B_3; B_4$.

16.45. Find the projective transformation which carries $A_1, A_2, A_3; A_4$ of Example 16.7, page 176, into $B_1, B_2, B_3; B_4$ respectively of Problem 16.43. Prove that the diagonal points of the complete quadrangle $A_1 A_2 A_3 A_4$ are carried into the diagonal points of $B_1 B_2 B_3 B_4$, without computations.

$$Ans. \quad \rho X' = \begin{bmatrix} 243 & -42 & 27 \\ -61 & 134 & 11 \\ 157 & -38 & 133 \end{bmatrix} X$$

16.46. For the transformation $\rho \begin{bmatrix} x_1' \\ x_2' \\ x_3' \end{bmatrix} = \begin{bmatrix} 1 & 3 & 5 \\ 2 & -1 & 3 \\ 3 & 2 & 8 \end{bmatrix} \cdot \begin{bmatrix} x_1 \\ x_2 \\ x_3 \end{bmatrix}$, show that

(a) It is singular, i.e. $|\mathbf{C}| = 0$.

(b) The point $P: (-2, -1, 1)$ has no image, i.e. is transformed into $(0, 0, 0)$.

(c) Every other point of the plane is carried into a point on the line $l': x_1 + x_2 - x_3 = 0$.

(d) The points of the line $l: x_1 - x_2 + x_3 = 0$, excepting P, are carried into the point $(4, 1, 5)$ on l'.

(e) Every other line in the plane is carried into the line l'.

16.47. Prove: The transformation $\rho X' = \mathbf{E} X$, $|\mathbf{E}| \neq 0$, preserves cross ratio.

16.48. For each of the following collineations

(a) $\rho X' = \begin{bmatrix} 1 & 1 & -2 \\ -1 & 2 & 1 \\ 0 & 1 & -1 \end{bmatrix} X$ (c) $\rho X' = \begin{bmatrix} 6 & 6 & -2 \\ -2 & -1 & 1 \\ -1 & -2 & 3 \end{bmatrix} X$

(b) $\rho X' = \begin{bmatrix} 0 & 1 & 1 \\ -1 & 2 & 1 \\ -2 & 2 & 3 \end{bmatrix} X$ (d) $\rho X' = \begin{bmatrix} 2 & 2 & 3 \\ -3 & -3 & -5 \\ 1 & 2 & 4 \end{bmatrix} X$

(e) $\rho X' = \begin{bmatrix} -2 & 2 & 1 \\ -10 & 7 & 2 \\ -5 & 2 & 4 \end{bmatrix} X$

obtain all double elements and a canonical form.

Ans. (a) D.P. $(3, 2, 1)$, $(1, 3, 1)$, $(1, 0, 1)$.

D.L. $x_1 - x_2 - x_3 = 0$, $x_1 - x_3 = 0$, $x_1 + 2x_2 - 7x_3 = 0$.

C.F. $\rho X' = \begin{bmatrix} 1 & 0 & 0 \\ 0 & 2 & 0 \\ 0 & 0 & -1 \end{bmatrix} X$

(b) D.P. All points on $x_1 - x_2 - x_3 = 0$; $(1,1,2)$. D.L. All lines on $(1,1,2)$; $x_1 - x_2 + x_3 = 0$.

$$\text{C.F.} \quad \rho X' = \begin{bmatrix} 3 & 0 & 0 \\ 0 & 1 & 0 \\ 0 & 0 & 1 \end{bmatrix} X$$

(c) D.P. $(2,-1,0)$, $(1,-1,-1)$.   D.L. $x_1 + 2x_2 = 0$, $x_1 + 2x_2 - x_3 = 0$.

$$\text{C.F.} \quad \rho X' = \begin{bmatrix} 2 & 0 & 0 \\ 0 & 3 & -1 \\ 0 & 0 & 3 \end{bmatrix} X$$

(d) D.P. $(1,-2,1)$.   D.L. $x_1 - x_3 = 0$. $\text{C.F.} \quad \rho X' = \begin{bmatrix} 1 & 2 & 2 \\ 0 & 1 & 1 \\ 0 & 0 & 1 \end{bmatrix} X$

(e) D.P. All points on $5x_1 - 2x_2 - x_3 = 0$.   D.L. All lines on $(1,2,1)$.

$$\text{C.F.} \quad \rho X' = \begin{bmatrix} 3 & 0 & 1 \\ 0 & 3 & 0 \\ 0 & 0 & 3 \end{bmatrix} X$$

16.49. Obtain the collineation which carries:

(a) $A:(1,0,0)$ into $P:(-1,1,1)$, $B:(0,1,0)$ into $Q:(1,-1,1)$, $C:(0,0,1)$ into $R:(1,1,-1)$, $D:(1,1,1)$ into $S:(1,2,3)$.

(b) A into $T:(-3,1,1)$, B into $U:(1,-2,1)$, C into R and leaves D fixed.

(c) A into P, B into Q, C into R and leaves D fixed.

(d) A into B, P into Q, $F:(1,0,-1)$ into $G:(0,1,-1)$ and leaves $E:(1,-1,0)$ fixed.

(e) B into P, G into Q, E into $H:(2,0,-1)$ and leaves R fixed.

Ans.

(a) $\rho X' = \begin{bmatrix} -5 & 4 & 3 \\ 5 & -4 & 3 \\ 5 & 4 & -3 \end{bmatrix} X$ (c) $\rho X' = \begin{bmatrix} -1 & 1 & 1 \\ 1 & -1 & 1 \\ 1 & 1 & -1 \end{bmatrix} X$

(b) $\rho X' = \begin{bmatrix} -9 & 4 & 6 \\ 3 & -8 & 6 \\ 3 & 4 & -6 \end{bmatrix} X$ (d) $\rho X' = \begin{bmatrix} 0 & 1 & 0 \\ 1 & 0 & 0 \\ 0 & 0 & 1 \end{bmatrix} X$ (e) $\rho X' = \begin{bmatrix} 3 & -1 & 0 \\ 1 & 1 & 0 \\ -1 & 1 & 2 \end{bmatrix} X$

16.50. In each of Problem 16.49(b)-(e), the two triangles are perspective from the fixed point (for example, in (d) the triangles APF and BQG are perspective from E). The collineations are called *perspective collineations*. Show that the double elements consist of a pencil of double lines on the double (fixed) point and a line of double points of equation (b) $3x_1 + 4x_2 + 6x_3 = 0$, (c) $x_1 + x_2 + x_3 = 0$, (d) $x_1 - x_2 = 0$, (e) $x_1 - x_2 = 0$. Verify that each line of double points is the axis of the corresponding pair of perspective triangles.

16.51. In Problem 16.49(b)-(d), verify that the fixed point is not on the line of double points, i.e. that the collineations are homologies.

In (b) take any two distinct points V, W on the axis of homology. On the double line $p = DV$ take another point Y, and on the double line $q = DW$ take another point Z. Locate the correspondents Y', Z' of Y, Z in the homology and verify: $(D, V; Y, Y') = (D, W; Z, Z')$.

Repeat for the homologies of (c) and (d). In (d), each set of four points is a harmonic set. The collineation is called a *harmonic homology*.

16.52. In Problem 16.49(e), verify that the collineation is an elation.

16.53. Let

$$S: \rho \begin{bmatrix} x_1' \\ x_2' \\ x_3' \end{bmatrix} = \mathbf{A} \begin{bmatrix} x_1 \\ x_2 \\ x_3 \end{bmatrix}; \quad \mathbf{A} = [a_{ij}], \ (i, j = 1, 2, 3); \ |\mathbf{A}| \neq 0$$

be a collineation which carries $P: (x_1, x_2, x_3)$ into $P': (x_1', x_2', x_3')$ and

$$T: \sigma \begin{bmatrix} x_1'' \\ x_2'' \\ x_3'' \end{bmatrix} = \mathbf{B} \begin{bmatrix} x_1' \\ x_2' \\ x_3' \end{bmatrix}; \quad \mathbf{B} = [b_{ij}], \ (i, j = 1, 2, 3); \ |\mathbf{B}| \neq 0$$

be a collineation which carries P' into $P'': (x_1'', x_2'', x_3'')$.   Then

$$TS: \tau \begin{bmatrix} x_1'' \\ x_2'' \\ x_3'' \end{bmatrix} = \mathbf{B} \cdot \mathbf{A} \begin{bmatrix} x_1 \\ x_2 \\ x_3 \end{bmatrix} = \mathbf{C} \begin{bmatrix} x_1 \\ x_2 \\ x_3 \end{bmatrix}, \quad \mathbf{B} \cdot \mathbf{A} = \mathbf{C}$$

is a collineation which carries P into P''. TS is called the *resultant* or *product* of S and T. We have established:

The product of two collineations is a collineation.

(a) Rewrite the initial collineations so that T carries $Q: (x_1, x_2, x_3)$ into $Q': (x_1', x_2', x_3')$ and S carries Q' into $Q'': (x_1'', x_2'', x_3'')$ and obtain

$$ST: \tau \begin{bmatrix} x_1'' \\ x_2'' \\ x_3'' \end{bmatrix} = \mathbf{A} \cdot \mathbf{B} \begin{bmatrix} x_1 \\ x_2 \\ x_3 \end{bmatrix}$$

(b) Verify that when $P = Q$, $P'' \neq Q''$ generally.

16.54. Consider the correlation (i):

$$\rho \begin{bmatrix} X_1 \\ X_2 \\ X_3 \end{bmatrix} = \mathbf{C} \begin{bmatrix} x_1 \\ x_2 \\ x_3 \end{bmatrix} = \begin{bmatrix} 1 & 3 & 5 \\ 1 & -2 & 3 \\ 2 & 4 & -3 \end{bmatrix} \begin{bmatrix} x_1 \\ x_2 \\ x_3 \end{bmatrix}$$

(a) Show that (i) carries the points $P: (1, 2, 1)$, $Q: (3, -1, 2)$, $R: (4, 1, 3)$ on the line $s: 5x_1 + x_2 - 7x_3 = 0$ into the lines $p: 12x_1 + 7x_3 = 0$, $q: 10x_1 + 11x_2 - 4x_3 = 0$, $r: 22x_1 + 11x_2 + 3x_3 = 0$ respectively on the point $S: (77, -118, -132)$.

(b) Verify that S and s are not correspondents under (i).

(c) Verify that S is the correspondent of s under the induced transformation (ii):

$$\sigma \begin{bmatrix} x_1 \\ x_2 \\ x_3 \end{bmatrix} = (\mathbf{C}^{-1})^T \cdot \begin{bmatrix} X_1 \\ X_2 \\ X_3 \end{bmatrix}$$

(d) Infer that if a correlation of matrix $\mathbf{C}$ carries the points on a line s into the lines on a point S, then it will carry s into S if and only if

$$(\mathbf{C}^{-1})^T = \mathbf{C}^{-1}$$

that is, if and only if $\mathbf{C}$ is symmetric.

Chapter 17

The Conic

THE POINT CONIC

Consider in Fig. 17-1 the pencils of lines

$$\{Ax\} + \lambda\{Bx\} = 0 \qquad (1)$$

and

$$\{Dx\} + \lambda\{Ex\} = 0 \qquad (2)$$

on distinct centers R and S. The correspondence which associates the line of pencil (1) determined by a given value of λ with the line of (2) determined by the same value of λ is clearly one-to-one. Moreover, since the cross ratio of any four lines of (1) is then equal to the cross ratio of their correspondents of (2), this correspondence is a projectivity. The locus of intersections of corresponding lines of the two pencils has the equation

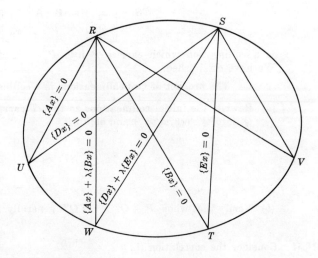

Fig. 17-1

$$\{Ax\} \cdot \{Ex\} - \{Dx\} \cdot \{Bx\} = 0 \qquad (3)$$

obtained by eliminating λ between (1) and (2). Multiplying out in (3) and collecting terms, we obtain

$$c_{11}x_1^2 + c_{22}x_2^2 + c_{33}x_3^2 + 2c_{12}x_1x_2 + 2c_{31}x_3x_1 + 2c_{23}x_2x_3 = 0 \qquad (4)$$

where $c_{11} = A_1E_1 - D_1B_1$, $c_{12} = \frac{1}{2}(A_1E_2 + A_2E_1 - D_1B_2 - D_2B_1)$, etc. By defining $c_{21} = c_{12}$, $c_{13} = c_{31}$, $c_{32} = c_{23}$, (4) may be written as

$$\sum c_{ij}x_ix_j = \sum_{i=1}^{3}\sum_{j=1}^{3} c_{ij}x_ix_j = 0, \qquad c_{ij} = c_{ji} \qquad (5)$$

and, in matrix notation,

$$[x_1, x_2, x_3] \cdot \begin{bmatrix} c_{11} & c_{12} & c_{13} \\ c_{21} & c_{22} & c_{23} \\ c_{31} & c_{32} & c_{33} \end{bmatrix} \cdot \begin{bmatrix} x_1 \\ x_2 \\ x_3 \end{bmatrix} = 0, \qquad c_{ij} = c_{ji} \qquad (6)$$

It is easy to verify that the locus (5) is on each of R and S. In Problem 17.1, we prove

Theorem 17.1. When the projectivity generating (5) is in reality a perspectivity, the left member of (5) is the product of two linear factors.

In the case of the theorem, the locus (5) is called a *degenerate point conic*; otherwise, the locus is called a *proper point conic*.

Let $P:(p)$ and $Q:(q)$ be two distinct points of the plane; then a point $X:(p + \alpha q)$ on PQ is also on (5) provided α is a root of

$$\sum c_{ij}(p_i + \alpha q_i)(p_j + \alpha q_j) \;=\; \sum c_{ij}p_i p_j + \alpha\left(\sum c_{ij}p_i q_j + \sum c_{ij}q_i p_j\right) + \alpha^2 \sum c_{ij}q_i q_j \;=\; 0 \qquad (7)$$

which since

$$\sum c_{ij}p_i q_j \;=\; \sum c_{ij}q_i p_j \qquad\qquad (8)$$

becomes

$$\sum c_{ij}p_i p_j + 2\alpha \sum c_{ij}p_i q_j + \alpha^2 \sum c_{ij}q_i q_j \;=\; 0 \qquad\qquad (9)$$

Suppose now that PQ is contained in (5), that is, (9) holds for every value of α; then

$$\sum c_{ij}p_i p_j \;=\; \sum c_{ij}p_i q_j \;=\; \sum c_{ij}q_i p_j \;=\; \sum c_{ij}q_i q_j \;=\; 0$$

Take $A:(a)$ where $a_1 = c_{11}p_1 + c_{12}p_2 + c_{13}p_3 = \sum_j c_{1j}p_j$, $a_2 = \sum_j c_{2j}p_j$, $a_3 = \sum_j c_{3j}p_j$ and $B:(b)$ where $b_1 = \sum_j c_{1j}q_j$, $b_2 = \sum_j c_{2j}q_j$, $b_3 = \sum_j c_{3j}q_j$. Since $\sum c_{ij}p_i p_j = \sum c_{ij}q_i p_j = 0$, both A and B are on the line

$$p_1 x_1 + p_2 x_2 + p_3 x_3 \;=\; 0 \qquad\qquad (10_1)$$

and since $\sum c_{ij}q_i p_j = \sum c_{ij}q_i q_j = 0$, both A and B are on the line

$$q_1 x_1 + q_2 x_2 + q_3 x_3 \;=\; 0 \qquad\qquad (10_2)$$

But these are distinct lines; hence there exists $\beta \neq 0$ such that

$$a_1 + \beta b_1 \;=\; c_{11}(p_1 + \beta q_1) + c_{12}(p_2 + \beta q_2) + c_{13}(p_3 + \beta q_3) \;=\; 0$$

$$a_2 + \beta b_2 \;=\; c_{21}(p_1 + \beta q_1) + c_{22}(p_2 + \beta q_2) + c_{23}(p_3 + \beta q_3) \;=\; 0 \qquad (11)$$

$$a_3 + \beta b_3 \;=\; c_{31}(p_1 + \beta q_1) + c_{32}(p_2 + \beta q_2) + c_{33}(p_3 + \beta q_3) \;=\; 0$$

From (11), we conclude $[c_{ij}] = 0$. We have proved

Theorem 17.2. If the locus (5) is degenerate, then $[c_{ij}] = 0$.

Returning to Fig. 17-1, let us assume T, U, V, W are four fixed points on a given conic and R, S two positions of a variable point on the same conic. By definition, $RT, RU,$ RV, RW and ST, SU, SV, SW are four pairs of corresponding lines of two projective pencils (1) and (2). Taking RT and ST as given by $\lambda = \infty$, RU and SU by $\lambda = 0$, RV and SV by $\lambda = 1$, then

$$(RT, RU; RV, RW) \;=\; (ST, SU; SV, SW) \;=\; (\infty, 0; 1, \lambda) \;=\; \lambda$$

and we have

Theorem 17.3. If four fixed points on a conic are joined to a variable fifth point, the cross ratio of the four lines is constant.

As a consequence, the points on a conic may be given by a single coordinate (parameter) as are the points on a line. Moreover, it is possible to establish a one-to-one correspondence between the points of a conic and the points of any line of the plane (see Fig. 17-2).

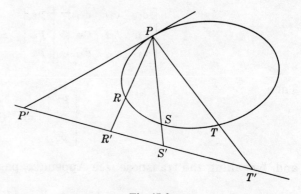

Fig. 17-2

POINT CONICS AND THEIR TANGENT LINES

Consider the point conic

$$\sum c_{ij}x_i x_j = 0, \quad c_{ij} = c_{ji}, \quad |c_{ij}| \neq 0 \tag{12}$$

Let $P:(p)$ and $Q:(q)$ be distinct points, neither of which is on (12), and $X:(p+\alpha q)$ be a variable point on the line PQ. It follows from (9) that PQ either

(i)   meets (12) in two distinct points $(p+\alpha_1 q)$, $(p+\alpha_2 q)$ and is a *secant*,

(ii)   meets (12) in two coincident points and is a *tangent*,

or   (iii) does not meet (12) and is a *non-secant*.

Suppose PQ is a tangent; then (9) is a perfect square and we have

Theorem 17.4.   A necessary and sufficient condition that the line joining two fixed points $P:(p)$ and $Q:(q)$ be tangent to the point conic (12) is

$$\left(\sum c_{ij}p_i p_j\right)\left(\sum c_{ij}q_i q_j\right) - \left(\sum c_{ij}p_i q_j\right)^2 = 0 \tag{13}$$

Consider again the point conic (12). Let $P:(p)$ be a fixed point, not on (12), and $Q:(q)$ be a variable point in the plane. If (13) can be written as the product of two linear factors in the q's with *real* coefficients, these factors equated to zero are the equations of the two lines on P tangent to the conic; if (13) cannot be so factored, no tangents on P can be drawn to the conic. In the first case, P is said to be outside the conic; in the second case, P is said to be inside the conic.

Finally, suppose $P:(p)$ is a fixed point on the conic (12) and $Q:(q)$ is a variable point of the plane. Then (13) reduces to $\left(\sum c_{ij}p_i q_j\right)^2 = 0$. Upon replacing Q by the more familiar $X:(x)$, we have

$$\sum c_{ij}p_i x_j = \sum c_{ij}p_j x_i = 0 \tag{14}$$

as the equation of the tangent to (12) at (with point of contact) P.

Suppose the line $[X_1, X_2, X_3]$ is tangent to (12), the point of contact being $P:(p)$. Equating these with the line coordinates from (14), we have

$$\sum_j c_{1j}p_j = X_1, \quad \sum_j c_{2j}p_j = X_2, \quad \sum_j c_{3j}p_j = X_3$$

or, more compactly,

$$\begin{bmatrix} c_{11} & c_{12} & c_{13} \\ c_{21} & c_{22} & c_{23} \\ c_{31} & c_{32} & c_{33} \end{bmatrix} \cdot \begin{bmatrix} p_1 \\ p_2 \\ p_3 \end{bmatrix} = \mathbf{C}\begin{bmatrix} p_1 \\ p_2 \\ p_3 \end{bmatrix} = \begin{bmatrix} X_1 \\ X_2 \\ X_3 \end{bmatrix}$$

Then

$$\begin{bmatrix} p_1 \\ p_2 \\ p_3 \end{bmatrix} = \mathbf{C}^{-1}\begin{bmatrix} X_1 \\ X_2 \\ X_3 \end{bmatrix} \tag{15}$$

and, by taking the transpose (see Appendix, page 237),

$$[p_1, p_2, p_3] = [X_1, X_2, X_3]\,\mathbf{C}^{-1} \tag{16}$$

Upon substituting from (*15*) and (*16*) into

$$[p_1, p_2, p_3] \cdot \mathbf{C} \cdot \begin{bmatrix} p_1 \\ p_2 \\ p_3 \end{bmatrix} = 0$$

the condition that P be on (*12*), we obtain

$$[X_1, X_2, X_3] \cdot \mathbf{C}^{-1} \cdot \begin{bmatrix} X_1 \\ X_2 \\ X_3 \end{bmatrix} = 0 \qquad (17)$$

the equation of the point conic in line coordinates.

Example 17.1.

Given the projective pencils of lines

$$x_1 - x_2 - x_3 + \lambda(2x_1 + x_2 + 4x_3) = 0$$

and

$$x_1 + 2x_2 - x_3 + \lambda(x_1 - 4x_2 + x_3) = 0$$

(*a*) Obtain the equation of the locus generated by them.

Eliminating λ, we have

$$(x_1 - x_2 - x_3)(x_1 - 4x_2 + x_3) - (x_1 + 2x_2 - x_3)(2x_1 + x_2 + 4x_3) = 0$$

The equation of the locus is

(i) $$\sum c_{ij} x_i x_j = x_1^2 - 2x_2^2 - 3x_3^2 + 10x_1 x_2 + 2x_3 x_1 + 4x_2 x_3 = 0$$

or   (ii) $$[x_1, x_2, x_3] \cdot \begin{bmatrix} 1 & 5 & 1 \\ 5 & -2 & 2 \\ 1 & 2 & -3 \end{bmatrix} \cdot \begin{bmatrix} x_1 \\ x_2 \\ x_3 \end{bmatrix} = 0$$

(*b*) Show that the locus is a point conic.

$$|\mathbf{C}| = \begin{vmatrix} 1 & 5 & 1 \\ 5 & -2 & 2 \\ 1 & 2 & -3 \end{vmatrix} = \begin{vmatrix} 1 & 0 & 0 \\ 5 & -27 & -3 \\ 1 & -3 & -4 \end{vmatrix} \neq 0; \quad \text{the locus is a point conic.}$$

(*c*) Find the intersections, if any, of (i) and the line joining $P: (2, 0, -1)$ and $Q: (0, -4, 1)$.

$X: (2, 0, -1) + \alpha(0, -4, 1) = (2, -4\alpha, -1+\alpha)$ is an arbitrary point on PQ. If it is also on (i) then

$$4 - 32\alpha^2 - 3 + 6\alpha - 3\alpha^2 - 80\alpha - 4 + 4\alpha + 16\alpha - 16\alpha^2 = -51\alpha^2 - 54\alpha - 3 = 0$$

and $\alpha = -1, -1/17$. There are two distinct points of intersection:

$$X_1: (2, 0, -1) - (0, -4, 1) = (2, 4, -2) \text{ or } (1, 2, -1)$$

and

$$X_2: (2, 0, -1) - \tfrac{1}{17}(0, -4, 1) = (2, 4/17, -18/17) \text{ or } (17, 2, -9)$$

(*d*) Show that $T: (3, 0, -1)$ is on (i) and write the equation of the tangent there.

$$[3, 0, -1] \begin{bmatrix} 1 & 5 & 1 \\ 5 & -2 & 2 \\ 1 & 2 & -3 \end{bmatrix} \begin{bmatrix} 3 \\ 0 \\ -1 \end{bmatrix} = [3, 0, -1] \begin{bmatrix} 2 \\ 13 \\ 6 \end{bmatrix} = 0$$

and T is on the conic. The equation of the tangent is

$$[x_1, x_2, x_3] \begin{bmatrix} 1 & 5 & 1 \\ 5 & -2 & 2 \\ 1 & 2 & -3 \end{bmatrix} \begin{bmatrix} 3 \\ 0 \\ -1 \end{bmatrix} = 2x_1 + 13x_2 + 6x_3 = 0$$

(e) Write the equation of the point conic in line coordinates.

$$\mathbf{C} = \begin{bmatrix} 1 & 5 & 1 \\ 5 & -2 & 2 \\ 1 & 2 & -3 \end{bmatrix} \quad \text{and} \quad \mathbf{C}^{-1} = \frac{1}{99}\begin{bmatrix} 2 & 17 & 12 \\ 17 & -4 & 3 \\ 12 & 3 & -27 \end{bmatrix}$$

The required equation is

(iii) $$[X_1, X_2, X_3]\begin{bmatrix} 2 & 17 & 12 \\ 17 & -4 & 3 \\ 12 & 3 & -27 \end{bmatrix}\begin{bmatrix} X_1 \\ X_2 \\ X_3 \end{bmatrix} = 0$$

Note. Most authors give the line equation of a conic whose point equation is (12) as

$$[X_1, X_2, X_3]\begin{bmatrix} b_{11} & b_{12} & b_{13} \\ b_{21} & b_{22} & b_{23} \\ b_{31} & b_{32} & b_{33} \end{bmatrix}\begin{bmatrix} X_1 \\ X_2 \\ X_3 \end{bmatrix} = 0, \quad b_{ij} = b_{ji}, \quad |b_{ij}| \neq 0 \qquad (18)$$

where b_{ij} is the cofactor C_{ij} of c_{ij} in $\mathbf{C}$. This is equivalent to replacing $\mathbf{C}^{-1}$ in (17) by adj $\mathbf{C}$, a natural reduction [see Example 17.1(e)] in numerical problems.

THE LINE CONIC

Consider in Fig. 17-3 the pencils of points

$$\{aX\} + \lambda\{cX\} = 0 \qquad (1')$$

and

$$\{dX\} + \lambda\{eX\} = 0 \qquad (2')$$

on respective lines r and s. The correspondence which associates the point of pencil $(1')$, determined by a given value of λ, with the point of $(2')$, determined by the same value of λ, is a projectivity. The envelope of joins of corresponding points has the equation

$$\{aX\}\cdot\{eX\} - \{dX\}\cdot\{cX\} = 0 \qquad (3')$$

which may be written as $$\sum b_{ij}X_iX_j = 0 \qquad (5')$$

where $b_{11} = a_1e_1 - d_1c_1$, $b_{12} = b_{21} = \frac{1}{2}(a_1e_2 + a_2e_1 - d_1c_2 - d_2c_1)$, etc., and in matrix notation as

$$[X_1, X_2, X_3]\begin{bmatrix} b_{11} & b_{12} & b_{13} \\ b_{21} & b_{22} & b_{23} \\ b_{31} & b_{32} & b_{33} \end{bmatrix}\begin{bmatrix} X_1 \\ X_2 \\ X_3 \end{bmatrix} = 0, \quad b_{ij} = b_{ji} \qquad (6')$$

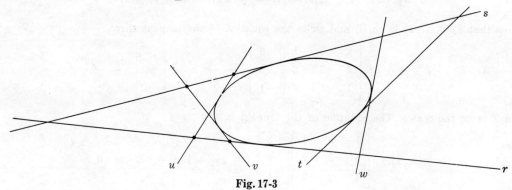

Fig. 17-3

The dual of Problem 17.1 shows that when the projectivity is, in reality, a perspectivity, (*3'*) is the product of two real factors and the locus degenerates into a pair of points, distinct or coincident. Otherwise, the locus is called a *line conic*. We leave for the reader to complete the dual of the sections above. Having done so, it will be noted that for each point conic (*12*) there corresponds a line conic (*18*) and, dually, for each line conic there corresponds a point conic. In other words, a point conic (*12*) together with its tangents and the corresponding line conic together with its points of contact are identical configurations. Each pair of identical configurations will be called a *non-singular conic* or, simply, a *conic*. A conic, then, has two equations: (*12*) in point coordinates and (*18*) in line coordinates.

POLAR LINES AND POLES. POINT CONICS

Denote by C the conic whose equation is (*12*) in point coordinates and take $P:(p)$ an

arbitrary point of the plane.   Since $\mathbf{C} = [c_{ij}]$ is non-singular, $\mathbf{C} \cdot \begin{bmatrix} p_1 \\ p_2 \\ p_3 \end{bmatrix} \neq \begin{bmatrix} 0 \\ 0 \\ 0 \end{bmatrix}$ and so

$$\sum c_{ij} p_i x_j = [x_1, x_2, x_3] \cdot \mathbf{C} \cdot \begin{bmatrix} p_1 \\ p_2 \\ p_3 \end{bmatrix} = 0 \qquad (19)$$

defines a unique line p of the plane. Call this line the *polar line* of P with respect to C. We now proceed to show that the polar line as here defined is identical with that defined in Chapter 9.

First, note that (*14*) and (*19*) have the same form; thus when P is on C, (*19*) is the equation of the tangent to C at P. Next, let $Q:(q)$ be an arbitrary point on (*19*); then

$$\sum c_{ij} p_i q_j = \sum c_{ij} q_i p_j = 0 \qquad (20)$$

Now the polar line of Q with respect to C is $\sum c_{ij} q_i x_j = 0$ and, by (*20*) P is on this line. We have proved

Theorem 17.5.   If Q is on the polar line of P with respect to C, then P is on the polar line of Q with respect to C.

There follows

Theorem 17.6.   As Q varies over the polar line p of P with respect to C, its polar line q with respect to C varies over the pencil of lines on P.

Call the point P, whose polar line with respect to C is p, the *pole* of p with respect to C. Now let p be any line of the plane. It follows from Theorem 17.6 that the pole of p with respect to C is uniquely determined as the point of intersection of the polar lines with respect to C of any two distinct points on p. (Where is the pole if p is tangent to C at P?)

The points P and Q of Theorem 17.5 are said to be *conjugate* to one another with respect to C or to constitute a pair of *conjugate points* with respect to C. When P is on C, (*19*) becomes (*14*); thus all points Q conjugate to P are on the tangent to C at P. When P is not on C, its polar line p is either a secant or non-secant of C. Suppose p is a secant and that it meets C in the points R and S. Now the polar line of R (S) with respect to C is tangent to C at R (S) and, by Theorem 17.5, is on P. Thus if P is outside C, the points of contact of the two tangents to C from P lie on the polar line of P. Finally, if P is inside C, the polar line of P, i.e., the locus of points conjugate to P, is a non-secant.

Let $P:(p)$ be any point not on C and p be its polar line with respect to C. If P is outside C, take $Q:(q)$ on p such that PQ meets C in distinct points R and S; if P is inside C take $Q:(q)$ on p and again denote the intersections of PQ and C as R and S. In either case, $R:(p+\lambda q)$ and $S:(p+\mu q)$ where λ, μ are the roots of (9). Since P and Q are conjugate points, $\sum c_{ij}p_iq_j = 0$ and so $\lambda + \mu = 0$. We have proved

Theorem 17.7.   Let P, Q be a pair of conjugate points with respect to C. If one of the points is on C, the other is on the tangent to C at that point. If neither of the points is on C but the line PQ is a secant meeting C in R and S, then $H(P, Q; R, S)$.

There follows

Theorem 17.8.   The pairs of conjugate points on a secant p of C form an involution whose double points are the intersections of p and C.

A triangle is said to be *self-polar* with respect to a conic C if each vertex and opposite side are pole and polar line with respect to C. Since, then, each vertex is conjugate to the other two vertices and, dually, each side is conjugate to the other two sides, we have

Theorem 17.9.   A triangle is self-polar with respect to a conic provided each two vertices are conjugate points and each two sides are conjugate lines with respect to the conic.

POLES AND POLAR LINES, LINE CONICS

Denote by C the conic whose equation in line coordinates is (18) and take $p:[P]$ an

arbitrary line of the plane. Since $\mathbf{B} = [b_{ij}]$ is non-singular, $\mathbf{B} \cdot \begin{bmatrix} P_1 \\ P_2 \\ P_3 \end{bmatrix} \neq \begin{bmatrix} 0 \\ 0 \\ 0 \end{bmatrix}$ and so

$$\sum b_{ij}P_iX_j = [X_1, X_2, X_3] \cdot \mathbf{B} \cdot \begin{bmatrix} P_1 \\ P_2 \\ P_3 \end{bmatrix} = 0 \qquad (19')$$

defines a unique point P of the plane. Call this point the *pole* of p with respect to C. Also define:

The poles with respect to C of any two distinct lines on a point P determine a line p, called the *polar line* of P with respect to C.

Two lines p and q are said to be *conjugate* with respect to C provided one is on the pole of the other.

We leave for the reader to demonstrate the duals of Theorems 17.5-17.9. The dual of Theorem 17.7 is

Theorem 17.7'.   Let p, q be a pair of conjugate lines with respect to C. If one of the lines is on C, the other is on the point of contact of that line. If neither of the lines is on C while two of the lines r, s of C are on $p \cdot q$, then $H(p, q; r, s)$.

From this and the previous section it follows that a fixed conic $C: \sum c_{ij}x_ix_j = 0$ or $\sum b_{ij}X_iX_j = 0$ establishes a one-to-one correspondence between the points (x_1, x_2, x_3) and the lines $[X_1, X_2, X_3]$ of the plane given by

$$\rho \begin{bmatrix} X_1 \\ X_2 \\ X_3 \end{bmatrix} = \mathbf{C} \cdot \begin{bmatrix} x_1 \\ x_2 \\ x_3 \end{bmatrix} \quad \text{and} \quad \sigma \begin{bmatrix} x_1 \\ x_2 \\ x_3 \end{bmatrix} = \mathbf{B} \cdot \begin{bmatrix} X_1 \\ X_2 \\ X_3 \end{bmatrix}$$

The correspondence preserves cross ratio (show this) and, hence, is a polar correlation or polarity (see Chapter 16).

EQUATIONS OF A CONIC

Up to now the conic

$$C: \sum c_{ij} x_i x_j = 0 \tag{12}$$

under consideration has been assumed in general position with the coordinate system in the plane. We now obtain the equation of this conic when referred to a triangle of reference whose vertices and sides are intimately connected with it.

Suppose we take $A_1: (a)$, $A_2: (b)$, $A_3: (c)$ on C as the vertices of a new triangle of reference and suppose the equation of the conic when referred to the new triangle is

$$\text{(i)} \qquad \sum c'_{ij} x'_i x'_j = 0$$

Since $A_1: (1, 0, 0)$, $A_2: (0, 1, 0)$, $A_3: (0, 0, 1)$ are on (i), $c'_{11} = c'_{22} = c'_{33} = 0$, and the equation (i) has the form

$$c'_{12} x'_1 x'_2 + c'_{31} x'_3 x'_1 + c'_{23} x'_2 x'_3 = 0 \tag{21}$$

or $\qquad [x'_1, x'_2, x'_3] \cdot \begin{bmatrix} 0 & c'_{12} & c'_{13} \\ c'_{21} & 0 & c'_{23} \\ c'_{31} & c'_{32} & 0 \end{bmatrix} \cdot \begin{bmatrix} x'_1 \\ x'_2 \\ x'_3 \end{bmatrix} = 0, \qquad c'_{12} \cdot c'_{31} \cdot c'_{23} \neq 0$

If the unit point $A_4: (1, 1, 1)$ is also taken on (21), we have the further condition $c'_{12} + c'_{31} + c'_{23} = 0$ on the coefficients.

Next, let $A_1: (a)$, $A_2: (b)$, $A_3: (c)$ be the vertices of a self-polar triangle with respect to (12). Let the equation of the conic, when referred to this triangle as reference triangle, be

$$\text{(ii)} \qquad \sum c'_{ij} x'_i x'_j = 0$$

Now the polar line of $A_1: (1, 0, 0)$ with respect to (ii), $c'_{11} x'_1 + c'_{21} x'_2 + c'_{31} x'_3 = 0$, is $x'_1 = 0$ so that $c'_{21} = c'_{31} = 0$. Similarly, since the polar line of $A_2: (0, 1, 0)$ is $x'_2 = 0$, we have $c'_{23} = 0$ and, so, the polar line of $A_3: (0, 0, 1)$ is $x'_3 = 0$ as required. Thus the equation of the conic has the form

$$c'_{11} x'^2_1 + c'_{22} x'^2_2 + c'_{33} x'_3 = 0 \tag{22}$$

or $\qquad [x'_1, x'_2, x'_3] \cdot \begin{bmatrix} c'_{11} & 0 & 0 \\ 0 & c'_{22} & 0 \\ 0 & 0 & c'_{33} \end{bmatrix} \cdot \begin{bmatrix} x'_1 \\ x'_2 \\ x'_3 \end{bmatrix} = 0$

where no one of the coefficients is zero and, since the conic is real, not all have the same sign.

By renaming the vertices of the self-polar triangle, if necessary, we can always obtain (22) with $c'_{11} > 0$, $c'_{22} > 0$, $c'_{33} < 0$. The transformation

$$(iii) \qquad \rho X'' = \begin{bmatrix} \sqrt{c'_{11}} & 0 & 0 \\ 0 & \sqrt{c'_{22}} & 0 \\ 0 & 0 & \sqrt{-c'_{33}} \end{bmatrix} X$$

then reduces (22) to $\qquad x_1''^2 + x_{22}''^2 - x_{33}''^2 = 0$ $\qquad\qquad$ (23)

Thus all (real, proper) conics are projectively equivalent.

The transformation $\rho X'' = BX$ which carries (22) directly into (23) preserves the triangle of reference but generally associates with it another unit point. (See Chapter 16.)

DEGENERATE CONICS

A degenerate point conic C having as equation

$$\sum c_{ij} x_i x_j = 0, \quad c_{ij} = c_{ji}, \quad |c_{ij}| = 0 \qquad\qquad (24)$$

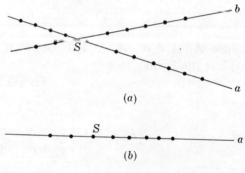

consists (see Fig. 17-4) either of two distinct lines or of one line counted twice. A point $S:(s)$ is called a *singular point* of C provided every line determined by it and any other point $Y:(y)$ on C is contained in C. Thus, when C consists of a pair of distinct lines a, b it has just one singular point $S = a \cdot b$; when C consists of a line a counted twice, every point of a is a singular point. When S is a singular point and Y is any other point of the plane, a point $(s) + \lambda(y)$ on SY is also on C provided λ is a root [see (9)] of

Fig. 17-4

$$\sum c_{ij}(s_i + \lambda y_i)(s_j + \lambda y_j) = \sum c_{ij} s_i s_j + 2\lambda \sum c_{ij} s_i y_j + \lambda^2 \sum c_{ij} y_i y_j = 0 \qquad (25)$$

The line SY is contained in C when Y is on C, and so all coefficients in (25) vanish; the line SY meets C only in S when Y is not on C, and then $\sum c_{ij} s_i s_j = \sum c_{ij} s_i y_j = 0$ in (25). Then $\sum c_{ij} s_i y_j = 0$ when Y is *any* point of the plane; hence,

$$\sum_i c_{i1} s_i = 0, \quad \sum_i c_{i2} s_i = 0, \quad \sum_i c_{i3} s_i = 0 \qquad\qquad (26)$$

Conversely, if $S:(s)$ is a point on C for which (26) holds, then, for every value of λ, the point $(s) + \lambda(y)$ is on C. We have proved

Theorem 17.10. The point $S:(s)$ is a singular point of the degenerate conic $C: \sum c_{ij} x_i x_j = 0$ if and only if (26) holds.

Suppose (24) consists of the pair of distinct lines

$$a: a_1 x_1 + b_1 x_2 + c_1 x_3 = 0 \quad \text{and} \quad b: a_2 x_1 + b_2 x_2 + c_2 x_3 = 0$$

on the point $S:(s)$. Let $P:(p)$ be any point of the plane distinct from S; then

$$(a_1 p_1 + b_1 p_2 + c_1 p_3)(a_2 x_1 + b_2 x_2 + c_2 x_3) + (a_2 p_1 + b_2 p_2 + c_2 p_3)(a_1 x_1 + b_1 x_2 + x_1 x_3)$$

$$= 2 \sum c_{ij} p_i x_j = 0 \qquad\qquad (27)$$

the polar line of P with respect to C is on S. If P is on C, i.e. if P is on either the line a or b, the polar line of P is that line. In keeping with the case of a non-degenerate conic, we shall call (27) the tangent line to C at P. If P is not on C, then (27) and

$$(a_1p_1 + b_1p_2 + c_1p_3)(a_2x_1 + b_2x_2 + c_2x_3) - (a_2p_1 + b_2p_2 + c_2p_3)(a_1x_1 + b_1x_2 + c_1x_3) = 0 \qquad (28)$$

the line PS, separate the lines a and b harmonically.

Two points P and Q, neither of which is on $C = a, b$, are conjugate with respect to C provided $H(a, b; PS, QS)$ where $S = a \cdot b$. If, however, $P \neq S$ is on C, then Q and P are conjugate if and only if Q is also on a or on b.

When $C = a, b$ the polar line of $S = a \cdot b$ is undefined. We may then define the lines a, b as the tangent lines to C at S. Moreover, since (26) assures $\sum c_{ij}s_ix_j$ identically zero for any triple (x_1, x_2, x_3), every point of the plane is conjugate to S.

A triangle, each two of whose vertices are a pair of conjugate points with respect to C, is a self-polar triangle of C. Every self-polar triangle of C must have S as one of its vertices while the sides joining S to the other two vertices must separate harmonically the two lines which constitute C.

Dually, a degenerate line conic C having as equation

$$\sum b_{ij}X_iX_j = 0, \qquad b_{ij} = b_{ji}, \qquad |b_{ij}| = 0 \qquad (29)$$

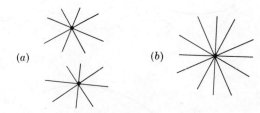

consists (see Fig. 17-5) either of two distinct points or of a point counted twice. A line $s: [S]$ is a *singular line* of a line conic if every point determined by it and any other line on C belongs to C. We leave for the reader to complete the study of degenerate line conics.

(a)　　　　　　　　(b)

Fig. 17-5

Earlier in this chapter it was found that a non-degenerate conic has two equations, one in point coordinates and one in line coordinates. A degenerate point (line) conic, on the other hand, has no equation in line (point) coordinates.

PAIRS OF CONICS

From our experience in the metric plane, we conclude that two distinct conics intersect in four points, not all of which are necessarily real. To settle the matter in the projective plane, let C and C' be two distinct point conics of the plane and suppose the reference triangle and unit point be chosen so that (see Problem 17.24)

$$C: x_1^2 - x_2x_3 = 0 \qquad (30)$$

while

$$C': \sum c_{ij}'x_ix_j = 0, \qquad c_{22}' \neq 0 \qquad (31)$$

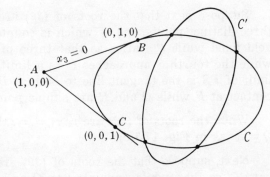

Fig. 17-6

The restriction in (31) merely requires that the vertex $B: (0, 1, 0)$ of the reference triangle not be a point of intersection of C and C'. Since the side $x_3 = 0$ of the reference triangle is tangent to C, it cannot be on any of the intersections of the two conics. Hence there will be no loss in intersections if we take $x_3 \neq 0$ in (30) and solve to obtain $x_2 = x_1^2/x_3$. Substituting this value for x_2 in (31), we obtain

$$c_{22}'r^4 + 2c_{12}'r^3 + (c_{11}' + 2c_{23}')r^2 + 2c_{13}'r + c_{33}' = 0 \qquad (32)$$

where, for convenience, we have written $x_1/x_3 = r$. Now (32) has four roots r_1, r_2, r_3, r_4; hence C and C' have four points of intersection: $(r_1, r_1^2, 1), (r_2, r_2^2, 1), (r_3, r_3^2, 1), (r_4, r_4^2, 1)$.

We now restrict attention exclusively to pairs of conics for which (32) has four *real* roots. Consider, first, the case in which the four roots are distinct. Then [see Fig. 17-7(a)] C and C' intersect in the distinct points P, Q, R, S. Denote by p the tangent line to C and by p' the tangent line to C' at their intersection P. Since there is just one conic on four general points and tangent to a given line at one of them, the lines p and p' are distinct. Call any point of intersection of two distinct conics *simple* if the tangent lines to the conics at the point are distinct. Thus if two conics have four distinct points of intersection, each is simple.

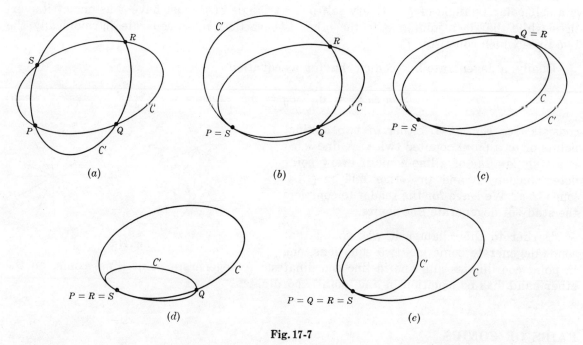

$$(a) \qquad\qquad\qquad (b) \qquad\qquad\qquad (c)$$

$$(d) \qquad\qquad\qquad (e)$$

Fig. 17-7

Suppose next that the roots of (32) are r_1, r_1, r_2, r_3 so that the two conics intersect in three distinct points, one of which is counted twice. Think of the conic C of Fig. 17-7(a) held fixed while C' varies so that three of the intersections P, Q, R with C remain fixed while the fourth S approaches P as a limit. Since S approaches P along both conics, the limit of PS is the tangent line to both at P. The conics [see Fig. 17-7(b)] have two-point contact at P while Q and R are simple points of intersection.

When the roots of (32) are r_1, r_1, r_2, r_2, the conics have two-point contact at both P and $Q = R$ as in Fig. 17-7(c).

Next, suppose that the roots of (32) are r_1, r_1, r_1, r_2. Keep C in Fig. 17-7(b) fixed and vary C' so that, while remaining on Q and maintaining two-point contact at P, the intersection R approaches P as a limit. Then Q remains a simple point of intersection [see Fig. 17-7(d)] while the conics have three-point contact at P.

Finally, when the roots of (32) are r_1, r_1, r_1, r_1, the four intersections of C and C' coincide [see Fig. 17-7(e)] and the conics have four-point contact at P.

The dual of the above paragraphs will again be left largely for the reader. Two sides b, c of the reference triangle and the unit line are chosen among the lines of C, with b not also a line of C', and the remaining side of the reference triangle is the line joining the points of contact of b and c. Restricting attention exclusively to pairs of conics for which $(32')$ has only real roots, we obtain the five cases illustrated in Fig. 17-8 (a), (b), (c), (d), (e). In Fig. 17-8(a), the two conics have p, q, r, s as simple common lines; in Fig. 17-8(b), the conics have two-line contact on p and have q and r as simple common lines; etc.

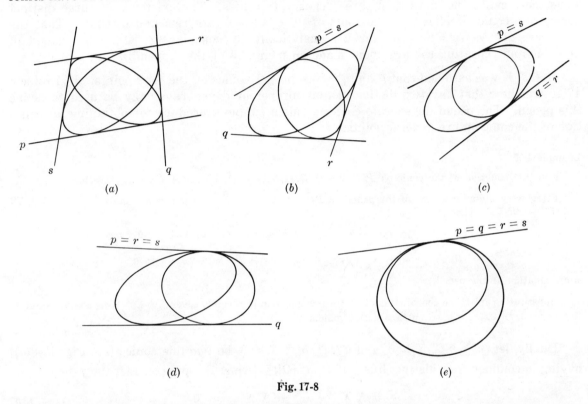

Fig. 17-8

QUADRANGULAR PENCIL OF CONICS

In Chapter 11 a quadrangular pencil of point conics was defined to be the set of all conics (proper or degenerate) on four distinct points, no three of which are collinear. Denote these points as P, Q, R, S and let

$$C: \sum c_{ij} x_i x_j = 0, \quad |c_{ij}| \neq 0 \qquad \text{and} \qquad C': \sum c'_{ij} x_i x_j = 0, \quad |c'_{ij}| \neq 0$$

be two distinct conics on them. We show that the totality of conics (including degenerate conics) represented by

$$\sum c_{ij} x_i x_j + \lambda \sum c'_{ij} x_i x_j = \left(\sum c_{ij} + \lambda \sum c'_{ij}\right) x_i x_j = 0, \qquad (33)$$

when λ varies over the extended real number system, is a quadrangular pencil. First, note that every locus given by (33) is on the four points. Conversely, let C'' be a conic (proper or degenerate) on the base points P, Q, R, S and let $T: (t)$ be any fifth point on C''. Now

$$\sum c'_{ij} t_i t_j \cdot \sum c_{ij} x_i x_j - \sum c_{ij} t_i t_j \cdot \sum c'_{ij} x_i x_j = 0 \qquad (34)$$

is a conic on the five points and, since a unique conic is determined by these points, must be C''. Finally, C'' is a member of the pencil (33), obtained when

$$\lambda = -\sum c_{ij} t_i t_j \Big/ \sum c'_{ij} t_i t_j$$

Clearly, the pairs of lines $PQ, RS; PR, QS; PS, QR$ are three degenerate conics of the pencil. The degenerate conics of this pencil are given by those values and only those values of λ for which

$$|\mathbf{C} + \lambda \mathbf{C}'| = |[c_{ij}] + \lambda[c'_{ij}]| = 0$$

Thus there can be no more than three degenerate conics. The existence of three distinct degenerate conics implies that the roots of $|\mathbf{C} + \lambda \mathbf{C}'| = 0$ are real and distinct. That the roots are always real may be verified analytically by considering $|\mathbf{C} + \lambda \mathbf{C}'|$ instead of $|\mathbf{A} - \lambda \mathbf{I}|$ and repeating the argument made on page 239 of the Appendix.

In (33) it was assumed that C and C' were proper conics on the four points. In Problem 17.10 we show that any two distinct conics (proper or degenerate) may be used to define this pencil. The use of degenerate conics yields a simple procedure for obtaining an equation of the conic on five general points.

Example 17.2.

Find the equation of the conic on $P: (1, 0, -1)$, $Q: (1, 0, 1)$, $R: (1, 2, 1)$, $S: (1, 2, -1)$, $T: (1, 3, 0)$.

First, we write an equation of the pencil on P, Q, R, S. Two of the degenerate conics are $C' = PQ, RS$ and $C'' = PR, QS$. Then

$$C': \; x_0(2x_1 - x_0) = 0, \qquad C'': \; (x_1 - x_0 + x_0)(x_1 - x_0 - x_0) = 0$$

and   (i) $C' + \lambda C'': 2x_1 x_2 - x_2^2 + \lambda(x_1^2 + x_2^2 - x_3^2 - 2x_1 x_2) = 0$

is an equation of the pencil.

Substituting in (i) the coordinates of T, we find the required conic, given by $\lambda = 3/4$, has as equation $3x_1^2 - x_2^2 - 3x_3^2 + 2x_1 x_2 = 0$. (Compare with Problem 17.14.)

Dually, let $C: \sum b_{ij} X_i X_j = 0$ and $C': \sum b'_{ij} X_i X_j = 0$ be two line conics [see Fig. 17-8(a)] having in common four distinct lines $p: [P]$, $q: [Q]$, $r: [R]$, $s: [S]$ and consider the pencil

$$\sum b_{ij} X_i X_j + \lambda \sum b'_{ij} X_i X_j = 0 \tag{35}$$

The degenerate conics of the pencil consist of the pairs of opposite vertices of the complete quadrilateral $pqrs$. It is easily shown that the equation of the pencil (35) may be replaced by

$$\sum b''_{ij} X_i X_j + \lambda \sum b'''_{ij} X_i X_j = 0$$

where $\sum b''_{ij} X_i X_j = 0$ and $\sum b'''_{ij} X_i X_j = 0$ are any two of the degenerate conics.

PENCILS OF CONICS

Let $C: \sum c_{ij} x_i x_j = 0$ and $C': \sum c'_{ij} x_i x_j = 0$ be two distinct (proper) conics whose points of intersection are all real, and consider the pencil

$$\sum c_{ij} x_i x_j + \lambda \sum c'_{ij} x_i x_j = 0 \tag{36}$$

We will be concerned here with those cases which arise when the points of intersection of C and C' are not distinct.

First, suppose C and C' have two-point contact at P and simple intersections at the distinct points Q and R. Denote the tangent to C at P by p. Then p is tangent to both C and C' and, hence, to every conic of the pencil at P. The degenerate conics of this pencil are

$C'' = p, QR$: $\sum c''_{ij} x_i x_j = 0$ and $C''' = PQ, PR$: $\sum c'''_{ij} x_i x_j = 0$,   the latter counted twice. The pencil may then be defined by

$$\sum c''_{ij} x_i x_j + \lambda \sum c'''_{ij} x_i x_j = 0$$

We have

Theorem 17.11.   There is a unique conic having two-point contact with a given conic C at a given point P, simple intersections at two other points, and on any given point T, neither on C nor on its tangent at P.

See Problem 17.12.

Next, suppose that C and C' have two-point contact at P and also at $Q \neq P$, the common tangents being p and q respectively. The pencil determined by C and C' contains two distinct degenerate conics $C'' = p, q$ and $C''' = PQ, PQ$, the latter counted twice. Hence,

Theorem 17.12.   There is a unique conic having two-point contact with a given conic C at the distinct points P and Q, and on any other point T, neither on C nor on its tangents at P and Q.

When C and C' have three-point contact at P and a simple intersection at Q, the pencil of conics determined by these conics has just one degenerate conic

$$C'' = p, PQ: \sum c''_{ij} x_i x_j = 0$$

counted three times.   Suppose

$$\sum c'_{ij} x_i x_j + \alpha \sum c_{ij} x_i x_j = \sum c''_{ij} x_i x_j;$$

then

$$\sum c'_{ij} x_i x_j = \sum c''_{ij} x_i x_j - \alpha \sum c_{ij} x_i x_j$$

and we have

Theorem 17.13.   There is a unique conic C' having three-point contact with a given conic C at P, a simple intersection at $Q \neq P$, and on any other point T, neither on C nor on its tangent p at P. This conic is a member of the pencil determined by C and the degenerate conic $C'' = p, PQ$.

We leave for the reader to prove

Theorem 17.14.   There is a unique conic C' having four-point contact with a given conic C at P and on any point, neither on C nor its tangent p at P. This conic is a member of the pencil determined by C and the degenerate conic $C'' = p, p$.

See Problem 17.13.

Solved Problems

17.1.   Obtain the locus generated by the perspective pencils of lines (1) and (2) of page 198.

Since the two pencils are perspective, the line RS joining their centers is self-corresponding. Assume this line to be given by $\lambda = 0$; then $\{Ax\} = \{Dx\}$ and equation (3) becomes

$$\{Ax\} \cdot [\{Ex\} - \{Bx\}] = 0$$

The locus consists of the line of centers $\{Ax\} = 0$ and the axis of perspectivity $\{Ex\} - \{Bx\} = 0$.

17.2. For the (non-perspective) projective pencils of lines (*1*) and (*2*), page 198, show that the correspondent in pencil (*2*) of the line RS of pencil (*1*) is tangent to the conic (*3*) at S.

Generally, a line of either pencil meets the conic (*3*), page 198, in distinct points — the center of the pencil, R or S, and the point of intersection T with its correspondent in the other pencil. Consider the line $p_1 = RS$ with equation $\{Ax\} + \lambda_1\{Bx\} = 0$ and its correspondent q_1 in pencil (*2*) with equation $\{Dx\} + \lambda_1\{Ex\} = 0$. Now $T = p_1 \cdot q_1 = S$ and so q_1 is tangent to (*3*) at S.

Similarly, the tangent to (*3*) at R is the correspondent of RS considered as a line of the pencil (*2*).

17.3. Obtain the equations of the tangents, if any, to the conic of Example 17.1, page 201, on the point $P: (-29, 1, 24)$.

The polar line p of P is

$$[x_1, x_2, x_3] \begin{bmatrix} 1 & 5 & 1 \\ 5 & -2 & 2 \\ 1 & 2 & -3 \end{bmatrix} \begin{bmatrix} -29 \\ 1 \\ 24 \end{bmatrix} = -99x_2 - 99x_3 = 0 \quad \text{or} \quad x_2 + x_3 = 0$$

We may use either of two procedures: (i) Select two distinct points as $(0, 1, -1)$, $(1, 1, -1)$ on p and proceed as in Example 17.1(c) or (ii) Eliminate x_3 between the equations of p and the conic. Using (ii), we obtain

$$x_1^2 - 2x_2^2 - 3x_2^2 + 10x_1x_2 - 2x_1x_2 - 4x_2^2 = x_1^2 + 8x_1x_2 - 9x_2^2 = (x_1 + 9x_2)(x_1 - x_2) = 0$$

Then $(9, -1, 1)$ and $(1, 1, -1)$ are the points of intersection of p and the conic. At $(9, -1, 1)$ the equation of the tangent is

$$[x_1, x_2, x_3] \begin{bmatrix} 1 & 5 & 1 \\ 5 & -2 & 2 \\ 1 & 2 & -3 \end{bmatrix} \begin{bmatrix} 9 \\ -1 \\ 1 \end{bmatrix} = 0 \quad \text{or} \quad 5x_1 + 49x_2 + 4x_3 = 0$$

At $(1, 1, -1)$ the equation of the tangent is $5x_1 + x_2 + 6x_3 = 0$.

17.4. Prove: If the four tangents to C drawn from the points $P: (p)$ and $Q: (q)$ are distinct, then P, Q and the four points of contact of the tangents are on a conic C'.

Let $C: \sum c_{ij}x_ix_j = 0$; then two of the points of contact are on $\sum c_{ij}p_ix_j = 0$ and the other two are on $\sum c_{ij}q_ix_j = 0$. Now

$$C': \sum c_{ij}x_ix_j \cdot \sum c_{ij}p_iq_j - \sum c_{ij}p_ix_j \cdot \sum c_{ij}q_ix_j = 0$$

is on each of the six points.

17.5. Prove: If a triangle is inscribed in a conic C, any line conjugate to one of the sides meets the other two sides in a pair of conjugate points. (See Problem 9.6, page 98.)

Take the inscribed triangle to be the reference triangle ABC with $A: (1, 0, 0)$, $B: (0, 1, 0)$, $C: (0, 0, 1)$; then C has equation of the form (*21*). Consider the side $c = AB: x_3 = 0$. The polar line of A has equation $c_{21}x_2 + c_{31}x_3 = 0$ and the polar line of B has equation $c_{12}x_1 + c_{32}x_3 = 0$; hence the pole of c is $C': (c_{23}, c_{13}, -c_{12})$. Take $P: (1, p, 0)$ on c; then

$$C'P: pc_{12}x_1 - c_{12}x_2 + (pc_{23} - c_{13})x_3 = 0$$

is conjugate to c.

The intersections of $C'P$ with $a: x_1 = 0$ and $b: x_2 = 0$ are respectively $Q: (0, pc_{23} - c_{13}, c_{12})$ and $R: (pc_{23} - c_{13}, 0, -pc_{12})$. The polar line of Q with respect to C has equation

$$[x_1, x_2, x_3] \cdot \begin{bmatrix} 0 & c_{12} & c_{13} \\ c_{21} & 0 & c_{23} \\ c_{31} & c_{32} & 0 \end{bmatrix} \cdot \begin{bmatrix} 0 \\ pc_{23} - c_{13} \\ c_{12} \end{bmatrix} = 0 \quad \text{or} \quad pc_{12}x_1 + c_{12}x_2 + (pc_{23} - c_{13})x_3 = 0$$

and is clearly on R. Thus, Q and R are conjugate with respect to C as required.

17.6. Prove: The six vertices of two self-polar triangles with respect to a conic C are on another conic C'.

Take the reference triangle ABC as one of the two triangles; then $C: c_{11}x_1^2 + c_{22}x_2^2 + c_{33}x_3^2 = 0$. Let $P: (p)$, $Q: (q)$, $R: (r)$ be the vertices of the second triangle. (The reader will show that the proposed theorem collapses if any coordinate of P, Q, R is zero.)

Now the polar line of P, $c_{11}p_1x_1 + c_{22}p_2x_2 + c_{33}p_3x_3 = 0$, is on Q and R provided

(i) $\qquad\qquad c_{11}p_1q_1 + c_{22}p_2q_2 + c_{33}p_3q_3 = 0$

(ii) $\qquad\qquad c_{11}p_1r_1 + c_{22}p_2r_2 + c_{33}p_3r_3 = 0$

Thus PQR is a self polar triangle with respect to C provided (i), (ii) and

(iii) $\qquad\qquad c_{11}q_1r_1 + c_{22}q_2r_2 + c_{33}q_3r_3 = 0$

hold. These equations imply

$$\begin{vmatrix} q_1r_1 & q_2r_2 & q_3r_3 \\ p_1r_1 & p_2r_2 & p_3r_3 \\ p_1q_1 & p_2q_2 & p_3q_3 \end{vmatrix} = 0$$

or dividing the first column by $p_1q_1r_1$, the second by $p_2q_2r_2$, and the third by $p_3q_3r_3$,

(iv) $$\begin{vmatrix} \dfrac{1}{p_1} & \dfrac{1}{p_2} & \dfrac{1}{p_3} \\[2mm] \dfrac{1}{q_1} & \dfrac{1}{q_2} & \dfrac{1}{q_3} \\[2mm] \dfrac{1}{r_1} & \dfrac{1}{r_2} & \dfrac{1}{r_3} \end{vmatrix} = 0$$

In turn (iv) insures the existence of constants a, b, c not all zero, such that

(v) $$\begin{bmatrix} \dfrac{1}{p_1} & \dfrac{1}{p_2} & \dfrac{1}{p_3} \\[2mm] \dfrac{1}{q_1} & \dfrac{1}{q_2} & \dfrac{1}{q_3} \\[2mm] \dfrac{1}{r_1} & \dfrac{1}{r_2} & \dfrac{1}{r_3} \end{bmatrix} \cdot \begin{bmatrix} a \\ b \\ c \end{bmatrix} = \begin{bmatrix} 0 \\ 0 \\ 0 \end{bmatrix}$$

Clearly, $C': ax_2x_3 + bx_3x_1 + cx_1x_2 = 0$ is the required conic since by (v) it is on P, Q, R and by (21), page 205, it is on A, B, C.

17.7. Prove: If two pairs of opposite vertices of a complete quadrilateral are pairs of conjugate points with respect to a conic C, then the third pair is also a conjugate pair.

Let the sides of the quadrilateral be $a: [1, 0, 0]$, $b: [0, 1, 0]$, $c: [0, 0, 1]$, $e: [1, 1, 1]$ and $C: \sum c_{ij}x_ix_j = 0$. Suppose $C = a \cdot b: (0, 0, 1)$, $F = c \cdot e: (1, -1, 0)$ are a pair of conjugate points with respect to C; then F on the polar line of C requires $c_{13} = c_{23}$. Similarly, if $B = c \cdot a: (0, 1, 0)$ and $E = b \cdot e: (1, 0, -1)$ are a pair of conjugate points, then $c_{12} = c_{32}$. Now the polar line of $A = b \cdot c: (1, 0, 0)$ is $c_{11}x_1 + c_{21}(x_2 + x_3) = 0$ and is on $D = a \cdot e: (0, 1, -1)$ and so A and D are conjugates, as required.

17.8. Prove: If Q, S are two points on a conic C and A, C are a pair of conjugate points on any line conjugate to QS with regard to C, then $AQ \cdot CS$ and $AS \cdot CQ$ are on C. (See Problems 9.26 and 17.24.)

Take $C: x_1^2 = x_2 x_3$ with $Q: (0, 1, 0)$, $S: (0, 0, 1)$, $E: (1, 1, 1)$ on C and $R: (1, 0, 0)$ the pole of QS. Any line as $r: x_2 - a x_3 = 0$ on R is conjugate to QS. Take $C: (b, a, 1)$ on r; its polar line $2b x_1 - x_2 - a x_3 = 0$ meets r in $A: (a, ab, b)$.

Then $AQ: b x_1 - a x_3 = 0$ and $CS: a x_1 - b x_2 = 0$ intersect in $U: (ab, a^2, b^2)$ while $AS: b x_1 - x_2 = 0$ and $CQ: x_1 - b x_3 = 0$ intersect in $V: (b, b^2, 1)$. Clearly, U and V are on C.

17.9. Prove: If a point moves on a given line, its polar lines with respect to two given conics intersect on a third conic.

Let the given line be determined by the points $P: (p)$ and $Q: (q)$ and let $\sum a_{ij} x_i x_j = 0$ and $\sum b_{ij} x_i x_j = 0$ be the given conics. The polar line of the point $(p + \lambda q)$ with respect to the given conics are $\sum a_{ij} p_i x_j + \lambda \sum a_{ij} q_i x_j = 0$ and $\sum b_{ij} p_i x_j + \lambda \sum b_{ij} q_i x_j = 0$ respectively. Eliminating λ, we have

$$\sum a_{ij} p_i x_j \cdot \sum b_{ij} q_i x_j - \sum a_{ij} q_i x_j \cdot \sum b_{ij} p_i x_j = 0$$

as the equation of the third conic.

17.10. Show that the quadrangular pencil (33), page 209, is given by

$$\sum c_{ij}'' x_i x_j + \lambda \sum c_{ij}''' x_i x_j = 0$$

where $\sum c_{ij}'' x_i x_j = 0$ and $\sum c_{ij}''' x_i x_j = 0$ are any two distinct conics (proper or degenerate) on the base points.

Suppose $\sum c_{ij}'' x_i x_j = 0$ and $\sum c_{ij}''' x_i x_j = 0$ are given by (33) when $\lambda = \lambda_1$ and $\lambda = \lambda_2 \neq \lambda_1$ respectively. Consider

$$\text{(i)} \qquad \sum c_{ij}'' x_i x_j + \lambda \sum c_{ij}''' x_i x_j$$

$$= \left(\sum c_{ij} x_i x_j + \lambda_1 \sum c_{ij}' x_i x_j \right) + \lambda \left(\sum c_{ij} x_i x_j + \lambda_2 \sum c_{ij}' x_i x_j \right)$$

$$= (1 + \lambda) \sum c_{ij} x_i x_j + (\lambda_1 + \lambda \lambda_2) \sum c_{ij}' x_i x_j = 0$$

or $\qquad$ (ii) $\qquad \sum c_{ij} x_i x_j + \mu \sum c_{ij}' x_i x_j = 0$

where $\mu = \dfrac{\lambda_1 + \lambda \lambda_2}{1 + \lambda}$ when $\lambda \neq -1$ and $\mu = \infty$ when $\lambda = -1$. Since $(1 + \lambda)\mu = \lambda_1 + \lambda \lambda_2$ is linear in both λ and μ, it follows that (ii) and (33) are equations of the same pencil.

17.11. Find the equation of the conic on the points $Q: (1, 0, 1)$, $R: (1, 1, 0)$, $T: (1, 1, 1)$ and tangent to $p: x_1 + x_2 + x_3 = 0$ at $P: (-1, 0, 1)$.

In the pencil of conics on Q, R and tangent to p at P, there are two degenerate conics — p, QR and PQ, PR. An equation of the pencil is then

$$(x_1 + x_2 + x_3)(x_1 - x_2 - x_3) + \lambda x_2 (x_1 - x_2 + x_3) = 0$$

The equation of the conic of the pencil on T, given by $\lambda = 3$, is

$$x_1^2 - 4x_2^2 - x_3^2 + 3x_1 x_2 + x_2 x_3 = 0$$

17.12. Find the equation of the conic having two-point contact with $C: x_1^2 + x_2^2 - 4x_3^2 - 2x_1x_2 + 3x_1x_3 = 0$ at $P:(1,1,0)$, simple intersections with C at $Q:(1,0,1)$ and $R:(1,2,1)$, and on $T:(1,1,2)$.

The tangent to C at P is $p: x_3 = 0$. The required conic is a member of the pencil having $C'' = p, QR: x_3(x_1 - x_3)$ and $C''' = PQ, PR: (x_1 - x_2 - x_3)(x_1 - x_2 + x_3)$ as degenerate conics. The pencil is given by

$$x_3(x_1 - x_3) + \lambda(x_1 - x_2 - x_3)(x_1 - x_2 + x_3) = 0$$

and the equation of the required conic, given when $\lambda = -\frac{1}{2}$, is

$$x_1^2 + x_2^2 + x_3^2 - 2x_1x_2 - 2x_1x_3 = 0$$

17.13. Find the equation of the conic on $T:(0,1,0)$ and having four-point contact with $C: x_1^2 + x_2^2 - x_3^2 + 2x_1x_2 + 2x_2x_3 = 0$ at $P:(1,0,1)$.

The tangent to C at P is $p: x_1 + 2x_2 - x_3 = 0$. Each conic of the pencil, having four-point contact with C at P, has p as tangent at P. The only degenerate conic of this pencil is the line p counted twice. An equation of the pencil is then

$$x_1^2 + x_2^2 - x_3^2 + 2x_1x_2 + 2x_2x_3 + \lambda(x_1 + 2x_2 - x_3)^2 = 0$$

The equation of the required conic, obtained when $\lambda = -\frac{1}{4}$, is

$$3x_1^2 - 5x_3^2 + 4x_1x_2 + 2x_3x_1 + 12x_2x_3 = 0$$

Supplementary Problems

17.14. Find the equation of the conic C on the points $R:(1,0,-1)$, $S:(1,0,1)$, $T:(1,2,1)$, $U:(1,2,-1)$, $V:(1,3,0)$.

Hint. Take R and S as the centers of the projective pencils

$$\{Ax\} + \lambda\{Bx\} = 0 \quad \text{and} \quad \{Dx\} + \lambda\{Ex\} = 0$$

by which C is generated. Let the points T, U, V be given when $\lambda = \infty, 0, 1$ respectively. Now $RU: x_1 + x_3 = 0$, $RT: x_1 - x_2 + x_3 = 0$ and $RV: 3x_1 - x_2 + 3x_3 = 0$; hence, $\{Ax\} = 2x_1 + 2x_3 = 0$ and $\{Bx\} = x_1 - x_2 + x_3 = 0$. Similarly, $\{Dx\} = x_1 - x_2 - x_3 = 0$ and $\{Ex\} = 2x_1 - 2x_3 = 0$. *Ans.* $3x_1^2 + 2x_1x_2 - x_2^2 - 3x_3^2 = 0$

17.15. Find the equation of the conic tangent at $S:(0,3,1)$ to $x_2 - 3x_3 = 0$ and on the points $R:(1,2,1)$, $T:(-1,2,1)$, $U:(2,0,1)$.

Hint. Choose representations $\{Ax\} = 0$ and $\{Bx\} = 0$ of RU and RT so that $RS: \{Ax\} + \{Bx\} = 0$, and representations $\{Dx\} = 0$ and $\{Ex\} = 0$ of SU and ST so that $\{Dx\} + \{Ex\} = x_2 - 3x_3 = 0$. *Ans.* $6x_1^2 - x_2^2 + 11x_2x_3 - 24x_3^2 = 0$

17.16. Find the equation of the conic on the lines $r:[0,1,1]$, $s:[1,0,1]$, $t:[1,1,1]$, $u:[1,1,0]$, $v:[2,6,5]$. *Ans.* $3X_1^2 + X_2^2 - 4X_3^2 - 4X_1X_2 + X_1X_3 + 3X_2X_3 = 0$

17.17. Find the equation of the conic tangent to $2x_1 + x_2 + x_3 = 0$ at $(-1,1,1)$, tangent to $2x_1 - x_2 - 2x_3 = 0$ at $(1,0,1)$, and tangent to $x_2 + 2x_3 = 0$. *Ans.* $5X_1^2 - 38X_1X_2 + 24X_1X_3 + 108X_2^2 - 146X_2X_3 + 46X_3^2 = 0$

17.18. Show that $10x_1 - x_2 + 8x_3 = 0$ is tangent to the conic of Example 17.1, page 201, first by eliminating x_2 between the equation of the line and (i) and second by showing that it is a line of (ii).

17.19. Find the intersections, if any, of the conic

$$[x_1, x_2, x_3] \cdot \begin{bmatrix} 1 & 3 & 2 \\ 3 & -1 & 1 \\ 2 & 1 & 3 \end{bmatrix} \cdot \begin{bmatrix} x_1 \\ x_2 \\ x_3 \end{bmatrix} = 0$$

and the line joining the points

(a) $(2, 1, 0)$ and $(1, 2, 1)$,   (b) $(0, 3, 2)$ and $(4, 1, -4)$,   (c) $(2, 1, 0)$ and $(1, -1, -1)$

Ans. (a) $(3, 0, -1)$, $(1, -4, -3)$; (b) $(2, -1, -3)$

17.20. (a) Write the equation of the conic of Problem 17.19 in line coordinates.

(b) Obtain the lines of this conic on the point $(2, -1, 2)$.

Hint. An arbitrary line on $(2, -1, 2)$ has equation $ax_1 + 2(a + c)x_2 + cx_3 = 0$.

Ans. (a) $[X_1, X_2, X_3] \begin{bmatrix} 4 & 7 & -5 \\ 7 & 1 & -5 \\ -5 & -5 & 10 \end{bmatrix} \begin{bmatrix} X_1 \\ X_2 \\ X_3 \end{bmatrix} = 0$ (b) $x_1 - 2x_2 - 2x_3 = 0$, $x_1 + 8x_2 + 3x_3 = 0$

17.21. Find the equation of the conic, five of whose lines are: $p\colon [1, 0, -1]$, $q\colon [1, 2, 1]$, $r\colon [1, 2, -1]$, $s\colon [1, 0, 1]$, $t\colon [1, 1, 2]$.   *Ans.* $X_1^2 - 3X_2^2 - X_3^2 + 6X_1X_2 = 0$

17.22. (a) Show that $P\colon (p)$ is outside the conic $x_1^2 + x_2^2 - x_3^2 = 0$ provided $p_1^2 + p_2^2 - p_3^2 > 0$.

(b) Show that if $P\colon (p)$ is inside the conic of (a), every line on P meets the conic in two distinct points.

17.23. Show that every self-polar triangle with respect to a conic has just one vertex inside the conic.

17.24. Show that the equation of any conic C may be reduced to $x_1^2 - x_2x_3 = 0$.

Hint. Take as reference triangle $A_1A_2A_3$ with $A_2\colon (0, 1, 0)$, $A_3\colon (0, 0, 1)$ on C and A_1 the pole of A_2A_3; also take $E\colon (1, 1, 1)$ on C.

17.25. Obtain the equation of a conic inscribed in the reference triangle.

Hint. If $x_1 = 0$ is tangent to $\sum c_{ij}x_ix_j = 0$, then $c_{23}^2 = c_{22} \cdot c_{33}$.

Ans. $c_{11}x_1^2 + c_{22}x_2^2 + c_{33}x_3^2 + 2\sqrt{c_{11} \cdot c_{22}}\, x_1x_2 + 2\sqrt{c_{33} \cdot c_{11}}\, x_3x_1 + 2\sqrt{c_{22} \cdot c_{33}}\, x_2x_3 = 0$

17.26. Let P be a point not on C, p its polar line with respect to C, and Q an arbitrary point on C. Show that $p \cdot PQ$ and $p \cdot q$, where q is the tangent to C at Q, are conjugate points.

17.27. Let A, B, P be distinct points on a conic. Show that PA and PB are separated harmonically by p, the tangent at P, and the join of P and the pole of AB.

17.28. Let A and B be conjugate points with respect to C. On A take a line r meeting C in P and Q. Let BP and BQ meet C again in R and S respectively. Show that A, R, S are collinear.

17.29. Prove: If a complete quadrangle is inscribed in a conic C, its diagonal triangle is self-polar with respect to C.

Hint. Take the vertices of the quadrangle as the vertices of the reference triangle and the unit point.

17.30. (a) Show that the triangle with vertices $A\colon (1, 0, 1)$, $B\colon (1, -1, 0)$, $C\colon (1, 1, -1)$ is self-polar with respect to the conic

$$C\colon 4x_1^2 + 16x_2^2 + 7x_3^2 - 16x_1x_2 - 2x_3x_1 + 22x_2x_3 = 0$$

(b) Obtain the equation of C as $9x_1'^2 + 36x_2'^2 - 9x_3'^2 = 0$ when referred to the self-polar triangle as triangle of reference.

Hint. Use $X = \begin{bmatrix} 1 & 1 & 1 \\ 0 & -1 & 1 \\ 1 & 0 & -1 \end{bmatrix} X' = \mathbf{A}^{-1}X'$.

(c) Obtain $X' = \mathbf{A}X$ and verify that it renames the self-polar triangle as reference triangle.

(d) Introduce in (b) the transformation $X' = \begin{bmatrix} 1/3 & 0 & 0 \\ 0 & 1/6 & 0 \\ 0 & 0 & 1/3 \end{bmatrix} X'' = \mathbf{B}^{-1}X''$ and verify that

$X = \mathbf{A}^{-1}\mathbf{B}^{-1}X''$ yields $x_1''^2 + x_2''^2 - x_3''^2 = 0$ as the equation of C.

(e) Show that $X'' = \mathbf{B}\mathbf{A}X$ preserves the triangle of reference and associates with it the unit point $E: (5, 1, 0)$.

17.31. (a) Show that the triangle whose vertices are $A: (1, -1, 0)$, $B: (7, -3, -4)$, $C: (1, 3, 4)$ is inscribed in the conic
$$C: x_1^2 + x_2^2 - 4x_3^2 + 2x_1x_2 + 4x_2x_3 = 0$$

(b) Show that the equation of C is $x_1'x_2' - x_3'x_1' + 4x_2'x_3' = 0$ when referred to this triangle as triangle of reference.

(c) Show that the transformation: $x_1' = 4x_1''$, $x_2' = x_2''$, $x_3' = x_3''$ reduces the equation of C to $x_1''x_2'' - x_3''x_1'' + x_2''x_3'' = 0$.

(d) Obtain the equation of transformation $X'' = \mathbf{A}X$ which effects the reduction; show that the unit point associated with triangle ABC is $E: (3, -1, 0)$.

17.32. Show that there is just one conic on the points $(1, 0, 0)$, $(0, 1, 0)$, $(0, 0, 1)$ and tangent to $ax_1 + bx_2 - (a + b)x_3 = 0$ at $(1, 1, 1)$.

17.33. Prove: If PQR is not self-polar with respect to C, the polar lines of its vertices meet the opposite sides in collinear points.

17.34. If the vertices of one triangle are the poles with respect to C of the sides of another triangle, show that the vertices of the second triangle are the poles of the sides of the first. Call either triangle the *polar triangle* of the other. What is the polar triangle of an inscribed triangle of C?

17.35. Show that polar triangles are perspective, the center and axis of perspectivity being pole and polar line.

17.36. C and E are conjugate points with respect to a conic C, and D is a point on C. If DC and DE meet the conic again in F and G respectively, show (a) CE and FG are conjugate lines with respect to C, (b) CG and EF meet on C.
Hint. Take the reference triangle ABC as in Problem 17.5, page 212, and $D: (1, 1, 1)$ on C. Then $E: (1, \alpha, 0)$ on AB is conjugate to C.

17.37. Prove the theorem of Problem 17.5, page 212, taking the inscribed triangle as ABD, where $A: (1, 0, 0)$, $B: (0, 1, 0)$, $D: (1, 1, 1)$, and $C: (0, 0, 1)$ as the pole of the side AB.

17.38. Let the distinct points $Y: (y)$ and $Z: (z)$ be distinct from the base points of the pencil of conics (33), page 209. Prove:

(a) The polar lines of Y with respect to the conics of (33) constitute a pencil of lines.

(b) The center of the pencil of lines and Y are a conjugate pair with respect to every conic of the pencil.

(c) The poles of the line YZ with respect to the conics of (33) are on
$$\sum c_{ij}y_ix_j \cdot \sum c_{ij}'z_ix_j - \sum c_{ij}z_ix_j \cdot \sum c_{ij}'y_ix_j = 0$$

17.39. Prove: The diagonal triangle of the complete quadrangle $PQRS$ is a self-polar triangle of every conic of the quadrangular pencil defined by P, Q, R, S.
Hint. Take as vertices of the quadrangle $(1, \pm1, \pm1)$ and $(1, \pm1, \mp1)$.

17.40. Prove: Any conic C circumscribing the quadrangle $PQRS$ of Problem 17.39 and any conic inscribed in the quadrilateral $pqrs$, where p, q, r, s are the tangents to C at P, Q, R, S respectively, have a common self-polar triangle.

17.41. Prove: Of the conics of a quadrangular pencil, those which meet a given line p, not on a vertex, do so in reciprocal pairs of the involution on p determined by its intersections with pairs of opposite sides of the quadrangle.

17.42. When will the involution of Problem 17.41 be hyperbolic? elliptic?

17.43. Let $Y: (y)$ be a point from which two tangents can be drawn to $C: \sum c_{ij} x_i x_j = 0$. Show that the equations of these tangents are given by $\sum c_{ij} y_i y_j \cdot \sum c_{ij} x_i x_j - \left(\sum c_{ij} y_i x_j \right)^2 = 0$.

17.44. Find the equation of the conic tangent to $x_1 - x_2 + x_3 = 0$ at $P: (1, 2, 1)$, tangent to $x_1 + x_2 - x_3 = 0$ at $R: (1, 0, 1)$, and on $T: (1, 1, -1)$.   *Ans.* $7x_1^2 - 4x_2^2 - x_3^2 - 6x_1 x_3 + 8x_2 x_3 = 0$

17.45. Find the equation of the conic on $T: (1, -1, 1)$ and having two-point contact with
$$x_1^2 + x_2^2 + 3x_3^2 - 2x_1 x_2 - 4x_3 x_1 + 6x_2 x_3 = 0$$
at $P: (1, 1, 0)$ and at $R: (1, 0, 1)$.   *Ans.* $2x_1^2 + 2x_2^2 + 3x_3^2 - 4x_1 x_2 - 5x_3 x_1 + 6x_2 x_3 = 0$

17.46. Find the equation of the conic having three-point contact with $x_1^2 - 2x_2^2 + x_1 x_2 - x_3 x_1 - x_2 x_3$ at $P: (1, 0, 1)$ and on the points $Q: (1, 1, 0)$ and $T: (1, 0, 0)$.
Ans. $2x_2^2 + x_3^2 - 2x_1 x_2 - x_3 x_1 + 2x_2 x_3 = 0$

17.47. Prove: If C and C' are distinct conics on four distinct points and Y is any point of the plane, then the polar lines of Y, with respect to the conics of the pencils determined by C and C', (a) are identical when Y is a singular point of a degenerate conic of the pencil, (b) constitute a pencil of lines otherwise.

Identify the center of the pencil in (b). (There are two cases to be considered.)

17.48. Prove: The conics of a quadrangular pencil have a unique common self-polar triangle.

17.49. Prove: If C and C' are distinct conics having two-point contact at the distinct points P and R, the conics of the pencil determined by them have an infinite number of common self-polar triangles. *Hint.* Begin with $A = p \cdot r$, the intersection of the common tangents at P and R.

17.50. State and prove the duals of Problems 17.47-17.49.

17.51. Call Y' *conjugate to Y with respect to a pencil of conics* provided Y' is conjugate to Y with respect to every conic of the pencil. For a quadrangular pencil show (a) the points not on the common self-polar triangle are conjugate in pairs, (b) each point on a side of the self-polar triangle is conjugate to the opposite vertex, (c) a vertex is conjugate to every point on the opposite side except the vertices on that side.

17.52. Prove Pascal's Theorem: If a simple hexagon $A_1 A_2 A_3 A_4 A_5 A_6$ is inscribed in a conic, the intersections $R = A_1 A_2 \cdot A_4 A_5$, $S = A_2 A_3 \cdot A_5 A_6$, $T = A_3 A_4 \cdot A_6 A_1$ of the three pairs of opposite sides are collinear.

Hint. Let the tangents to $C: \sum c_{ij} x_i x_j = 0$ at A_1 and A_2 meet at B. Take $A_1: (1, 0, 0)$, $A_2: (0, 1, 0)$, $B: (0, 0, 1)$ and $A_3: (1, 1, 1)$; then $C: x_3^2 = x_1 x_2$. Take $A_4: (1, a^2, a)$, $A_5: (1, b^2, b)$ and $A_5: (1, c^2, c)$.

17.53. Prove Pascal's Theorem taking three of the vertices of the hexagon as vertices of the reference triangle and (a_1, a_2, a_3), (b_1, b_2, b_3), (c_1, c_2, c_3) as the remaining vertices.

17.54. Using the conic and inscribed hexagon of Problem 17.52, show that the Brianchon point of the circumscribed hexagon whose sides are the tangents $a_1, a_2, a_3, a_4, a_5, a_6$ to C at $A_1, A_2, A_3, A_4, A_5, A_6$ respectively is the pole of the Pascal line RST.

Chapter 18

Projective, Affine and Euclidean Geometry

THE PROJECTIVE GROUP

The set G of all non-singular n-square matrices is said to form a *group* with respect to multiplication since, when $\mathbf{A}, \mathbf{B}, \mathbf{C}$ are any matrices in G,

  (i) $\mathbf{AB}$ and $\mathbf{BA}$ are in G,   (the *closure property*)

  (ii) $\mathbf{A}(\mathbf{BC}) = (\mathbf{AB})\mathbf{C}$,   (the *associative law*)

  (iii) there exists a matrix $\mathbf{I}$ (*the identity*) in G such that $\mathbf{IA} = \mathbf{AI} = \mathbf{A}$,

  (iv)   for each matrix $\mathbf{A}$ in G there exists a matrix $\mathbf{A}^{-1}$ (*the inverse* of $\mathbf{A}$) in G such that $\mathbf{AA}^{-1} = \mathbf{A}^{-1}\mathbf{A} = \mathbf{I}$.

A group G is called *commutative* or *Abelian* if the further property

  (v) $\mathbf{AB} = \mathbf{BA}$ for every $\mathbf{A}, \mathbf{B}$ in G (the *commutative law*)

holds.   The group G of non-singular n-square matrices is non-Abelian.

When expressed in homogeneous coordinates a projective transformation of a line onto itself is essentially a non-singular 2-square matrix and a projective transformation of the plane onto itself is essentially a non-singular 3-square matrix.   Interpreting the product of two matrices as the resultant of one transformation followed by another (see Problem 16.43), we have

Theorem 18.1.   The set of all projective transformations of a line onto itself forms a non-Abelian group.

and

Theorem 18.2.   The set of all projective transformations (collineations) of the form

$$\rho X' = \mathbf{E}X, \quad |\mathbf{E}| \neq 0 \tag{1}$$

of the plane onto itself forms a non-Abelian group, called the *general projective group* $\mathcal{T}$.

The content of projective geometry on a line consists of those properties of sets of points which remain unchanged (invariant) under the group of Theorem 18.1.   The content of plane projective geometry consists of those properties of configurations in the plane which remain unchanged under the group $\mathcal{T}$ of Theorem 18.2.   It will be noted that the study of projective geometry on a line may be made without considering the plane or may be obtained as a byproduct of a study of all collineations of the plane having the line as double line.

SUBGROUPS OF THE PROJECTIVE GROUP

Any non-empty subset G' of the elements of a group G, which in itself satisfies the properties of a group, is called a subgroup of G.   For example, the subset consisting of all matrices of G of determinant ± 1 forms a subgroup of G.   In turn, this subgroup (and hence

219

the group G) has a subgroup consisting of all matrices of G of determinant $+1$. The essential conditions that a subset G' of G be a subgroup are: (i) G' contain the inverse of each of its elements, (ii) G' contain AB for every pair A, B of G'. Thus the subset of all matrices of G of determinant -1 is not a subgroup. An exhaustive study of the subgroups of T will not be attempted here. Certain subgroups of interest in later sections will be considered next.

Select in the plane any line l and choose a coordinate system having this line as the side $x_3 = 0$ of the reference triangle. A collineation of the form

$$
\begin{aligned}
\rho x_1' &= e_{11}x_1 + e_{12}x_2 + e_{13}x_3 \\
\rho x_2' &= e_{21}x_1 + e_{22}x_2 + e_{23}x_3, \quad |\mathbf{E}| = e_{33}\begin{vmatrix} e_{11} & e_{12} \\ e_{21} & e_{22} \end{vmatrix} \neq 0 \\
\rho x_3' &= \phantom{e_{21}x_1 + e_{22}x_2 +} e_{33}x_3
\end{aligned}
\tag{2}
$$

has $x_3 = 0$ as a double (fixed) line. We leave for the reader to verify that the set of all collineation of the form (2) constitutes a subgroup T_a of T.

Consider an element of T_a having the form

$$
\begin{aligned}
\rho x_1' &= e_{11}x_1 \phantom{+ e_{11}x_2} + e_{13}x_3 \\
\rho x_2' &= \phantom{e_{11}x_1} e_{11}x_2 + e_{23}x_3, \quad e_{11} \cdot e_{33} \neq 0 \\
\rho x_3' &= \phantom{e_{11}x_1 + e_{11}x_2 +} e_{33}x_3
\end{aligned}
\tag{3}
$$

Now (3) induces on $x_3 = 0$ the identity transformation (show that $(1, 0, 0)$, $(0, 1, 0)$ and $(1, 1, 0)$ are fixed points) and hence has $x_3 = 0$ as a line of double points. Moreover, the subset of all elements of T_a having the form (3) constitutes a subgroup of T_a (verify this) and, hence, of T. This subgroup consists of homologies (when $e_{11} \neq e_{33}$) and elations (when $e_{11} = e_{33}$).

AFFINE GEOMETRY

In Chapter 13 the affine plane was obtained either by singling out a line l of the projective plane to be endowed with special properties and called the ideal line or by removing this line from the projective plane. Taking l to be the side $x_3 = 0$ of a triangle of reference, it follows that the collineations of the form (2) may be interpreted as those leaving the ideal line unchanged or as those carrying the affine plane (the totality of points (x_1, x_2, x_3) with $x_3 \neq 0$) onto itself.

In homogeneous coordinates each point of the affine plane may be given with coordinates of the form $(x_1, x_2, 1)$; in non-homogeneous coordinates each point may be given as (x_1, x_2) or, by a change of notation, as (x, y). In homogeneous coordinates, (2) may be replaced by

$$
\begin{aligned}
x_1' &= a_{11}x_1 + a_{12}x_2 + a_{13}x_3 \\
x_2' &= a_{21}x_1 + a_{22}x_2 + a_{23}x_3, \quad |\mathbf{A}| = \begin{vmatrix} a_{11} & a_{12} \\ a_{21} & a_{22} \end{vmatrix} \neq 0 \\
x_3' &= \phantom{a_{21}x_1 + a_{22}x_2 +} x_3
\end{aligned}
\tag{2a}
$$

where each coefficient in (2a) is the corresponding coefficient in (2) divided by $e_{33} \neq 0$, and in non-homogeneous coordinates by

$$
\begin{aligned}
x' &= a_{11}x + a_{12}y + a_{13} \\
y' &= a_{21}x + a_{22}y + a_{23}
\end{aligned}, \quad \begin{vmatrix} a_{11} & a_{12} \\ a_{21} & a_{22} \end{vmatrix} \neq 0
$$

The content of plane affine geometry consists of the properties of configurations in the affine plane which are unchanged under the group of affine transformations (2a).

COORDINATE SYSTEM

In passing from the projective plane to the affine plane, a triangle of reference (of the projective plane) may be made to lose the side $x_3 = 0$. The remaining sides $x_2 = 0$ and $x_1 = 0$ are then taken as coordinate axes — $x_2 = 0$, along which x_1 varies, as the x_1 *axis* (in non-homogeneous coordinates the x *axis*) and $x_1 = 0$, along which x_2 varies as the x_2 *axis* (in non-homogeneous coordinates the y *axis*). The point of intersection $(0, 0, 1)$ or $(0, 0)$ of the axes is called the *origin*. The ideal points on the x and y axes are $(1, 0, 0)$ and $(0, 1, 0)$ respectively.

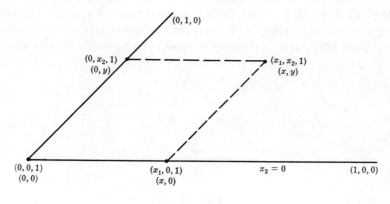

Fig. 18-1

In what is to follow, non-homogeneous coordinates will be used whenever more convenient. Homogeneous coordinates are necessary, however, in order to give an affine interpretation of a projective theorem. In such instances, the particular set of affine coordinates for which $x_3 = 1$ will be used.

THE AFFINE TRANSFORMATIONS

The affine transformations (*2a*) carry points into points and lines into lines; they also preserve collinearity and concurrency. Two affine lines are called parallel provided they do not intersect (their meet is on the ideal line). Since the affine transformations leave the ideal line unchanged, they carry parallel affine lines into parallel affine lines; thus the affine transformations preserve parallelism.

Consider next the distinct points $P: p$, $Q: q$, $R: r$ and $S: s$ on a projective line. Since (see Chapter 16), the cross ratio $(P, Q; R, S)$ is unchanged under a projective transformation, the same is then true under an affine transformation when these points are affine points. Suppose P, Q, R are affine points but $S = P_\infty$ is the ideal point on the line l. A general affine transformation carries these points into P', Q', R', P'_∞ respectively on a line l'. Now

$$(P, Q; R, P_\infty) \;=\; \frac{p - r}{q - r} \;=\; (P', Q'; R', P'_\infty)$$

Hence the ratio of the segments into which R divides the segment PQ is an affine invariant.

Finally, let $A: (a_1, a_2, 1)$, $B: (b_1, b_2, 1)$, $C: (c_1, c_2, 1)$ be three non-collinear points of the affine plane and define the *measure* μ of the triangle ABC as $\mu = \begin{vmatrix} a_1 & a_2 & 1 \\ b_1 & b_2 & 1 \\ c_1 & c_2 & 1 \end{vmatrix}$. (The reader will recognize this as twice the area, apart from possibly the sign, of the triangle in terms of a unit square. He will also recall that the sign of the measure depends upon the order in which the coordinates of the vertices are set down in the determinant, that is, by the sense in which the triangle is traversed.) As unit of measure, we take [see Fig. 18.2(a)] the triangle OHJ with vertices $O: (0,0,1)$, $H: (1,0,1)$, $J: (0,1,1)$ and measure

$$\begin{vmatrix} 0 & 0 & 1 \\ 1 & 0 & 1 \\ 0 & 1 & 1 \end{vmatrix} = +1.$$ Note that as we move around the triangle OHJ — from O to H to J and

back to O — the triangle, considered as a portion of the plane bounded by straight line segments, lies always on our left. The triangle ABC of Fig. 18-2(b) has positive measure while the triangle $A'B'C'$ of Fig. 18-2(c) has negative measure. An affine transformation $(2a)$ carries the unit triangle into a triangle of measure $|\mathbf{A}|$ and a triangle of measure μ into one of measure $\mu|\mathbf{A}|$. Thus affine transformations preserve the *ratio* of measures of triangles.

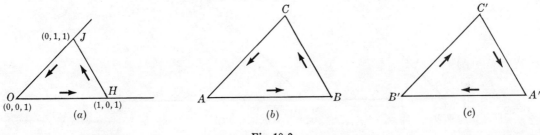

Fig. 18-2

An affine transformation for which $|\mathbf{A}| = +1$, preserves then both the measure and sense of a triangle; an affine transformation for which $|\mathbf{A}| = -1$ preserves the measure except for sign, that is, preserves the measure but changes the sense. The set of all affine transformations for which $|\mathbf{A}| = +1$ forms a group whose elements are called *equi-affine* or *special affine* transformations. The set of all affine transformations for which $|\mathbf{A}| = -1$ does not form a group since the product of two such transformations is an equi-affine transformation. Thus the set of all affine transformations for which $|\mathbf{A}| = \pm 1$ forms a group, each element of which is called an *equiareal* transformation.

An affine transformation, having the form

$$\begin{aligned} x_1' &= a_{11}x_1 && + a_{13}x_3 \\ x_2' &= && a_{11}x_2 + a_{23}x_3, && a_{11} \neq 0 \\ x_3' &= && x_3 \end{aligned}$$ $(3a)$

in homogeneous coordinates and

$$\begin{aligned} x' &= ax + b_1 \\ y' &= ay + b_2 \end{aligned}, \quad a \neq 0$$

in non-homogeneous coordinates, is called a *homothetic transformation*. The set of all homothetic transformations forms a group $\mathcal{T}_h$ (verify this), a subgroup of $\mathcal{T}_a$. Aside from the identity transformation, each element of $\mathcal{T}_h$ has a double line (*axis*) consisting entirely of double points — the ideal line $x_3 = 0$ — and a unique double point (*center*) given by $(a_{13}, a_{23}, 1 - a_{11})$.

Two affine lines p and p' which correspond under a homothetic transformation are parallel. In other words, any homothetic transformation carries any affine line p into an affine line p' parallel to p. Thus under a homothetic transformation a triangle is carried into another such that corresponding sides are parallel. The two triangles, being perspective from the ideal line, axis of the transformation, are perspective from a point. This point is the center of the transformation.

The homothetic transformations may be separated into two sets — *dilations*, having an affine point as center (i.e., $a_{11} \neq 1$) and *translations*, having an ideal point as center (i.e., $a_{11} = 1$). The set of all translations forms a group (verify); the set of all dilations does not.

Two triangles which correspond under a homothetic transformation are called *homothetic*; in more familiar terms, they are said to be similar and similarly placed.

EUCLIDEAN PLANE GEOMETRY

It will be recalled that Euclid's geometry (high school plane geometry) was concerned both with similar triangles and with congruent triangles, that is, was concerned both with affine and metric properties of configurations. In order to test for congruency, that is, to compare the lengths of the respective sides of two triangles ABC and $A'B'C'$, it is necessary to have a unit of measure which operates equally on all lines of the plane. This, in turn, requires (see Chapter 14) the definition of perpendicular lines. We are thus led to define a Euclidean transformation as an equiareal transformation which preserves perpendicularity.

Consider on the ideal line $x_3 = 0$ an elliptic involution

$$\begin{aligned} x_1' &= c_{11}x_1 + c_{12}x_2 \\ x_2' &= c_{21}x_1 - c_{11}x_2 \end{aligned}, \qquad c_{11}^2 + c_{12}c_{21} < 0 \qquad (4)$$

The effect of the projective transformation (alias)

$$x_1' = c_{21}x_1 - c_{11}x_2$$
$$x_2' = -c_{11}x_1 - c_{12}x_2$$
$$x_3' = (c_{11}^2 + c_{12}c_{21})x_3$$

is to reduce (4) to

$$x_1' = x_2, \quad x_2' = -x_1 \qquad (5)$$

in homogeneous coordinates or to

$$x' = -1/x \qquad (6)$$

in non-homogeneous coordinates of the points on $x_3 = 0$. Hereafter, we assume always on $x_3 = 0$ the elliptic involution (5) and call it the *absolute involution*.

Let any line p of the affine plane meet $x_3 = 0$ in $P_\infty : (x_1, x_2, 0)$. Under the absolute involution, the points P_∞ and $P_\infty' : (x_2, -x_1, 0)$ correspond, that is, P_∞ and P_∞' are a reciprocal pair of the involution. We define any affine line p' on P_∞' to be perpendicular to p. Under any projective transformation which leaves the absolute involution unchanged, any reciprocal pair P_∞, P_∞' of the absolute involution is carried into a reciprocal pair Q_∞, Q_∞' and, accordingly, any pair of perpendicular lines p, p' is carried into a pair of perpendicular lines q, q'.

The equiareal transformation

$$\rho X' = \begin{bmatrix} a_{11} & a_{12} & a_{13} \\ a_{21} & a_{22} & a_{23} \\ 0 & 0 & 1 \end{bmatrix} X, \qquad \begin{vmatrix} a_{11} & a_{12} \\ a_{21} & a_{22} \end{vmatrix} = \pm 1 \qquad (7)$$

being an affine transformation, leaves the ideal line $x_3 = 0$ unchanged. If (7) also leaves unchanged the absolute involution on $x_3 = 0$, it must carry any reciprocal pair $P_\infty : (x_1, x_2, 0)$ and $P_\infty' : (x_2, -x_1, 0)$ into a reciprocal pair

$$(a_{11}x_1 + a_{12}x_2,\ a_{21}x_1 + a_{22}x_2,\ 0) \quad \text{and} \quad (a_{11}x_2 - a_{12}x_1,\ a_{21}x_2 - a_{22}x_1,\ 0)$$

of the absolute involution. Thus there must exist a non-zero number k such that

$$a_{11}x_1 + a_{12}x_2 = k(a_{21}x_2 - a_{22}x_1)$$

$$a_{21}x_1 + a_{22}x_2 = -k(a_{11}x_2 - a_{12}x_1)$$

or
$$(a_{11} + ka_{22})x_1 + (a_{12} - ka_{21})x_2 = 0$$

$$(a_{21} - ka_{12})x_1 + (a_{22} + ka_{11})x_2 = 0$$

In order that these relations be independent of the choice of reciprocal pair, each coefficient must vanish. Thus

$$k = -\frac{a_{11}}{a_{22}} = -\frac{a_{22}}{a_{11}} = \frac{a_{12}}{a_{21}} = \frac{a_{21}}{a_{12}}$$

or
$$a_{11}^2 = a_{22}^2, \quad a_{12}^2 = a_{21}^2, \quad a_{11}a_{21} + a_{12}a_{22} = 0$$

Then (7) has either the form

$$\begin{aligned} x' &= ax - by + c \\ y' &= bx + ay + d \end{aligned}, \quad a^2 + b^2 = 1 \tag{7a}$$

or

$$\begin{aligned} x' &= ex + fy + g \\ y' &= fx - ey + h \end{aligned}, \quad e^2 + f^2 = 1 \tag{7b}$$

The totality of transformations (7a) and (7b) constitutes the group of *Euclidean transformations.*

LENGTHS

Let $A: (x_1, y_1)$ and $B: (x_2, y_2)$ be two distinct points. Any definition of the distance AB between the two points must satisfy the conditions:

(i) $AB = 0$ if and only if $A = B$.

(ii) $AB = BA$.   (We are defining distance, not directed distance.)

(iii) For any three points A, B, C then $AB + BC \geqq AC$ (the triangle property).

We leave for the reader to verify that

$$AB = (+)\sqrt{(x_2 - x_1)^2 + (y_2 - y_1)^2} \tag{8}$$

meets the above requirements.

We now show that (8) is invariant under the group of Euclidean transformations. Consider the transformation (7a) which carries

$$(x_1, y_1) \quad \text{into} \quad (x_1', y_1') = (ax_1 - by_1 + c,\ bx_1 + ay_1 + d)$$

and
$$(x_2, y_2) \quad \text{into} \quad (x_2', y_2') = (ax_2 - by_2 + c,\ bx_2 + ay_2 + d)$$

Then
$$x_2' - x_1' = a(x_2 - x_1) - b(y_2 - y_1)$$

$$y_2' - y_1' = b(x_2 - x_1) + a(y_2 - y_1)$$

and $(x_2' - x_1')^2 + (y_2' - y_1')^2 = (a^2 + b^2)[(x_2 - x_1)^2 + (y_2 - y_1)^2] = (x_2 - x_1)^2 + (y_2 - y_1)^2$

The verification for (7b) follows in a similar manner.

ANGLES

Consider in Fig. 18-3 two half-lines (rays) issuing from a point P. Assume for the moment the existence of a transformation which keeps P fixed and carries one ray point by point onto the other, that is, a transformation which rotates one ray onto the other. The inverse of such a transformation would then rotate the second ray back onto the first. We now restrict our attention to such transformations (rotations) whose sense is counterclockwise. Thus if the transformation τ carries r_1 counterclockwise onto r_2 then τ^{-1} carries r_2 *counterclockwise* onto r_1.

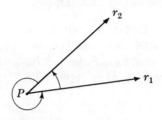

Fig. 18-3

To simplify matters, let the rays OA and OB (see Fig. 18-4) be such that their common point O is the origin, OA is the positive x axis, and OB is such that

$$x' = ax - by \\ y' = bx + ay \ , \qquad a^2 + b^2 = 1 \tag{9}$$

carries OA counterclockwise onto OB. Call the figure consisting of the two rays together with the associated rotation (9) the angle (AOB). As here defined, (AOB) is a directed angle and is always positive. While it differs from the directed angle of Chapter 14 in certain respects, the essential difference (as we shall see later) is that there is no restriction as to its size. Note that (9) carries $A:(1,0)$ into $B:(a,b)$.

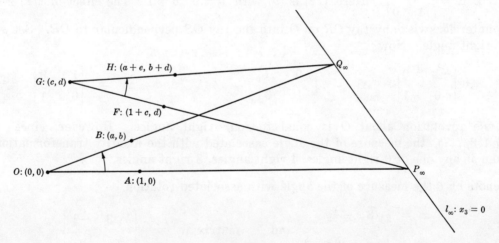

Fig. 18-4

Under a translation

$$x' = x + c \\ y' = y + d \tag{10}$$

angle (AOB) is carried into angle (FGH) for which the image of $O:(0,0)$ is the vertex $G:(c,d)$, the image of $A:(1,0)$ is $F:(1+c,d)$ and the image of $B:(a,b)$ is $H:(a+c, b+d)$. Associated with (FGH) is the rotation

$$x' - c = a(x-c) - b(y-d) \\ y' - d = b(x-c) + a(y-d) \ , \qquad a^2 + b^2 = 1 \tag{11}$$

The two rotations (9) and (11) have the same coefficients and are called *equivalent*; the two associated angles (AOB) and (FGH) have equivalent rotations and are called *equal*.

Conversely, given an angle (FGH) with associated rotation (11), we can always obtain an equal angle with vertex at the origin, initial side along the positive x axis and equivalent associated rotation (9).

Consider again the angle (AOB) with associated rotation (9). Under an equiareal transformation (7), the points A, O, B correspond respectively to $Q: (a_{11} + a_{13}, a_{21} + a_{23})$, $R: (a_{13}, a_{23})$, $S: (aa_{11} + ba_{12} + a_{13}, aa_{21} + ba_{22} + a_{23})$. If (AOB) and (QRS) are to be equal, the rotation associated with (QRS) must have the form

$$x' - a_{13} = a(x - a_{13}) - b(y - a_{23})$$

$$y' - a_{23} = b(x - a_{13}) + a(y - a_{23})$$

Now this rotation carries Q into $Q': (aa_{11} - ba_{21} + a_{13}, ba_{11} + aa_{21} + a_{23})$ and Q' will be on the side RS if and only if $a_{12} = -a_{21}$ and $a_{11} = a_{22}$. Thus any transformation (7) carries an angle into an equal angle if and only if it has the form $(7a)$.

MEASURE OF AN ANGLE

Let it be assumed for the moment that an angle (AOB) with associated rotation (9) has a measure. Denote by α the measure of the angle with associated rotation

$$x' = -y, \quad y' = x \tag{12}$$

of matrix $\Delta = \begin{bmatrix} 0 & -1 \\ 1 & 0 \end{bmatrix}$. Here, (12) is (9) with $a = 0$, $b = 1$. The effect of (12) is to rotate counterclockwise any ray OR on O into the ray OS perpendicular to OR. Set α equal to one right angle. Now

$$\Delta^2 = \begin{bmatrix} -1 & 0 \\ 0 & -1 \end{bmatrix}, \quad \Delta^3 = \begin{bmatrix} 0 & 1 \\ -1 & 0 \end{bmatrix} = \Delta^{-1} \quad \text{and} \quad \Delta^4 = \begin{bmatrix} 1 & 0 \\ 0 & 1 \end{bmatrix} = \mathbf{I}$$

Thus one revolution about O is equal to four right angles. However, since $\mathbf{I}^n = \mathbf{I}$, $(n = 0, 1, 2, \ldots)$, the measure of the angle associated with the identity transformation may be taken as any one of 0 right angles, 4 right angles, 8 right angles, $\ldots$.

Denote by β the measure of the angle with associated rotation

$$\begin{aligned} x' &= \tfrac{1}{2}\sqrt{3}\,x - \tfrac{1}{2}y \\ y' &= \tfrac{1}{2}x + \tfrac{1}{2}\sqrt{3}\,y \end{aligned} \qquad \text{and} \qquad \text{matrix } \Delta_1 = \begin{bmatrix} \tfrac{1}{2}\sqrt{3} & -\tfrac{1}{2} \\ \tfrac{1}{2} & \tfrac{1}{2}\sqrt{3} \end{bmatrix}$$

Now $\Delta_1^2 = \begin{bmatrix} \tfrac{1}{2} & -\tfrac{1}{2}\sqrt{3} \\ \tfrac{1}{2}\sqrt{3} & \tfrac{1}{2} \end{bmatrix}$, $\Delta_1^3 = \begin{bmatrix} 0 & -1 \\ 1 & 0 \end{bmatrix} = \Delta$ and $\Delta_1^{12} = \mathbf{I}$. Hence, $\beta = \tfrac{1}{3}$ right angle, $\tfrac{13}{3}$ right angles, $\tfrac{25}{3}$ right angles, $\ldots$. From these examples $\Big($the reader may also examine

(i) $\begin{aligned} x' &= -\tfrac{1}{2}x - \tfrac{1}{2}\sqrt{3}\,y \\ y' &= \tfrac{1}{2}\sqrt{3}\,x - \tfrac{1}{2}y \end{aligned}$ and   (ii) $\begin{aligned} \sqrt{2}\,x' &= -x + y \\ \sqrt{2}\,y' &= -x - y \end{aligned}\Big)$ we conclude that the measure in

right angles of an angle has one value γ satisfying $0 \leqq \gamma < 4$ (call this the *principal value* of the measure) from which all other values may be obtained by adding positive multiples of 4 right angles. In practice, of course, the principal value, θ degrees, is obtained from a suitable table of natural trigonometric functions by the use of any two of $\cos\theta = a$, $\sin\theta = b$, $\tan\theta = b/a$.

RESUME

In this chapter, we began with the general projective group $\mathcal{T}$, that is, all collineations

$$\rho X' = \mathbf{E}X, \quad |\mathbf{E}| \neq 0 \tag{1}$$

of the projective plane onto itself. These collineations carry collinear points (concurrent lines) into collinear points (concurrent lines) and preserve cross ratios.

Then, selecting a (any) line of the plane and a coordinate system having this line as the side $x_3 = 0$ of the reference triangle, we obtained a subgroup $\mathcal{T}_a$ of $\mathcal{T}$ whose elements (collineations) left $x_3 = 0$ invariant, that is, all collineations having $x_3 = 0$ as a double (fixed) line. The group $\mathcal{T}_a$ — called the general affine group — preserves in the affine plane parallelism of lines, the ratio of the segments into which a point R divides a segment PQ, and the ratio of the measures of two triangles. The subgroup of $\mathcal{T}_a$ for which $|\mathbf{A}| = \pm 1$, called the equiareal group, preserves the measure of a triangle but not necessarily its sense; the subgroup of the equiareal group for which $|\mathbf{A}| = +1$, called the equi-affine group, preserves both the measure and the sense of a triangle.

The subgroup $\mathcal{T}_h$ of $\mathcal{T}_a$, each element of which has the form $(3a)$, has the line $x_3 = 0$ as a line of double points. The elements of this group carry lines into parallel lines and, hence, triangles into similar triangles.

Finally, taking the affine plane as the Euclidean plane and establishing on $x_3 = 0$ an elliptic involution as absolute involution, two lines p and p' which meet $x_3 = 0$ in a reciprocal pair of the absolute involution were defined as mutually perpendicular. All equiareal transformations which preserve the absolute involution were defined to be Euclidean transformations. These transformations preserve distances and angles but, again, not always the sense of the angle.

Consider a Euclidean transformation of the type

$$\begin{matrix} x' = ax - by + c \\ y' = bx + ay + d \end{matrix}, \quad a^2 + b^2 = 1 \tag{7a}$$

If $a = 1$, $b = c = d = 0$, $(7a)$ is the identity transformation; if $a = 1$, $b = 0$ but not both c and d are zero, $(7a)$ is a translation $\begin{matrix} x' = x + c \\ y' = y + d \end{matrix}$; if $c = d = 0$, $(7a)$ is a rotation $\begin{matrix} x' = x\cos\theta - y\sin\theta \\ y' = x\sin\theta + y\cos\theta \end{matrix}$; otherwise, $(7a)$ effects a rotation $\begin{matrix} \bar{x} = x\cos\theta - y\sin\theta \\ \bar{y} = x\sin\theta + y\cos\theta \end{matrix}$ followed by a translation $\begin{matrix} x' = \bar{x} + c \\ y' = \bar{y} + d \end{matrix}$. Clearly, the set of all transformations of the type $(7a)$ forms a group, a subgroup of the Euclidean group.

Consider next a Euclidean transformation of the type

$$\begin{matrix} x' = ex + fy + g \\ y' = fx - ey + h \end{matrix}, \quad e^2 + f^2 = 1 \tag{7b}$$

The double points are given by

$$\begin{matrix} (e-1)x + fy = -g \\ fx - (e+1)y = -h \end{matrix} \tag{13}$$

Now $\begin{vmatrix} e-1 & f \\ f & -(e+1) \end{vmatrix} = 0$; hence the two equations in (13) represent either the same line or distinct parallel lines.

In the first case, $(7b)$ has the form

$$x' = ex + fy + g$$
$$y' = fx - ey + fg/(e-1)$$
, $\quad e^2 + f^2 = 1 \qquad (7b_1)$

Its double elements consist of a line of double points $p: (e-1)x + fy + g = 0$ and a double point $P'_\infty: (e-1, f, 0)$ not on p. Since the ideal point $P_\infty: (-f, e-1, 0)$ on p and P'_∞ are a reciprocal pair in the absolute involution, every line on P'_∞ is perpendicular to p. Let $R: (r_1, r_2)$ be an arbitrary point in the plane and denote by R' its correspondent under the transformation $(7b_1)$. Now R, R', P'_∞ are collinear and the midpoint of the segment RR' is on p. (The reader will verify this.) Thus $(7b_1)$ is an *orthogonal reflection in the line p*. (See Fig. 18-5 where the joins $AA', BB', CC', \ldots$ of corresponding points are perpendicular to p and the segments $AA', BB', CC', \ldots$ are bisected by p.)

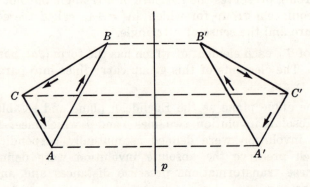

Fig. 18-5

In the second case, $(7b)$ is an orthogonal reflection $\begin{aligned} \bar{x} &= x\cos\theta + y\sin\theta \\ \bar{y} &= x\sin\theta - y\cos\theta \end{aligned}$ in the line $y = x \tan \tfrac{1}{2}\theta$ followed by a translation $\begin{aligned} x' &= \bar{x} + g \\ y' &= \bar{y} + h \end{aligned}$.

It is customary to interpret the Euclidean transformations as rigid motions whereby, for example, a triangle on the plane is moved as a rigid entity into another position on the plane. Two triangles which correspond under a rigid motion are said to be congruent. Any transformation of the type $(7b)$, considered as a rigid motion, is wholly or in part a displacement in ordinary three-space. For, an orthogonal reflection in a line is equivalent to (has the same effect as) a rotation about the line; that is, since the reflection changes the sense of a triangle, its effect (see Fig. 18-5) is simply to turn the triangle over and place it in a precisely defined position on the plane. On the contrary, any transformation of the type $(7a)$ when considered as a rigid motion is a displacement entirely on the plane. These transformations of the type $(7a)$ are, then, the transformations of Plane Analytic Geometry by which, for example, the equation of a conic is reduced to standard form. It is merely a matter of convenience that such transformations are given in the inverse form

$$x = x' \cos \phi - y' \sin \phi + c'$$
$$y = x' \sin \phi + y' \cos \phi + d'$$

Appendix

Matrix Algebra

DEFINITIONS

A rectangular array of numbers enclosed by a pair of brackets, such as

$$(a)\ [1,2,3],\quad (b)\begin{bmatrix}1&3&1\\2&1&4\\4&7&6\end{bmatrix},\quad (c)\begin{bmatrix}2&3&-1&-4\\3&-1&2&10\\-5&3&3&1\end{bmatrix},\quad (d)\begin{bmatrix}1\\2\\-3\end{bmatrix}$$

and subject to certain operations given below is called a *matrix*. A matrix consisting of m rows and n columns is said to be of *order* $m \times n$. When $m = n$, the matrix is called a square matrix of order n or an *n-square* matrix. For example, (a) is of order 1×3, (b) is 3-square, (d) is of order 3×1.

In the matrix $\mathbf{A} = \begin{bmatrix}a_{11}&a_{12}&a_{13}&a_{14}\\a_{21}&a_{22}&a_{23}&a_{24}\\a_{31}&a_{32}&a_{33}&a_{34}\end{bmatrix}$, the numbers a_{ij} are called its *elements*. In the double subscript notation adopted here, the first subscript indicates the row and the second subscript indicates the column in which the element stands. Thus, a_{24} is the element standing in the second row and fourth column; all elements in the third row have 3 as first subscript; all elements in the first column have 1 as second subscript. The matrix of this paragraph may also be given by writing

$$\mathbf{A} = [a_{ij}],\quad (i=1,2,3;\ j=1,2,3,4)$$

When the order has been established, we shall write "$\mathbf{A} = [a_{ij}]$" or "the matrix $\mathbf{A}$".

A matrix $\mathbf{A}$, all of whose elements are real numbers, is called a *real matrix*. It is to be understood that all matrices considered here are real matrices.

ADDITION OF MATRICES

Two matrices $\mathbf{A} = [a_{ij}]$ and $\mathbf{B} = [b_{ij}]$, of the same order, are said to be conformable for addition or subtraction. The sum (difference) of two $m \times n$ matrices is an $m \times n$ matrix whose elements are the sums (differences) of the correspondingly positioned elements of the given matrices. Thus, $\mathbf{A} + \mathbf{B} = [a_{ij} + b_{ij}] = \mathbf{B} + \mathbf{A}$; $\mathbf{A} - \mathbf{B} = [a_{ij} - b_{ij}]$; $\mathbf{B} - \mathbf{A} = [b_{ij} - a_{ij}]$.

Example 1. When $\mathbf{A} = \begin{bmatrix}1&3&1\\2&1&4\\4&7&6\end{bmatrix}$ and $\mathbf{B} = \begin{bmatrix}0&-2&3\\1&1&1\\4&0&2\end{bmatrix}$,

$$\mathbf{A} + \mathbf{B} = \begin{bmatrix}1+0&3-2&1+3\\2+1&1+1&4+1\\4+4&7+0&6+2\end{bmatrix} = \begin{bmatrix}1&1&4\\3&2&5\\8&7&8\end{bmatrix} = \mathbf{B} + \mathbf{A}$$

$$\mathbf{A} - \mathbf{B} = \begin{bmatrix} 1-0 & 3+2 & 1-3 \\ 2-1 & 1-1 & 4-1 \\ 4-4 & 7-0 & 6-2 \end{bmatrix} = \begin{bmatrix} 1 & 5 & -2 \\ 1 & 0 & 3 \\ 0 & 7 & 4 \end{bmatrix}$$

$$\mathbf{B} - \mathbf{A} = \begin{bmatrix} 0-1 & -2-3 & 3-1 \\ 1-2 & 1-1 & 1-4 \\ 4-4 & 0-7 & 2-6 \end{bmatrix} = \begin{bmatrix} -1 & -5 & 2 \\ -1 & 0 & -3 \\ 0 & -7 & -4 \end{bmatrix}$$

$$\mathbf{A} + \mathbf{A} + \mathbf{A} = \begin{bmatrix} 3 & 9 & 3 \\ 6 & 3 & 12 \\ 12 & 21 & 18 \end{bmatrix} = 3\mathbf{A}$$

We define: $p\mathbf{A} = [pa_{ij}]$, for p any real number.   In particular, $(-1)\mathbf{A} = [-a_{ij}] = -[a_{ij}]$. Thus, in Example 1, $\mathbf{B} - \mathbf{A} = -(\mathbf{A} - \mathbf{B})$.

MULTIPLICATION OF MATRICES

Two matrices $\mathbf{A}$ and $\mathbf{B}$ are said to be conformable for multiplication in the order $\mathbf{A} \cdot \mathbf{B}$ provided the number of columns of $\mathbf{A}$ is equal to the number of rows of $\mathbf{B}$, and in order $\mathbf{B} \cdot \mathbf{A}$ provided the number of columns of $\mathbf{B}$ is equal to the number of rows of $\mathbf{A}$.   Two n-square matrices $\mathbf{A}$ and $\mathbf{B}$ are conformable for multiplication in both the order $\mathbf{A} \cdot \mathbf{B}$ and $\mathbf{B} \cdot \mathbf{A}$.   In general, however, $\mathbf{A} \cdot \mathbf{B} \neq \mathbf{B} \cdot \mathbf{A}$.

When $\mathbf{A} = [a_{11} \ a_{12} \ a_{13}]$ and $\mathbf{B} = \begin{bmatrix} b_{11} \\ b_{21} \\ b_{31} \end{bmatrix}$,   we define

$$\mathbf{A} \cdot \mathbf{B} = a_{11}b_{11} + a_{12}b_{21} + a_{13}b_{31} = \sum_j a_{1j}b_{j1}$$

Let $\mathbf{A} = \begin{bmatrix} a_{11} & a_{12} & a_{13} \\ a_{21} & a_{22} & a_{23} \\ a_{31} & a_{32} & a_{33} \end{bmatrix}$ and $\mathbf{B} = \begin{bmatrix} b_{11} & b_{12} & b_{13} \\ b_{21} & b_{22} & b_{23} \\ b_{31} & b_{32} & b_{33} \end{bmatrix}$.   To obtain $\mathbf{A} \cdot \mathbf{B}$, think of each row of $\mathbf{A}$ as a 1×3 matrix and think of each column of $\mathbf{B}$ as a 3×1 matrix, i.e.,

$\mathbf{A} = \begin{bmatrix} \mathbf{A}_1 \\ \mathbf{A}_2 \\ \mathbf{A}_3 \end{bmatrix}$ and $\mathbf{B} = [\mathbf{B}_1 \ \mathbf{B}_2 \ \mathbf{B}_3]$.   Then

$$\mathbf{A} \cdot \mathbf{B} = \begin{bmatrix} \mathbf{A}_1 \cdot \mathbf{B}_1 & \mathbf{A}_1 \cdot \mathbf{B}_2 & \mathbf{A}_1 \cdot \mathbf{B}_3 \\ \mathbf{A}_2 \cdot \mathbf{B}_1 & \mathbf{A}_2 \cdot \mathbf{B}_2 & \mathbf{A}_2 \cdot \mathbf{B}_3 \\ \mathbf{A}_3 \cdot \mathbf{B}_1 & \mathbf{A}_3 \cdot \mathbf{B}_2 & \mathbf{A}_3 \cdot \mathbf{B}_3 \end{bmatrix} = \begin{bmatrix} \sum_j a_{1j}b_{j1} & \sum_j a_{1j}b_{j2} & \sum_j a_{1j}b_{j3} \\ \sum_j a_{2j}b_{j1} & \sum_j a_{2j}b_{j2} & \sum_j a_{2j}b_{j3} \\ \sum_j a_{3j}b_{j1} & \sum_j a_{3j}b_{j2} & \sum_j a_{3j}b_{j3} \end{bmatrix} \qquad (1)$$

$$= \begin{bmatrix} a_{11}b_{11} + a_{12}b_{21} + a_{13}b_{31} & a_{11}b_{12} + a_{12}b_{22} + a_{13}b_{32} & a_{11}b_{13} + a_{12}b_{23} + a_{13}b_{33} \\ a_{21}b_{11} + a_{22}b_{21} + a_{23}b_{31} & a_{21}b_{12} + a_{22}b_{22} + a_{23}b_{32} & a_{21}b_{13} + a_{22}b_{23} + a_{23}b_{33} \\ a_{31}b_{11} + a_{32}b_{21} + a_{33}b_{31} & a_{31}b_{12} + a_{32}b_{22} + a_{33}b_{32} & a_{31}b_{13} + a_{32}b_{23} + a_{33}b_{33} \end{bmatrix}$$

To obtain $\mathbf{B} \cdot \mathbf{A}$, interchange the roles of $\mathbf{A}$ and $\mathbf{B}$ above.

Example 2. For the matrices of Example 1,

$$\mathbf{A} \cdot \mathbf{B} = \begin{bmatrix} 1 & 3 & 1 \\ 2 & 1 & 4 \\ 4 & 7 & 6 \end{bmatrix} \cdot \begin{bmatrix} 0 & -2 & 3 \\ 1 & 1 & 1 \\ 4 & 0 & 2 \end{bmatrix} = \begin{bmatrix} 1 \cdot 0 + 3 \cdot 1 + 1 \cdot 4 & 1 \cdot (-2) + 3 \cdot 1 + 1 \cdot 0 & 1 \cdot 3 + 3 \cdot 1 + 1 \cdot 2 \\ 2 \cdot 0 + 1 \cdot 1 + 4 \cdot 4 & 2 \cdot (-2) + 1 \cdot 1 + 4 \cdot 0 & 2 \cdot 3 + 1 \cdot 1 + 4 \cdot 2 \\ 4 \cdot 0 + 7 \cdot 1 + 6 \cdot 4 & 4 \cdot (-2) + 7 \cdot 1 + 6 \cdot 0 & 4 \cdot 3 + 7 \cdot 1 + 6 \cdot 2 \end{bmatrix}$$

$$= \begin{bmatrix} 7 & 1 & 8 \\ 17 & -3 & 15 \\ 31 & -1 & 31 \end{bmatrix}$$

and

$$\mathbf{B} \cdot \mathbf{A} = \begin{bmatrix} 0 & -2 & 3 \\ 1 & 1 & 1 \\ 4 & 0 & 2 \end{bmatrix} \cdot \begin{bmatrix} 1 & 3 & 1 \\ 2 & 1 & 4 \\ 4 & 7 & 6 \end{bmatrix}$$

$$= \begin{bmatrix} 0 \cdot 1 + (-2) \cdot 2 + 3 \cdot 4 & 0 \cdot 3 + (-2) \cdot 1 + 3 \cdot 7 & 0 \cdot 1 + (-2) \cdot 4 + 3 \cdot 6 \\ 1 \cdot 1 + 1 \cdot 2 + 1 \cdot 4 & 1 \cdot 3 + 1 \cdot 1 + 1 \cdot 7 & 1 \cdot 1 + 1 \cdot 4 + 1 \cdot 6 \\ 4 \cdot 1 + 0 \cdot 2 + 2 \cdot 4 & 4 \cdot 3 + 0 \cdot 1 + 2 \cdot 7 & 4 \cdot 1 + 0 \cdot 4 + 2 \cdot 6 \end{bmatrix} = \begin{bmatrix} 8 & 19 & 10 \\ 7 & 11 & 11 \\ 12 & 26 & 16 \end{bmatrix}$$

Example 3. When $\mathbf{A} = \begin{bmatrix} 1 & 2 & 3 \\ 1 & 3 & 3 \\ 1 & 2 & 4 \end{bmatrix}$ and $\mathbf{B} = \begin{bmatrix} 6 & -2 & -3 \\ -1 & 1 & 0 \\ -1 & 0 & 1 \end{bmatrix}$, show

$$\mathbf{A} \cdot \mathbf{B} = \mathbf{B} \cdot \mathbf{A}$$

$$\mathbf{A} \cdot \mathbf{B} = \begin{bmatrix} 1 & 2 & 3 \\ 1 & 3 & 3 \\ 1 & 2 & 4 \end{bmatrix} \cdot \begin{bmatrix} 6 & -2 & -3 \\ -1 & 1 & 0 \\ -1 & 0 & 1 \end{bmatrix} = \begin{bmatrix} 6 - 2 - 3 & -2 + 2 + 0 & -3 + 0 + 3 \\ 6 - 3 - 3 & -2 + 3 + 0 & -3 + 0 + 3 \\ 6 - 2 - 4 & -2 + 2 + 0 & -3 + 0 + 4 \end{bmatrix} = \begin{bmatrix} 1 & 0 & 0 \\ 0 & 1 & 0 \\ 0 & 0 & 1 \end{bmatrix}$$

$$\mathbf{B} \cdot \mathbf{A} = \begin{bmatrix} 6 & -2 & -3 \\ -1 & 1 & 0 \\ -1 & 0 & 1 \end{bmatrix} \cdot \begin{bmatrix} 1 & 2 & 3 \\ 1 & 3 & 3 \\ 1 & 2 & 4 \end{bmatrix} = \begin{bmatrix} 6 - 2 - 3 & 12 - 6 - 6 & 18 - 6 - 12 \\ -1 + 1 + 0 & -2 + 3 + 0 & -3 + 3 + 0 \\ -1 + 0 + 1 & -2 + 0 + 2 & -3 + 0 + 4 \end{bmatrix} = \begin{bmatrix} 1 & 0 & 0 \\ 0 & 1 & 0 \\ 0 & 0 & 1 \end{bmatrix}$$

Suppose $\mathbf{A}, \mathbf{B}, \mathbf{C}$ are matrices conformable for multiplication in the order $\mathbf{A} \cdot \mathbf{B} \cdot \mathbf{C}$. We shall assume (the verification for 3-square matrices is not difficult but somewhat tedious) that

$$\mathbf{A} \cdot \mathbf{B} \cdot \mathbf{C} = (\mathbf{A} \cdot \mathbf{B}) \cdot \mathbf{C} = \mathbf{A} \cdot (\mathbf{B} \cdot \mathbf{C})$$

that is, matrix multiplication is associative.

Let $\mathbf{A} = [a_{ij}]$ and $\mathbf{B} = [b_{ij}]$ be matrices conformable for addition. If $\mathbf{A}, \mathbf{B}, \mathbf{C} = [c_{ij}]$ are conformable for multiplication in the orders $\mathbf{A} \cdot \mathbf{C}$ and $\mathbf{B} \cdot \mathbf{C}$, then $(\mathbf{A} + \mathbf{B}) \cdot \mathbf{C} = \mathbf{A} \cdot \mathbf{C} + \mathbf{B} \cdot \mathbf{C}$. For, the element standing in the ith row and jth column of $(\mathbf{A} + \mathbf{B}) \cdot \mathbf{C}$

$$(a_{i1} + b_{i1})c_{1j} + (a_{i2} + b_{i2})c_{2j} + \cdots + (a_{in} + b_{in})c_{nj}$$
$$= (a_{i1}c_{1j} + a_{i2}c_{2j} + \cdots + a_{in}c_{nj}) + (b_{i1}c_{1j} + b_{i2}c_{2j} + \cdots + b_{in}c_{nj})$$

is the sum of the elements standing in the ith row and jth column of $\mathbf{A} \cdot \mathbf{C}$ and $\mathbf{B} \cdot \mathbf{C}$. Thus multiplication of matrices is distributive with respect to addition.

SOME TYPES OF MATRICES

In any square matrix, the elements $a_{11}, a_{22}, a_{33}, \ldots$ are said to lie in the *principal diagonal* of $\mathbf{A}$. If all the elements below the principal diagonal are zeros, $\mathbf{A}$ is said to be *upper triangular*; if all of the elements above the principal diagonal are zeros, $\mathbf{A}$ is said to be *lower triangular*. If $\mathbf{A}$ is both upper triangular and lower triangular, $\mathbf{A}$ is called a *diagonal* matrix. If the diagonal elements of a diagonal matrix are all equal, $\mathbf{A}$ is called a *scalar* matrix. If the diagonal elements of a scalar matrix are all equal to 1, $\mathbf{A}$ is called the *identity* matrix and is denoted by $\mathbf{I}$. For example, $\begin{bmatrix} 1 & 3 & 2 \\ 0 & 2 & 1 \\ 0 & 0 & 3 \end{bmatrix}$ is upper triangular,

$\begin{bmatrix} 1 & 0 & 0 \\ 0 & 2 & 0 \\ 3 & 1 & 0 \end{bmatrix}$ is lower triangular, $\begin{bmatrix} 1 & 0 & 0 \\ 0 & 2 & 0 \\ 0 & 0 & -3 \end{bmatrix}$ is a diagonal matrix, $\begin{bmatrix} 3 & 0 & 0 \\ 0 & 3 & 0 \\ 0 & 0 & 3 \end{bmatrix}$ is a scalar

matrix and $\begin{bmatrix} 1 & 0 & 0 \\ 0 & 1 & 0 \\ 0 & 0 & 1 \end{bmatrix}$ is the 3-square identity matrix $\mathbf{I}$.

In matrix multiplication, $\mathbf{I}$ plays the role of 1 in ordinary multiplication, that is, $\mathbf{I} \cdot \mathbf{A} = \mathbf{A}$ and $\mathbf{B} \cdot \mathbf{I} = \mathbf{B}$. In any product of two or more matrices, $\mathbf{I}$ may be inserted as a factor or dropped as a factor at will.

DETERMINANT OF A SQUARE MATRIX

The determinant of a square matrix $\mathbf{A}$ is denoted by $|\mathbf{A}|$. When $\mathbf{A} = [a]$, $|\mathbf{A}| = a$; when $\mathbf{A} = \begin{bmatrix} a & b \\ c & d \end{bmatrix}$, $|\mathbf{A}| = ad - bc$. When $\mathbf{A}$ is n-square, $n \geqq 3$, there are a number of procedures for evaluating $|\mathbf{A}|$. We consider two of these for the 3-square matrix $\mathbf{A} = \begin{bmatrix} a_{11} & a_{12} & a_{13} \\ a_{21} & a_{22} & a_{23} \\ a_{31} & a_{32} & a_{33} \end{bmatrix}$:

$$\text{(i)} \quad |\mathbf{A}| = a_{11} \cdot \begin{vmatrix} a_{22} & a_{23} \\ a_{32} & a_{33} \end{vmatrix} - a_{12} \cdot \begin{vmatrix} a_{21} & a_{23} \\ a_{31} & a_{33} \end{vmatrix} + a_{13} \cdot \begin{vmatrix} a_{21} & a_{22} \\ a_{31} & a_{32} \end{vmatrix}$$

$$= -a_{21} \cdot \begin{vmatrix} a_{12} & a_{13} \\ a_{32} & a_{33} \end{vmatrix} + a_{22} \cdot \begin{vmatrix} a_{11} & a_{13} \\ a_{31} & a_{33} \end{vmatrix} - a_{23} \cdot \begin{vmatrix} a_{11} & a_{12} \\ a_{31} & a_{32} \end{vmatrix}$$

$$= a_{31} \cdot \begin{vmatrix} a_{12} & a_{13} \\ a_{22} & a_{23} \end{vmatrix} - a_{32} \cdot \begin{vmatrix} a_{11} & a_{13} \\ a_{21} & a_{23} \end{vmatrix} + a_{33} \cdot \begin{vmatrix} a_{11} & a_{12} \\ a_{21} & a_{22} \end{vmatrix}$$

$$= a_{11} \cdot \begin{vmatrix} a_{22} & a_{23} \\ a_{32} & a_{33} \end{vmatrix} - a_{21} \cdot \begin{vmatrix} a_{12} & a_{13} \\ a_{32} & a_{33} \end{vmatrix} + a_{31} \cdot \begin{vmatrix} a_{12} & a_{13} \\ a_{22} & a_{23} \end{vmatrix}$$

$$= -a_{12} \cdot \begin{vmatrix} a_{21} & a_{23} \\ a_{31} & a_{33} \end{vmatrix} + a_{22} \cdot \begin{vmatrix} a_{11} & a_{13} \\ a_{31} & a_{33} \end{vmatrix} - a_{32} \cdot \begin{vmatrix} a_{11} & a_{13} \\ a_{21} & a_{23} \end{vmatrix}$$

$$= a_{13} \cdot \begin{vmatrix} a_{21} & a_{22} \\ a_{31} & a_{32} \end{vmatrix} - a_{23} \cdot \begin{vmatrix} a_{11} & a_{12} \\ a_{31} & a_{32} \end{vmatrix} + a_{33} \cdot \begin{vmatrix} a_{11} & a_{12} \\ a_{21} & a_{22} \end{vmatrix}$$

Each of the above expansions is the sum of three terms; in turn, each term is the signed product of an element of $\mathbf{A}$ and the determinant of a 2-square matrix. In any expansion the elements are those of a row or column of $\mathbf{A}$; the 2-square matrix whose determinant multiplies the element a_{pq} is the matrix $\mathbf{M}_{pq}$ which remains when the row and column in which a_{pq} stands is removed from $\mathbf{A}$; the sign associated with the product $a_{pq} \cdot |\mathbf{M}_{pq}|$ is $(-1)^{p+q}$. $|\mathbf{M}_{pq}|$ is called the *minor* of a_{pq}; $(-1)^{p+q} |\mathbf{M}_{pq}|$ is called the *cofactor* of a_{pq} and is denoted by $\mathbf{A}_{pq}$.

Example 4.

(a) $\begin{vmatrix} 1 & 3 & 1 \\ 2 & 1 & 4 \\ 4 & 7 & 8 \end{vmatrix} = 1 \cdot \begin{vmatrix} 1 & 4 \\ 7 & 8 \end{vmatrix} - 3 \cdot \begin{vmatrix} 2 & 4 \\ 4 & 8 \end{vmatrix} + 1 \cdot \begin{vmatrix} 2 & 1 \\ 4 & 7 \end{vmatrix} = -20 + 0 + 10 = -10$

Here, the elements are from the first row.

(b) $\begin{vmatrix} 2 & 1 & -1 \\ 4 & 3 & 0 \\ 1 & 2 & 3 \end{vmatrix} = -4 \cdot \begin{vmatrix} 1 & -1 \\ 2 & 3 \end{vmatrix} + 3 \cdot \begin{vmatrix} 2 & -1 \\ 1 & 3 \end{vmatrix} - 0 \cdot \begin{vmatrix} 2 & 1 \\ 1 & 2 \end{vmatrix} = -20 + 21 = 1$

Here, to take advantage of the element 0, the elements are from the second row. It is suggested that the reader write the expansion taking the elements from the third column.

(c) $\begin{vmatrix} 2 & 3 & 5 \\ 0 & -1 & 2 \\ 0 & 0 & -3 \end{vmatrix} = 2 \cdot \begin{vmatrix} -1 & 2 \\ 0 & -3 \end{vmatrix} = 6$

(ii) In any text on Determinants it is proved that the following operations on a determinant $|\mathbf{A}|$ have the stated effects.

O_{ij}:   The interchange of the ith and jth rows (columns) changes the sign of $|\mathbf{A}|$.

$O_{p(i)}$:   The multiplication of the elements of the ith row (column) by the non-zero constant p multiplies $|\mathbf{A}|$ by p.

$O_{(i)+p(j)}$:   The addition to the elements of the ith row (column) of p times the corresponding elements of the jth row (column) does not change the value of $|\mathbf{A}|$.

Example 4(c) suggests that these operations be used initially to reduce $\mathbf{A}$ to triangular form.

Example 5.

(a) Using row operations $O_{(2)-2(1)}, O_{(3)-4(1)}; O_{(3)-(2)}$:

$$\begin{vmatrix} 1 & 3 & 1 \\ 2 & 1 & 4 \\ 4 & 7 & 8 \end{vmatrix} = \begin{vmatrix} 1 & 3 & 1 \\ 0 & -5 & 2 \\ 0 & -5 & 4 \end{vmatrix} = \begin{vmatrix} 1 & 3 & 1 \\ 0 & -5 & 2 \\ 0 & 0 & 2 \end{vmatrix} = -10$$

(b) Using column operations $O_{(2)-3(1)}, O_{(3)-(1)}; O_{(3)+\frac{2}{5}(2)}$:

$$\begin{vmatrix} 1 & 3 & 1 \\ 2 & 1 & 4 \\ 4 & 7 & 8 \end{vmatrix} = \begin{vmatrix} 1 & 0 & 0 \\ 2 & -5 & 2 \\ 4 & -5 & 4 \end{vmatrix} = \begin{vmatrix} 1 & 0 & 0 \\ 2 & -5 & 0 \\ 4 & -5 & 2 \end{vmatrix} = -10$$

(c) Using row operations $O_{(1)-(2)}; O_{(2)+3(1)}, O_{(3)+5(1)}; O_{(3)-\frac{9}{4}(2)}$:

$$\begin{vmatrix} 2 & 4 & 3 \\ 3 & 4 & 0 \\ 5 & 9 & 6 \end{vmatrix} = \begin{vmatrix} -1 & 0 & 3 \\ 3 & 4 & 0 \\ 5 & 9 & 6 \end{vmatrix} = \begin{vmatrix} -1 & 0 & 3 \\ 0 & 4 & 9 \\ 0 & 9 & 21 \end{vmatrix} = \begin{vmatrix} -1 & 0 & 3 \\ 0 & 4 & 9 \\ 0 & 0 & \frac{3}{4} \end{vmatrix} = -3$$

We state, without proof,

If $\mathbf{A}, \mathbf{B}, \dots, \mathbf{S}$ are n-square matrices, then $|\mathbf{A} \cdot \mathbf{B} \cdots \cdots \mathbf{S}| = |\mathbf{A}| \cdot |\mathbf{B}| \cdots \cdots |\mathbf{S}|$.

ELEMENTARY TRANSFORMATIONS ON A MATRIX

The *elementary transformations* on a matrix are precisely the operations on determinants listed above. Of these, we will make use of only two:

the row operation $O_{p(i)}$ which will hereafter be indicated by writing simply $p(i)$,

the row operation $O_{(i)+p(j)}$ which will hereafter be indicated by writing $(i) + p(j)$.

The effect of one or more applications of these transformations on a matrix $\mathbf{A}$ is to replace it by another $\mathbf{A}'$ which is said to be *equivalent* to $\mathbf{A}$. The equivalence of two matrices $\mathbf{A}$ and $\mathbf{A}'$ will be indicated by writing $\mathbf{A} \sim \mathbf{A}'$.

SYSTEMS OF LINEAR EQUATIONS

Consider the system of three non-homogeneous equations in three unknowns

$$\begin{cases} 2x + 3y - z = -4 \\ 3x - y + 2z = 10 \\ -5x + 3y + 3z = 1 \end{cases}$$

The usual procedure for solving this system consists in combining the equations so as to replace the given system by a simpler one. The same result will be obtained by the use of the transformation $(i) + p(j)$ on the rows of

$$\begin{bmatrix} 2 & 3 & -1 & -4 \\ 3 & -1 & 2 & 10 \\ -5 & 3 & 3 & 1 \end{bmatrix} = [\mathbf{A} \ \mathbf{H}]$$

where $\mathbf{A}$ is the square matrix of the coefficients and $\mathbf{H}$ is the 3×1 matrix of constant terms, to reduce $\mathbf{A}$ to triangular form. We find

$$\begin{bmatrix} 2 & 3 & -1 & -4 \\ 3 & -1 & 2 & 10 \\ -5 & 3 & 3 & 1 \end{bmatrix} \overset{(1)-(2)}{\sim} \begin{bmatrix} -1 & 4 & -3 & -14 \\ 3 & -1 & 2 & 10 \\ -5 & 3 & 3 & 1 \end{bmatrix} \overset{\substack{(2)+3(1) \\ (3)-5(1)}}{\sim} \begin{bmatrix} -1 & 4 & -3 & -14 \\ 0 & 11 & -7 & -32 \\ 0 & -17 & 18 & 71 \end{bmatrix}$$

$$\overset{(3)+\frac{17}{11}(2)}{\sim} \begin{bmatrix} -1 & 4 & -3 & -14 \\ 0 & 11 & -7 & -32 \\ 0 & 0 & 79/11 & 237/11 \end{bmatrix}$$

Then $\frac{79}{11}z = \frac{237}{11}$ and $z = 3$. Next, $11y = 7z - 32 = 21 - 32 = -11$ and $y = -1$. Finally, $x = 4y - 3z + 14 = -4 - 9 + 14 = 1$. The required solution is $x = 1$, $y = -1$, $z = 3$.

SINGULAR AND NON-SINGULAR SQUARE MATRICES

A square matrix $\mathbf{C}$ is called *singular* when $|\mathbf{C}| = 0$ and is called *non-singular* when $|\mathbf{C}| \neq 0$. A well-known theorem of algebra may now be restated as follows:

The system of homogeneous linear equations

$$c_{11}x_1 + c_{12}x_2 + c_{13}x_3 = 0$$

$$c_{21}x_1 + c_{22}x_2 + c_{23}x_3 = 0$$

$$c_{31}x_1 + c_{32}x_2 + c_{33}x_3 = 0$$

will have non-trivial solutions if and only if the matrix of coefficients $\mathbf{C} = [c_{ij}]$ is singular.

Suppose the 3-square matrix $\mathbf{B} = [b_{ij}]$ is singular so that there exist constants p, q, r, not all zero, such that

$$b_{11}p + b_{12}q + b_{13}r = 0$$
$$b_{21}p + b_{22}q + b_{23}r = 0 \quad \text{or} \quad \mathbf{B}\begin{bmatrix} p \\ q \\ r \end{bmatrix} = \begin{bmatrix} 0 \\ 0 \\ 0 \end{bmatrix}$$
$$b_{31}p + b_{32}q + b_{33}r = 0$$

Let $\mathbf{A}$ be any 3-square matrix; then $\mathbf{A} \cdot \mathbf{B} \cdot \begin{bmatrix} p \\ q \\ r \end{bmatrix} = \mathbf{A} \cdot \begin{bmatrix} 0 \\ 0 \\ 0 \end{bmatrix} = \begin{bmatrix} 0 \\ 0 \\ 0 \end{bmatrix}$. Thus $[\mathbf{A} \cdot \mathbf{B}]\begin{bmatrix} p \\ q \\ r \end{bmatrix} = \begin{bmatrix} 0 \\ 0 \\ 0 \end{bmatrix}$ and we have proved

If $\mathbf{A}$ and $\mathbf{B}$ are 3-square matrices and if $\mathbf{B}$ is singular, so also is $\mathbf{A} \cdot \mathbf{B}$.

INVERSE OF A SQUARE MATRIX

If $\mathbf{A}$ and $\mathbf{B}$ are square matrices such that $\mathbf{A} \cdot \mathbf{B} = \mathbf{B} \cdot \mathbf{A} = \mathbf{I}$ (see Example 3), then $\mathbf{B}$ is called the *inverse* of $\mathbf{A}$ and we write $\mathbf{B} = \mathbf{A}^{-1}$. Similarly, $\mathbf{A}$ is called the *inverse* of $\mathbf{B}$ and we write $\mathbf{A} = \mathbf{B}^{-1}$.

Suppose $\mathbf{B}$ is the inverse of $\mathbf{A}$. Since $\mathbf{A} \cdot \mathbf{B} = \mathbf{I}$ and $|\mathbf{A} \cdot \mathbf{B}| = |\mathbf{I}| = 1 \neq 0$, it follows from the result proved in the preceding section that $\mathbf{B} = \mathbf{A}^{-1}$ is non-singular. Similarly, from $\mathbf{B} \cdot \mathbf{A} = \mathbf{I}$, it follows that $\mathbf{A}$ is non-singular. Thus

If the square matrix $\mathbf{A}$ has an inverse, then $\mathbf{A}$ is non-singular.

A proof of the converse

If $\mathbf{A}$ is non-singular, then $\mathbf{A}^{-1}$ exists.

will not be given. The interested reader is referred to any book on matrices.

The inverse of any non-singular matrix $\mathbf{A}$ may be computed as follows:

(i)   Write the matrix $[\mathbf{A} \ \mathbf{I}]$.

(ii)   By means of a suitable sequence of the two elementary transformations reduce $[\mathbf{A} \ \mathbf{I}]$ to $[\mathbf{I} \ \mathbf{B}]$. Clearly, this reduction is always possible unless somewhere along the way a row of $\mathbf{A}' \sim \mathbf{A}$ consists entirely of zeros, in which case $|\mathbf{A}'| = 0$. But this can never happen. For, since $\mathbf{A}$ is non-singular, $|\mathbf{A}| \neq 0$ and the elementary transformations are such that $|\mathbf{A}'| = p|\mathbf{A}|$, $p \neq 0$.

Example 6. Obtain the inverse of $\mathbf{A} = \begin{bmatrix} 1 & 2 & 3 \\ 1 & 3 & 3 \\ 1 & 2 & 4 \end{bmatrix}$.

Writing out the matrix $[\mathbf{A} \ \mathbf{I}]$, we proceed to reduce it to $[\mathbf{I} \ \mathbf{B}]$. We have

$$[\mathbf{A} \ \mathbf{I}] = \begin{bmatrix} 1 & 2 & 3 & 1 & 0 & 0 \\ 1 & 3 & 3 & 0 & 1 & 0 \\ 1 & 2 & 4 & 0 & 0 & 1 \end{bmatrix} \overset{\substack{(2)-(1) \\ (3)-(1)}}{\sim} \begin{bmatrix} 1 & 2 & 3 & 1 & 0 & 0 \\ 0 & 1 & 0 & -1 & 1 & 0 \\ 0 & 0 & 1 & -1 & 0 & 1 \end{bmatrix}$$

$$\overset{(1)-2(2)}{\sim} \begin{bmatrix} 1 & 0 & 3 & 3 & -2 & 0 \\ 0 & 1 & 0 & -1 & 1 & 0 \\ 0 & 0 & 1 & -1 & 0 & 1 \end{bmatrix} \overset{(1)-3(3)}{\sim} \begin{bmatrix} 1 & 0 & 0 & 6 & -2 & -3 \\ 0 & 1 & 0 & -1 & 1 & 0 \\ 0 & 0 & 1 & -1 & 0 & 1 \end{bmatrix}$$

$$= \ [\mathbf{I} \ \mathbf{B}]$$

That $\mathbf{B} = \mathbf{A}^{-1} = \begin{bmatrix} 6 & -2 & -3 \\ -1 & 1 & 0 \\ -1 & 0 & 1 \end{bmatrix}$ was verified in Example 3.

Example 7. Obtain the inverse of $\mathbf{A} = \begin{bmatrix} 2 & 3 & -1 \\ 3 & -1 & 2 \\ -5 & 3 & 3 \end{bmatrix}$.

$$[\mathbf{A}\ \mathbf{I}] = \begin{bmatrix} 2 & 3 & -1 & 1 & 0 & 0 \\ 3 & -1 & 2 & 0 & 1 & 0 \\ -5 & 3 & 3 & 0 & 0 & 1 \end{bmatrix} \overset{(1)-(2)}{\sim} \begin{bmatrix} -1 & 4 & -3 & 1 & -1 & 0 \\ 3 & -1 & 2 & 0 & 1 & 0 \\ -5 & 3 & 3 & 0 & 0 & 1 \end{bmatrix}$$

$$\overset{-(1)}{\sim} \begin{bmatrix} 1 & -4 & 3 & -1 & 1 & 0 \\ 3 & -1 & 2 & 0 & 1 & 0 \\ -5 & 3 & 3 & 0 & 0 & 1 \end{bmatrix} \overset{\substack{(2)-3(1) \\ (3)+5(1)}}{\sim} \begin{bmatrix} 1 & -4 & 3 & -1 & 1 & 0 \\ 0 & 11 & -7 & 3 & -2 & 0 \\ 0 & -17 & 18 & -5 & 5 & 1 \end{bmatrix}$$

$$\overset{\frac{1}{11}(2)}{\sim} \begin{bmatrix} 1 & -4 & 3 & -1 & 1 & 0 \\ 0 & 1 & -7/11 & 3/11 & -2/11 & 0 \\ 0 & -17 & 18 & -5 & 5 & 1 \end{bmatrix} \overset{\substack{(1)+4(2) \\ (3)+17(2)}}{\sim} \begin{bmatrix} 1 & 0 & 5/11 & 1/11 & 3/11 & 0 \\ 0 & 1 & -7/11 & 3/11 & -2/11 & 0 \\ 0 & 0 & 79/11 & -4/11 & 21/11 & 1 \end{bmatrix}$$

$$\overset{\frac{11}{79}(3)}{\sim} \begin{bmatrix} 1 & 0 & 5/11 & 1/11 & 3/11 & 0 \\ 0 & 1 & -7/11 & 3/11 & -2/11 & 0 \\ 0 & 0 & 1 & -4/79 & 21/79 & 11/79 \end{bmatrix}$$

$$\overset{\substack{(1)-\frac{5}{11}(3) \\ (2)+\frac{7}{11}(3)}}{\sim} \begin{bmatrix} 1 & 0 & 0 & 9/79 & 12/79 & -5/79 \\ 0 & 1 & 0 & 19/79 & -1/79 & 7/79 \\ 0 & 0 & 1 & -4/79 & 21/79 & 11/79 \end{bmatrix}$$

$$= [\mathbf{I}\ \mathbf{B}]$$

Now $\mathbf{B} \cdot \mathbf{A} = \dfrac{1}{79} \begin{bmatrix} 9 & 12 & -5 \\ 19 & -1 & 7 \\ -4 & 21 & 11 \end{bmatrix} \cdot \begin{bmatrix} 2 & 3 & -1 \\ 3 & -1 & 2 \\ -5 & 3 & 3 \end{bmatrix} = \begin{bmatrix} 1 & 0 & 0 \\ 0 & 1 & 0 \\ 0 & 0 & 1 \end{bmatrix} = \mathbf{I};$ hence,

$$\mathbf{B} = \mathbf{A}^{-1} = \frac{1}{79} \begin{bmatrix} 9 & 12 & -5 \\ 19 & -1 & 7 \\ -4 & 21 & 11 \end{bmatrix}$$

Example 8. Show that the inverse of a non-singular matrix is unique.

Suppose the contrary, that is, suppose for the non-singular matrix $\mathbf{A}$, two distinct inverses $\mathbf{B}_1$ and $\mathbf{B}_2$ are found. Then

$$\mathbf{A} \cdot \mathbf{B}_1 = \mathbf{I} \qquad (\mathbf{B}_1 = \mathbf{A}^{-1})$$

$$(\mathbf{B}_2 \cdot \mathbf{A}) \cdot \mathbf{B}_1 = \mathbf{B}_2 \cdot \mathbf{I} = \mathbf{B}_2$$

$$\mathbf{I} \cdot \mathbf{B}_1 = \mathbf{B}_2 \qquad (\mathbf{B}_2 = \mathbf{A}^{-1})$$

and

$$\mathbf{B}_1 = \mathbf{B}_2$$

Thus the assumption is false; the inverse is unique.

From $\mathbf{A}^{-1} \cdot \mathbf{A} = \mathbf{I}$, it follows that $|\mathbf{A}^{-1}| = 1/|\mathbf{A}|$.

Example 9. The system of linear equations on page 234 when written in matrix notation is

$$\begin{bmatrix} 2 & 3 & -1 \\ 3 & -1 & 2 \\ -5 & 3 & 3 \end{bmatrix} \cdot \begin{bmatrix} x \\ y \\ z \end{bmatrix} = \begin{bmatrix} -4 \\ 10 \\ 1 \end{bmatrix}$$

Multiplying this relation on the left by the inverse of the matrix **A** of coefficients (see Example 7), we have

$$\frac{1}{79}\begin{bmatrix} 9 & 12 & -5 \\ 19 & -1 & 7 \\ -4 & 21 & 11 \end{bmatrix} \cdot \begin{bmatrix} 2 & 3 & -1 \\ 3 & -1 & 2 \\ -5 & 3 & 3 \end{bmatrix} \cdot \begin{bmatrix} x \\ y \\ z \end{bmatrix} = \frac{1}{79}\begin{bmatrix} 9 & 12 & -5 \\ 19 & -1 & 7 \\ -4 & 21 & 11 \end{bmatrix}\begin{bmatrix} -4 \\ 10 \\ 1 \end{bmatrix}$$

$$\begin{bmatrix} 1 & 0 & 0 \\ 0 & 1 & 0 \\ 0 & 0 & 1 \end{bmatrix} \cdot \begin{bmatrix} x \\ y \\ z \end{bmatrix} = \begin{bmatrix} x \\ y \\ z \end{bmatrix} = \begin{bmatrix} 1 \\ -1 \\ 3 \end{bmatrix}$$

and $x = 1$, $y = -1$, $z = 3$ is the solution of the system.

THE TRANSPOSE OF A MATRIX

Let **A** be a matrix. By the transpose $\mathbf{A}^T$ or $\mathbf{A}'$ of **A** is meant the matrix obtained by writing in the same order the rows of **A** as columns. For example, if $\mathbf{A} = [a_{11}, a_{12}, a_{13}]$,

then $\mathbf{A}^T = \begin{bmatrix} a_{11} \\ a_{12} \\ a_{13} \end{bmatrix}$; if $\mathbf{A} = \begin{bmatrix} a_{11} & a_{12} & a_{13} \\ a_{21} & a_{22} & a_{23} \\ a_{31} & a_{32} & a_{33} \end{bmatrix}$, then $\mathbf{A}^T = \begin{bmatrix} a_{11} & a_{21} & a_{31} \\ a_{12} & a_{22} & a_{32} \\ a_{13} & a_{23} & a_{33} \end{bmatrix}$.

Concerning the transpose of a matrix, we state:

(i) If **A** is a square matrix, then $|\mathbf{A}^T| = |\mathbf{A}|$.

(ii) If **A** and **B** are conformable for multiplication in the order $\mathbf{A} \cdot \mathbf{B}$, then $(\mathbf{A} \cdot \mathbf{B})^T = \mathbf{B}^T \cdot \mathbf{A}^T$. In general, the transpose of the product of two or more matrices is the product of their transposes in reverse order. For example, $(\mathbf{A} \cdot \mathbf{B} \cdot \mathbf{C})^T = \mathbf{C}^T \cdot \mathbf{B}^T \cdot \mathbf{A}^T$.

The proof of (ii) is somewhat tedious. We shall need (ii) for the special cases when

$\mathbf{A} = [a_{11}, a_{12}, a_{13}]$, $\mathbf{B} = [b_{ij}]$, $(i, j = 1, 2, 3)$, $\mathbf{C} = \begin{bmatrix} c_{11} \\ c_{21} \\ c_{31} \end{bmatrix}$. We leave the verifications for

the reader.

Example 10. Prove: $(\mathbf{A}^{-1})^T = (\mathbf{A}^T)^{-1}$.

From $\mathbf{A}^{-1} \cdot \mathbf{A} = \mathbf{I}$ follows

$$(\mathbf{A}^{-1} \cdot \mathbf{A})^T = \mathbf{A}^T \cdot (\mathbf{A}^{-1})^T = \mathbf{I}^T = \mathbf{I}$$

Then

$$(\mathbf{A}^T)^{-1} \cdot \mathbf{A}^T \cdot (\mathbf{A}^{-1})^T = (\mathbf{A}^T)^{-1} \cdot \mathbf{I} = (\mathbf{A}^T)^{-1}$$

and so

$$\mathbf{I} \cdot (\mathbf{A}^{-1})^T = (\mathbf{A}^{-1})^T = (\mathbf{A}^T)^{-1}$$

as required.

RANK OF A MATRIX

Consider the matrix $\mathbf{A} = \begin{bmatrix} a_{11} & a_{12} & a_{13} \\ a_{21} & a_{22} & a_{23} \\ a_{31} & a_{32} & a_{33} \end{bmatrix} \neq \begin{bmatrix} 0 & 0 & 0 \\ 0 & 0 & 0 \\ 0 & 0 & 0 \end{bmatrix}$. The matrix which remains after any one row and any one column of **A** have been erased is called a *2-rowed submatrix* of **A**; the determinant of this submatrix is called a *2-rowed minor* of **A**. Similarly, when any two rows and any two columns of **A** are erased, the matrix which remains is called a 1-rowed submatrix of **A** and the determinant of this submatrix is called a 1-rowed minor of **A**.

There are nine 2-rowed minors of **A**; for example, $\begin{vmatrix} a_{11} & a_{12} \\ a_{21} & a_{22} \end{vmatrix}, \begin{vmatrix} a_{12} & a_{13} \\ a_{22} & a_{23} \end{vmatrix}, \begin{vmatrix} a_{21} & a_{23} \\ a_{31} & a_{33} \end{vmatrix}, \dots.$

There are nine 1-rowed minors of **A** — the nine elements of **A**.

When $|\mathbf{A}| \neq 0$, we say that the *rank r* of **A** is 3. When $|\mathbf{A}| = 0$, but not every 2-rowed minor is 0, we say that the rank of **A** is $r = 2$. When $|\mathbf{A}| = 0$ and all 2-rowed minors are 0, but not every element is 0, we say that the rank of **A** is $r = 1$. For example, the rank of

$\mathbf{A} = \begin{bmatrix} 1 & 2 & 3 \\ 2 & 3 & 4 \\ 3 & 5 & 7 \end{bmatrix}$ is $r = 2$ since $|\mathbf{A}| = 0$ but $\begin{vmatrix} 1 & 2 \\ 2 & 3 \end{vmatrix} \neq 0$; the rank of $\mathbf{A} = \begin{bmatrix} 1 & 2 & 3 \\ 2 & 4 & 6 \\ 3 & 6 & 9 \end{bmatrix}$ is

$r = 1$ since $|\mathbf{A}| = 0$, every 2-rowed minor is 0 (check this completely) but $|4| = 4 \neq 0$.

The matrix $\mathbf{A} = \begin{bmatrix} 0 & 0 & 0 \\ 0 & 0 & 0 \\ 0 & 0 & 0 \end{bmatrix}$ is of rank $r = 0$.

It can be shown that the elementary transformations on a matrix preserve its rank.

THE ADJOINT OF A SQUARE MATRIX

Let $\mathbf{A} = [a_{ij}]$ be a 3-square matrix and denote by $\mathbf{A}_{ij}$ the cofactor of a_{ij} in **A**. We define

$$\text{adjoint } \mathbf{A} = \text{adj } \mathbf{A} = \begin{bmatrix} \mathbf{A}_{11} & \mathbf{A}_{21} & \mathbf{A}_{31} \\ \mathbf{A}_{12} & \mathbf{A}_{22} & \mathbf{A}_{32} \\ \mathbf{A}_{13} & \mathbf{A}_{23} & \mathbf{A}_{33} \end{bmatrix}$$

(Note carefully that the cofactors of the elements of the *i*th row (column) of **A** are the elements of the *i*th column (row) of adj **A**.)

We state, without proof,

If **A** is non-singular, then $|\mathbf{A}| \cdot \mathbf{A}^{-1} = \text{adj } \mathbf{A}$.

Example 11. From Example 7, the inverse of

$$\mathbf{A} = \begin{bmatrix} 2 & 3 & -1 \\ 3 & -1 & 2 \\ -5 & 3 & 3 \end{bmatrix} \text{ is } \mathbf{A}^{-1} = \frac{1}{79} \cdot \begin{bmatrix} 9 & 12 & -5 \\ 19 & -1 & 7 \\ -4 & 21 & 11 \end{bmatrix}$$

Now $|\mathbf{A}| = -79$; hence

$$\text{adj } \mathbf{A} = |\mathbf{A}| \cdot \mathbf{A}^{-1} = \begin{bmatrix} -9 & -12 & 5 \\ -19 & 1 & -7 \\ 4 & -21 & -11 \end{bmatrix}$$

We leave for the reader to verify that this agrees with the definition of adj **A**.

CHARACTERISTIC EQUATIONS AND ROOTS

Associated with any 3-square matrix $\mathbf{A} = [a_{ij}]$ is an equation

$$\phi(\rho) = \begin{vmatrix} a_{11} - \rho & a_{12} & a_{13} \\ a_{21} & a_{22} - \rho & a_{23} \\ a_{31} & a_{32} & a_{33} - \rho \end{vmatrix} = 0$$

called the *characteristic equation* of **A**. When $|\mathbf{A}| \neq 0$, the only case which concerns us here, $\phi(\rho) = 0$ is of degree three and has three roots ρ_1, ρ_2, ρ_3 called the characteristic roots of **A**, no one of which is zero. Since the a_{ij} are real numbers, $\phi(\rho)$ has real coefficients and $\phi(\rho) = 0$ has either three real roots or one real and two imaginary roots.

Clearly, the characteristic equations of **A** and $\mathbf{A}^T$ are identical and, hence, have the same roots. We state without proof:

If ρ_i is a characteristic root of the non-singular matrix **A**, then $1/\rho_i$ is a characteristic root of $\mathbf{A}^{-1}$ and, hence, of $(\mathbf{A}^{-1})^T$.

If ρ_i is a characteristic root of the non-singular matrix **A**, then $|\mathbf{A}|/\rho_i$ is a characteristic root of adj **A**.

SYMMETRIC MATRICES

A square matrix **A** is called *symmetric* provided $\mathbf{A}^T = \mathbf{A}$.

Suppose **A** is a non-singular symmetric matrix. Then $(\mathbf{A}^T)^{-1} = \mathbf{A}^{-1}$ and, by Example 10, $(\mathbf{A}^{-1})^T = (\mathbf{A}^T)^{-1} = \mathbf{A}^{-1}$. Thus,

The inverse of a non-singular symmetric matrix is symmetric; also, the adjoint of a symmetric matrix is symmetric.

We prove

The characteristic roots of a (real) symmetric matrix **A** are all real.

Suppose the contrary, that is, suppose that $h + ik$, $i = \sqrt{-1}$, $k \neq 0$, is a root of $|\mathbf{A} - \lambda\mathbf{I}| = 0$. Then, by a theorem of algebra, $h - ik$ is also a root. Consider

$$\mathbf{C} = [\mathbf{A} - (h + ik)\mathbf{I}] \cdot [\mathbf{A} - (h - ik)\mathbf{I}] = (\mathbf{A} - h\mathbf{I})^2 + k^2\mathbf{I}$$

Since $|\mathbf{C}| = 0$, there exists a non-zero 3×1 matrix $X = \begin{bmatrix} a \\ b \\ c \end{bmatrix}$ such that $\mathbf{C}X = \begin{bmatrix} 0 \\ 0 \\ 0 \end{bmatrix}$. Then

$X^T\mathbf{C}X = X^T[\mathbf{A} - h\mathbf{I}]^2X + k^2X^TX = [(\mathbf{A} - h\mathbf{I})X]^T \cdot [(\mathbf{A} - h\mathbf{I})X] + k^2X^TX = 0$. Now $(\mathbf{A} - h\mathbf{I})X$ is real and so $[(\mathbf{A} - h\mathbf{I})X]^T \cdot [(\mathbf{A} - h\mathbf{I})X] \geqq 0$ while $X^TX > 0$. Thus, $k = 0$ and $|\mathbf{A} - \lambda\mathbf{I}| = 0$ has only real roots.

SIMILAR MATRICES

Two n-square matrices **A** and **B** are called *similar* provided there exists an n-square non-singular matrix **R** such that

$$\mathbf{B} = \mathbf{R}^{-1}\mathbf{A}\mathbf{R}$$

Let **A** and **B** be similar. Then

$$\mathbf{B} - \lambda\mathbf{I} = \mathbf{R}^{-1}\mathbf{A}\mathbf{R} - \lambda\mathbf{I} = \mathbf{R}^{-1}\mathbf{A}\mathbf{R} - \mathbf{R}^{-1}\lambda\mathbf{I}\mathbf{R} = \mathbf{R}^{-1}(\mathbf{A} - \lambda\mathbf{I})\mathbf{R}$$

and $$|\mathbf{B} - \lambda\mathbf{I}| = |\mathbf{R}^{-1}(\mathbf{A} - \lambda\mathbf{I})\mathbf{R}| = |\mathbf{R}^{-1}| \cdot |\mathbf{A} - \lambda\mathbf{I}| \cdot |\mathbf{R}| = |\mathbf{A} - \lambda\mathbf{I}|$$

Thus,

Two similar matrices have the same characteristic equation and the same characteristic roots.

The converse of the above theorem is not true.

INDEX